Advances in Human Genetics

Advances in Human Genetics

Edited by David Rhodes

SYRAWOOD
PUBLISHING HOUSE
New York

Published by Syrawood Publishing House,
750 Third Avenue, 9th Floor,
New York, NY 10017, USA
www.syrawoodpublishinghouse.com

Advances in Human Genetics
Edited by David Rhodes

International Standard Book Number: 978-1-68286-591-0 (Hardback)

Cataloging-in-Publication Data

Advances in human genetics / edited by David Rhodes.
p. cm.
Includes bibliographical references and index.
ISBN 978-1-68286-591-0
1. Human genetics. 2. Human biology. I. Rhodes, David.
QH431 .A38 2018
599.935--dc23

Printed in China.

TABLE OF CONTENTS

PREFACE

This book has been a concerted effort by a group of academicians, researchers and scientists, who have contributed their research works for the realization of the book. This book has materialized in the wake of emerging advancements and innovations in this field. Therefore, the need of the hour was to compile all the required researches and disseminate the knowledge to a broad spectrum of people comprising of students, researchers and specialists of the field.

Human genetics studies the structure of genes and the inheritance of certain characteristics, diseases and physical appearance caused by them in humans. Human genetics has a number of allied fields like molecular genetics, population genetics, clinical genetics, cytogenetics, etc. Studying human genetics helps us in understanding and treating a number of neural as well as physical disorders. This book explores all the important aspects of human genetics in the present day scenario. Scientists and students actively engaged in this field will find this book full of crucial and unexplored concepts.

At the end of the preface, I would like to thank the authors for their brilliant chapters and the publisher for guiding us all-through the making of the book till its final stage. Also, I would like to thank my family for providing the support and encouragement throughout my academic career and research projects.

Editor

1

Possible Regulatory Roles of Promoter G-Quadruplexes in Cardiac Function-Related Genes – Human *TnIc* as a Model

Wenhua Zhou[1], Kogularamanan Suntharalingam[2], Nigel J. Brand[3], Paul J. R. Barton[3,4], Ramon Vilar[2], Liming Ying[1]*

1 Molecular Medicine, National Heart and Lung Institute, Imperial College London, London, United Kingdom, **2** Department of Chemistry, Imperial College London, London, United Kingdom, **3** Harefield Heart Science Centre, National Heart and Lung Institute, Imperial College London, Middlesex, United Kingdom, **4** NIHR Cardiovascular Biomedical Research Unit, Royal Brompton and Harefield NHS Trust, London, United Kingdom

Abstract

G-quadruplexes (G4s) are four-stranded DNA secondary structures, which are involved in a diverse range of biological processes. Although the anti-cancer potential of G4s in oncogene promoters has been thoroughly investigated, the functions of promoter G4s in non-cancer-related genes are not well understood. We have explored the possible regulatory roles of promoter G4s in cardiac function-related genes using both computational and a wide range of experimental approaches. According to our bioinformatics results, it was found that potential G4-forming sequences are particularly enriched in the transcription regulatory regions (TRRs) of cardiac function-related genes. Subsequently, the promoter of human cardiac troponin I (*TnIc*) was chosen as a model, and G4s found in this region were subjected to biophysical characterisations. The chromosome 19 specific minisatellite G4 sequence (MNSG4) and near transcription start site (TSS) G4 sequence (−80 G4) adopt anti-parallel and parallel structures respectively in 100 mM KCl, with stabilities comparable to those of oncogene G4s. It was also found that TnIc G4s act cooperatively as enhancers in gene expression regulation in HEK293 cells, when stabilised by a synthetic G4-binding ligand. This study provides the first evidence of the biological significance of promoter G4s in cardiac function-related genes. The feasibility of using a single ligand to target multiple G4s in a particular gene has also been discussed.

Editor: Rafael Josef Najmanovich, Universite de Sherbrooke, Canada

Funding: This work was funded by a National Heart and Lung Institute Foundation PhD studentship (to WZ) and the British Heart Foundation (grant No.PG/11/81/29130). KS and RV thank the UK's Engineering and Physical Sciences Research Council (EPSRC) for funding (grant number: EP/H005285/1). PB is supported by the National Institutes of Health Research (NIHR) Cardiovascular Biomedical Research Unit at the Royal Brompton and Harefield NHS Foundation Trust and Imperial College London. The funders had no role in study design, data collection and analysis, decision to publish, or preparation of the manuscript.

Competing Interests: The authors have declared that no competing interests exist.

* E-mail: l.ying@imperial.ac.uk

Introduction

Although the majority of human genomic DNA adopts the iconic double helical structure (also known as B-form or duplex DNA) proposed by Watson and Crick [1], other non-B-form DNA secondary structures formed within certain sequences in the human genome have also been revealed in many *in vivo* and *in vitro* studies, such as the G-quadruplex DNA (G4 DNA) and i-motifs (i-tetraplexes) formed in oligonucleotides with guanine and cytosine tracts respectively (reviewed in [2,3]). As 98% of the human genome is non-coding, the possible biological functions of these transient and dynamic non-B-form DNA structures have been a subject of increasing interest.

As compared to other non-B-form DNA, G4s are relatively stable in solution under near-physiological conditions (thermodynamic and kinetic stability of G4s are reviewed in [4] and [5], respectively), which enables them to compete with adjacent duplex DNA and therefore participate in certain biological processes. The accumulated distribution of G4s in gene promoter regions throughout the human genome also makes them particularly important as compared to other non-B-form DNA structures [6]. Unsurprisingly, there have been a growing number of studies on the biological functions of these DNA structures, since the discovery that G4s can form in the G-rich regions from human telomeric oligonucleotides in the late 1980s [7].

Early studies about the regulatory roles of promoter G4s have focused on a few specific gene loci. For example, the formation of G4 structures (mostly stabilised by certain G4-binding ligands) in promoters of the human insulin [8,9], *MYC* [10,11], *KRAS* [12,13], and *PDGF-A* [14] genes were shown to have significant influences on downstream gene transcription activity. Subsequently, promoter G4s have been identified in important functional regions (promoters, NHE regions, enhancers, and so on) of many other cancer-related genes, and their biophysical properties in solution have been thoroughly characterised. Examples include the G4s found in promoters of *KIT* [15–19], *RET* [20], *VEGF* [21–24], *RB1* [25–27], *BCL-2* [28–30], and *HIF-1α* [31] genes. These isolated examples have led to the hypothesis that G4s may act as a group of common *cis*-regulatory elements via various mechanisms *in vitro* and *in vivo*. Indeed, accumulating bioinformatics studies in the past decade supported this hypothesis as described below.

Firstly, the enrichment and biased distribution of G4s are found in gene transcription regulatory regions in the human genome, such as in promoter regions, 5′-ends of 5′-UTRs, within the first intron and within regions immediately downstream of the 3′-ends of genes. These findings indicate that these non-canonical DNA structures are favoured by natural selection in the process of evolution. Secondly, potential G4-forming motifs closely correlate with functional regions in gene promoters in the human genome, such as the gene regulatory elements [32] and G-rich TF (transcription factor) binding sites [33,34], which further supports the hypothesis that G4s are involved in gene transcription regulation. Furthermore, the existence of potential G4-forming motifs is also related with downstream gene expression levels, indirectly demonstrating the regulatory roles of G4s in gene transcription regulatory regions (TRRs) [35,36]. Finally, correlation between prevalence of G4s and gene functions also indicates potential roles of G4s in the human genome [37].

Due to its diverse biological functions and wide distribution in the human genome, G4 DNA has attracted exceptional attention from nucleic acid chemists to design and synthesise novel G4-binding ligands that can interact efficiently and selectively with this peculiar DNA structure. So far, most G4-binding ligands are researched as potential anti-cancer agents by targeting cancer-related G4s, especially the human telomeric G4 and oncogene promoter G4s. The affinity and selectivity of G4-binding ligands to G4s rather than to other DNA forms are mainly achieved by π–π stacking and electrostatic interaction, which are also the two main criteria in G4-binding ligand design [38,39].

To date, almost all G4 studies have focussed on their involvement in cancer and therefore their potential as targets for anticancer agents. Thus, most studies have centred on telomeric G4 and promoter G4s in proto-oncogenes. Importantly, extensive studies of these cancer-related G4s have contributed greatly to forward our knowledge in this field, which have established the fundamentals to investigate the role of G4s in other genes, such as those involved in regulating cardiac functions. Since possible biological functions of G4s have been identified in some heart-related genes, such as *PDGF-A*, *VEGF-A* and *HIF-1α*, G4 research also has relevance to cardiovascular science.

Taking into account the instability of G4s and their much lower abundance as compared to adjacent duplex DNA in the human genome *in vivo*, and the structural similarity between different G4s, it makes the design of selective G4-binding ligands extremely challenging, and off-target effects are practically inevitable. However off-target effects may lead to multi-targeting effects in some cases, which in turn become advantageous for G4-binding ligands, and may enhance their therapeutic potential, such as the ligand naphthalene diimide derivative 1 [40], 12459 [41–43], and CX-3543 [44].

In this article, we first evaluated, by bioinformatics approaches, the importance of G4s in TRRs of key cardiac genes as compared to other genes, especially those involved in cancer. We found that potential G4-forming motifs are particularly enriched in TRRs of genes active in the heart, and TRRs of genes involved in cardiac function-related pathways are also highly G4-abundant. Although very basic, these findings clearly supported the hypothesis that G4s are important in regulating cardiac function-related genes. Furthermore, due to the abundance of G4-forming motifs in TRRs of cardiac function-related genes, potentially, multi-targeting strategy may have similar therapeutic potential as those observed in anti-cancer research.

To validate the bioinformatics results (i.e. that G4s are involved in regulating cardiac function-related genes) and to initially test the multi-targeting strategy within a single gene, subsequently, the human cardiac troponin I (*TnIc*) gene was selected as a model system. Human TnIc is the only sarcomeric protein found to be exclusively expressed in the cardiac myocyte, which makes it an ideal model to investigate cardiac-specific gene expression and in turn to provide new approaches to the construction of cardiac-specific vectors in gene therapy. Although TnIc has been regarded as a preferred biomarker in MI (myocardial infarction) diagnosis for more than a decade [45,46], the molecular mechanisms of human *TnIc* transcription regulation are still poorly understood, and only a few TFs have been found to be involved in regulating the transcription of *TnIc* [47–49]. This is partially due to the difficulties in obtaining suitable cell models, and possibly due to the extremely G-rich in its promoter. Indeed, TRR of *TnIc* is highly G4-enriched according to the bioinformatics analysis. Multiple G4-forming sequences have been identified in the proximal promoter of human *TnIc*, and two types of G4-forming motifs were selected for detailed studies, which are designated as "MNSG4" (between −528 and −319 bp upstream to the TSS, transcription start site), and "−80 G4" (centred at −80 bp to the TSS). The MNSG4 motif is composed of 6 repeats of chromosome 19 specific G-rich minisatellite (MNS), which are highly conserved in primates. Although conserved chromosome 19 specific MNS repeats are also found in many other genes, such as the neuronally expressed Shc adaptor homolog *SCK1/SLI* [50], the EFG-like module-containing mucin-like receptor, *EMR3* [51], the human apolipoprotein C-II gene [52], and the human *TnT* [53], the actual functions of this widely distributed G4-forming sequence are still unclear. On the other hand, similar as the KIT1 G4-forming sequence, the −80 G4 is found to be unique in the human genome [18]. More importantly, multiple putative or proven TF binding sites are clustered within or around the −80 G4-forming region, such as a CACC-box, an Sp1 binding site, an E-box, and a GATA biding site [48].

We have used a combination of biophysical and biochemical approaches to characterise the formation, stability and possible regulatory roles of the G4s (MNSG4 and −80 G4) in the promoter of the human *TnIc*. The methods include CD (circular dichroism) spectroscopy [3], EMSA (electrophoresis mobility shift assay) [7], FRET (fluorescence resonance energy transfer) [54], DMS (dimethyl sulfate) footprinting [31], FID (fluorescent intercalator displacement) experiments [55], and dual luciferase reporter assays [10] in HEK293 cell line. These methods are reviewed in [56] and described in detail in [57]. It was found that these G4s have comparable stability as that of proto-oncogene G4s, and they may act cooperatively as enhancers in gene expression regulation in HEK293 cells when they are stabilised by a synthetic G4-binding ligand. This study provides the first evidence about the biological significance of cardiac promoter G4s.

Materials and Methods

Enrichment of G4-forming Motifs in Different Tissues

Sequences of the human genome (version 37.64) were downloaded from ENSEMBL FTP, and gene/transcripts features were exported from BioMart (detailed protocols are described in Protocols S1). TRR of individual transcripts were defined as −2,000 bp to +1,000 bp from TSS of the transcript. Although the actual mechanisms and final consequences of downstream and upstream (relative to TSS) G4s are different, we are only interested in whether or not G4s are involved in regulating a particular gene expression, and thus down- and upstream G4s are investigated together in most cases in this study. G4-forming motifs were identified as $G_{\geq 3}(N_{1\sim 7}G_{\geq 3})_{\geq 3}$. Tissue-specific enrichment of G4-forming motifs was defined as the percentage of genes active in certain tissue (GNF/Atlas organism part) containing at least one G4-forming motif in a certain region of the TRR (Equation 1)

$$\text{Tissue-specific Enrichment of G4} = \frac{\text{No. of gene containing at least one G4}}{\text{Total No. of gene active in the tissue}} \times 100\% \quad (1)$$

Besides the full TRRs, enrichment of G4s was also analysed and compared in the distal promoter region (−2,000∼−501 bp), proximal promoter region (−500∼−1 bp), and downstream region (TSS∼+1,000 bp). The analysis was carried out in 10 iterations to consider transcript redundancy. Hence the tissue-specific enrichment of G4-forming motifs was presented as mean ± standard deviation (SD). In analysing specific tissue, differences among iterations were evaluated by one-way ANOVA test, in which $P \geq 0.05$ represents that no significant difference was found.

Importance of G4s in Different Pathways

To evaluate the importance of G4s in individual transcript, two scores, *F* and *Q*, were calculated respectively to reflect the abundance and location significance of G4s identified in each transcript. In the TRR of each transcript, the percentage of nucleotides involved in G4 formation is defined as *F* values according to Equation 2

$$F = \frac{\sum_{i=1}^{n} (ge_i - gs_i + 1)}{T} \times 100\% \quad (2)$$

where n is the number of G4s identified in the TRR ($F = 0$ when $n = 0$), gs_i and ge_i are the start and end positions of ith G4 identified in the TRR relative to the TSS in 5′ to 3′ direction, and T, the length of TRR being searched, is fixed at 3,000 in this study.

Abundance of G4s on the coding strand, and template strand were separately calculated and labelled as F_{cd}, and F_{ncd}, (subscripts cd: coding strand; ncd: noncoding strand). Overall abundance of G4s, the F_o (o: overall), were calculated as the sum of F_{cd}, and F_{ncd}. To genome-widely compare the abundance of G4s in TRRs of different transcripts, *F* scores of all transcripts without redundancy were calculated (one transcript was randomly picked when multiple transcripts were reported from the same gene), and the cumulative frequencies of *F* scores of all transcript were calculated and denoted as CF_{cd}, CF_{ncd}, and CF_o. To reach a more stringent criterion, transcripts with *F* scores equal to 0 were excluded when calculating *CF*. Genes with transcript *CF* values higher than 50% were regarded as G4-rich in their TRR, while those with transcript *CF* values lower than 50% were regarded as G4-scarce.

Except for the G4 abundance, the location of individual G4 in a particular TRR may also correlate with its biological significance. As reported in previous genome-wide analyses, G4 was found to be enriched in the promoter region between −200 bp and TSS, and peaked at around −50 bp to TSS in the human genome [58]. This positional bias of G4 distribution in human gene promoters is believed to be a result of evolutionary pressure [59]. Most promoter G4s with confirmed biological functions localize in this region, such as *KIT* and *MYC* promoter G4s. Thus, another indicator, the *Q* score, is introduced to evaluate the potential location significance of G4s in particular TRR.

Following on the previously reported method [58], the probability distribution of each TRR position relative to TSS involved in G4 formation along coding and template (noncoding) strands of all TRRs was calculated according to Equation 3

$$W_k = \frac{\sum_{j=1}^{m} \alpha_{j,k}}{m} \times 100\% \quad (3)$$

where W_k is the normalized probability of nucleotide in position k of all TRRs involved in G4 formation, m is the total number of transcripts analyzed. $\alpha_{j,k}$ is the logical value of nucleotide in position k of jth TRR; $a_{j,k} = 1$ when the nucleotide is involved in G4 formation, otherwise $a_{j,k} = 0$. G4 location significance score Q (Q_{cd}, Q_{ncd} and Q_o for coding and template strand, and overall, respectively; $Q_o = Q_{cd} + Q_{ncd}$) for the TRR of a particular transcript was calculated according to Equation 4

$$Q = \sum_{i=1}^{n} \frac{\sum_{k=gs}^{ge} W_k}{ge_i - gs_i + 1} \quad (4)$$

The cumulative frequencies of *Q* scores, *CQ*, were also calculated. Similarly, transcripts with the zero *Q* scores were excluded. Genes with both *CF* and *CQ* values higher than 50% were regarded as G4-important genes, while those with both values lower than 50% were regarded as G4-less-important.

Correlation between the *CF* and *CQ* scores on the coding, template strand and both strands were also investigated. The correlation coefficient was calculated according to Equation 5

$$r_{xy} = \frac{\sum_{i=1}^{n} (x_i - \bar{x})(y_i - \bar{y})}{\sqrt{\sum_{i=1}^{n} (x_i - \bar{x})^2 \sum_{i=1}^{n} (y_i - \bar{y})^2}} \quad (5)$$

in which x_i and y_i are the *CF* and *CQ* scores of the coding, template strand or both strands at genome level, $\bar{x}$ and $\bar{y}$ are the corresponding mean values *CF* and *CQ* scores at genome level.

CF and *CQ* were calculated for all transcripts without redundancy. To evaluate bias introduced in random-picking when multiple transcripts exist in one gene, calculation was carried out for 10 iterations. Gene list for all pathways was downloaded from KEGG database. All pathways including more than 50 genes were subject to pathway analysis. When more than 50% component genes in a pathway were calculated as G4-important (excluding those with both *CF* and *CQ* equal to zero), the pathway was regarded as G4-important. To evaluate the G4-importance of individual pathway, distribution of *CF* and *CQ* values of pathways were compared to the distribution of all transcripts in humane genome by Wilcoxon rank sum test. Differences between iterations of individual pathway were evaluated by one-way ANOVA test.

The source code of bioinformatics studies is available upon request.

DNA Oligos

All labelled and unlabelled DNA oligos used in biophysical studies were purchased from IBA Biotechnology (Göttingon, Germany) and HPLC purified (double HPLC purification for labelled oligos). Fluorophores (Cy3 or Cy5) were coupled to nucleotides via NHS (N-hydroxysuccinimide) esters. Oligo concentration was determined by absorption at 260 nm. Oligo **Tr$_{MNS}$-I** is the consensus sequence of G4-forming minisatellites between −528 and −319 bp to the TSS of human *TnIc* gene.

Oligo **Tr$_{MNS}$-II** is the C-rich complimentary strand of oligo **Tr$_{MNS}$-I**. Oligo **Tr$_{MNS}$-III** is the Cy3-labeled **Tr$_{MNS}$-I** attached with a 34-mer artificial linker on its 3′-end. **Tr$_{-80}$-I** is a G4-forming sequence from the noncoding strand fragment of human *TnIc* promoter localized between −97 and −73 bp to the TSS, while oligo **Tr$_{-80}$-II** is its C-rich complimentary strand. Oligo **Tr$_{-80}$-III** is the Cy3-labeled **Tr$_{-80}$-I** attached with a 34-mer artificial linker on its 3′-end. Oligo **Comp-Cy5** is a Cy5-labeled oligo complimentary to the 34-mer artificial linker in oligos **Tr$_{MNS}$-III** and **Tr$_{-80}$-III**. Oligos and primers used in plasmid construction were purchased from Eurofins MWG Operon (Ebersberg, Germany) and were subjected to HPLC purification. The sequences and labeling sites of these oligos are listed in Table 1.

Formation of G4s in Solution

Formation of G4s was determined by EMSA, CD spectroscopy, and DMS footprinting in solution. Detailed protocols are described in Protocols S1.

G4 Unfolding Thermodynamics and Kinetics

Unfolding thermodynamics of G4s and the stabilising ability of G4-binding compounds to these G4s were investigated by CD melting experiments. Unfolding kinetics of these G4s were investigated via a pseudo-first order approach as described previously [60]. Briefly, the fluorophores-labeled unfolded G4s were trapped by excess C-rich oligos, and hence resulted in decreased FRET signals. Detailed protocols are described in Protocols S1.

G4-ligand Interaction in Solution by FID (Fluorescent Intercalator Displacement) Assay

The corresponding oligonucleotides were first dissolved in MilliQ water to yield a 20 μM stock solution, and then were diluted in 10 mM potassium cacodylate (pH 7.4)/50 mM potassium chloride (60 mM K^+) buffer to the appropriate concentrations. Prior to use in the FID assay, the DNA strands were incubated to allow the formation of G4s as stated above. The compounds to be analysed and thiazole orange (TO) were dissolved in DMSO to give 1 mM stock solutions. The corresponding solution was then diluted using 10 mM potassium cacodylate (pH 7.4)/50 mM potassium chloride (60 mM K^+) buffer to the appropriate concentrations. The FID assay was carried out according to the protocol as previously reported [55]. Briefly, to a mixture of DNA sequence (0.25 μM) and TO (0.50 μM) in 10 mM potassium cacodylate (pH 7.4)/50 mM potassium chloride (60 mM K^+) buffer an increasing amount of the molecule under study was added (0.125 to 5 μM, which corresponds to 0.5 to 20 equiv). After an equilibration time of 3 min the emission spectrum was recorded between 510 and 750 nm with an excitation wavelength of 501 nm. This was recorded using a Varian Cary Eclipse Spectrometer (Agilent Technologies, Yarnton, UK). The fluorescence area was calculated using the "trapezium rule" method. The area was converted into percentage TO displacement by the following formula: % TO displacement = 100 − [(fluorescence area of sample/fluorescence area of standard) × 100]. The standard fluorescence spectrum was obtained in the absence of any G4-binding ligands. % TO displacement was then plotted against each of the compound concentrations to give the respective FID curves, from which the DC_{50} values were determined.

Table 1. Details of oligos used in biophysical studies in solution.

Name	Sequence (5′–3′)
Synthetic oligos used in biophysical studies	
Tr$_{MNS}$-I	TGG GTC TGA GGG AGG AGG GGC TGG GGG C
Tr$_{MNS}$-II	GCC CCC AGC CCC TCC TCC CTC AGA CCC A
Tr$_{MNS}$-III	^{Cy3}TGG GTC TGA GGG AGG AGG GGC TGG GGG CGA GGT AAA AGG ATA ATG GCT ACG GTG CGG ACG GC
Tr$_{MNS}$-IV	^{Cy3}TGG GTC TGA GGG AGG AGG GGC TGG GGG ^{Cy5}C
Tr$_{-80}$-I	CGG GGG CGC GTG AGG GGC GGG GTG GGC
Tr$_{-80}$-II	GCC CAC CCC GCC CCT CAC GCG CCC CCG
Tr$_{-80}$-III	^{Cy3}CGG GGG CGC GTG AGG GGC GGG GTG GGC GAG GTA AAA GGA TAA TGG CTA CGG TGC GGA CGG C
Tr$_{-80}$-IV	^{Cy3}CGG GGG CGC GTG AGG GGC GGG GTG GG ^{Cy5}C
Comp-Cy5	GCC GTC CGC ACC GTA GCC ATT ATC CTT ^{Cy5}TTA CCT C
Synthetic oligos used in plasmid construction	
MNS_G	CCT GGA CTC TTG GGT CTG AGG GAG GAG GGG CTG GGG G
MNS_C	TCC AGG CCC CCA GCC CCT CCT CCC TCA GAC CCA AGA G
MNSM_G	CCT GGA CTC TTG AAT CTG AGA GAG GAG AGA CTG AGA G
MNSM_C	TCC AGG CTC TCA GTC TCT CCT CTC TCA GAT TCA AGA G
5′Linker_G	GGG GAC AAG TTT GTA CAA AAA AGC AGG CTG AAT TC
5′Linker_C	TCC AGG GAA TTC AGC CTG CTT TTT TGT ACA AAC TTG TCC CC
3′Linker_G	CCT GGA CAC TTG AGT CTG CAG ACC CAG CTT TCT TGT ACA AAG TGG TCC CC
3′Linker_C	GGG GAC CAC TTT GTA CAA GAA AGC TGG GTC TGC AGA CTC AAG TG

Note: $^{Cy3/5}$N indicates a Cy3 or Cy5-labeled nucleotide, which is coupled via NHS ester.

Construction of Wild-type/Mutated Minisatellite Concatemer

Engineered wild-type (E-WT, oligo **MNS_G/C**) and mutated (M, oligo **MNSM_G/C**) minisatellite concatemers were constructed by ligation and recombination processes. Oligos **MNS_G/C** and **MNSM_G/C** were first phosphorylated individually in 10 µl phosphorylation reaction system (DNA 250 pmol; 10× PNK buffer, 1 µl; 10 mM ATP, 1 µl; 10 U/µl PNK, 0.5 µl) at 37°C for 1 hour. Then, phosphorylated oligos **MNS_G/C** and **MNSM_G/C** were annealed to each other by a PCR thermocycler (Stratagene) at 95°C for 5 min, then gradually cooling to 25°C over 2 hours. Concatemers with random number of repeat were constructed by self-ligation (first round ligation) of annealed single repeat of E-WT/M minisatellites. 5′linker and 3′linker were phosphorylated and annealed from oligos **5′Linker_G/C** and **3′Linker_G/C** respectively through the same protocol. Then, annealed 5′linker was ligated to concatemers in a second round ligation, and the 3′linker was added in the third ligation subsequently. All three rounds ligation were carried out in 20 µl system at 16°C for 2 hours. Ligation products larger than 100 bp were extracted by QIAGEN Gel extraction kit after each round of ligation. All enzymes and buffers used in this step were purchased from NEB (Hitchin, UK).

Construction of Luciferase Vector

Entry vector containing minisatellite concatemers was made by recombinating the third round ligation products into pDONR221 using Gateway® BP reaction kit (Invitrogen). The minisatellite concatemers were further recombinated into a pGL3-Basic-Gateway vector using Gateway® LR reaction kit (Invitrogen). The insertions were confirmed by sequencing using M13 primer pair. The −299 to +51 fragment of human *TnIc* promoter was PCR amplified from plasmid pGL3-Basic_−553 containing the −553 to +69 fragment of human *TnIc.* Plasmid pGL3-Basic_−553 is a kind gift from Dr. Nigel Brand. Vectors hTnIc-6MNS(E)-WT, hTnIc-3MNS(E)-WT, and hTnIc-1MNS(E)-WT plasmids were constructed by inserting the −299 to +51 fragment into pGL3-Basic-Gateway vectors downstream to the engineered WT minisatellite concatemers with 6 repeats, 3 repeats, and 1 repeat via *Pst*I and *Xho*I restriction sites. By using similar approach, the vector containing 6 repeats of mutated MNSG4 was constructed as hTnIc-6MNS-M. Vector hTnIc-299-WT was constructed by inserting the −299 to +51 fragment into empty pGL3-Basic-Gateway vector via *EcoR*I and *Pst*I restriction sites. Vectors hTnIc-80 G4-M, hTnIc-80Sp1-M, and hTnIc-80 G4/Sp1-M were derived from hTnIc-6MNSG(E)-WT by *Dpn*I-mediated site-direct mutagenesis, in which the −80 G4-forming sequence, the Sp1 binding site in the −80 G4 motif, and both of them were mutated respectively. An overall mutant hTnIc-6MNS/80 G4-M was generated from hTnIc-6MNS-M by mutating both −80 G4-forming sequence and the Sp1 binding site in it. Vectors are schematically illustrated in Figure 1.

Transfection of Cultured Human Cells and G4-binding Ligands Treatment

HEK293 cells were inoculated in 75 cm^2 flasks and grown to 70–80% confluence at 37°C and under 5% CO_2 atmosphere in complete medium (Dulbecco Modified Eagle's Medium (DMEM) supplemented with 4500 mg/L D-glucose, 5.0 mM L-glutamine, 10% fetal calf serum, and 83.3 U/ml of penicillin and streptomycin (Invitrogen)). The cells were harvested by Trypsin-EDTA, and reseeded 2×10^4 cells/well in 24 well plates with a volume of 300 µl/well complete medium, and the cells were cultured overnight. The medium was replaced by 300 µl fresh medium at least 2 hours before transfection. Transfection was carried out by following calcium phosphate precipitation. Briefly, 20 ng pRL-TK control plasmid (Renilla luciferase reporter vector driven by herpes simplex virus thymidine kinase promoter $P_{HSV\text{-}TK}$) and 150 ng testing pGL3 luciferase plasmid was transfected in each well, and cells were incubated with precipitated DNA for another 24 hours before adding G4-binding ligands. G4-binding ligands were diluted in DMSO to different concentrations as 1000× stock solution. 0.3 µl G4-binding ligand stock solution was diluted to 300 µl by fresh medium to replace the transfection medium in each well. Cells were then incubated with G4-binding ligands for 24 hours before cell lysis for luciferase assays. Experiments were carried out in triplicates.

Luciferase Assay

Transfected cells were firstly washed by ice-cold PBS to reduce background signals from the medium, and luciferase assays were performed subsequently according to manufacturer's instruction (Promega). Generally, cells were lysed by 100 µl passive lysis buffer per well, and 10 µl cell lysate were added to 50 µl luciferase assay reagent. After 3 s' delay, luciferase signals were collected for 10 s in a Glomax 20/20 luminometer (Promega). After adding 50 µl of Stop and Glo reagent, Renilla signals were collected for 10 s as internal control.

Results

Enrichment of G4 DNA in the Heart

The enrichment of G4s in the transcriptional regulatory regions (TRRs: −2,000 bp to +1,000 bp around Ensembl Transcript Start) of all known protein-coding genes was first analysed as a reference point. In total, of 19,633 genes that were analysed, 67.79±0.01% of them were found to contain at least one G4 on both coding strand and template strand, with 52.42±0.02% and 46.91±0.02% containing at least one G4 on the coding strand (G4) and template strand (C4), respectively. Compared with the widely accepted results that around 40% human gene promoters contain G4s [6], the much higher percentage (67.79±0.01%) identified here can be attributed to the much longer region around the TSS that was analysed in this study. Similar results have been previously reported in the chicken genome when regions of similar length were analysed [61].

The enrichment of conserved G4s in different tissues has been previously investigated by Verma *et al.* [35]. Following a similar method, we focused on investigating the enrichment of G4s in different regions of gene TRR in different tissues (different GNF/Atlas organism) in the human genome. As shown in Figure 2 and Table S1, G4s are highly enriched in the whole TRR of genes active in the lung, heart and brain including cerebellum peduncles and caudate nucleus. The high G4 enrichment in these tissues may indicate that these tissues have evolved certain mechanisms to utilize G4s to regulate gene transcription (i.e. providing binding sites for tissue-restricted TFs, maintain the separation of coding and template strands, and so on), or at least to avoid possible negative effects caused by the formation of G4s in TRR (i.e. to resolve folding of G4 by helicases). For tissues with the lowest G4 enrichment at the whole TRR level, corresponding G4 enrichment in different regions of TRR is also among the lowest. Interestingly, for some tissues, although moderate or relatively low G4 enrichment was found at whole TRR level, G4s were found to be enriched in either distal or proximal promoters in the TRRs.

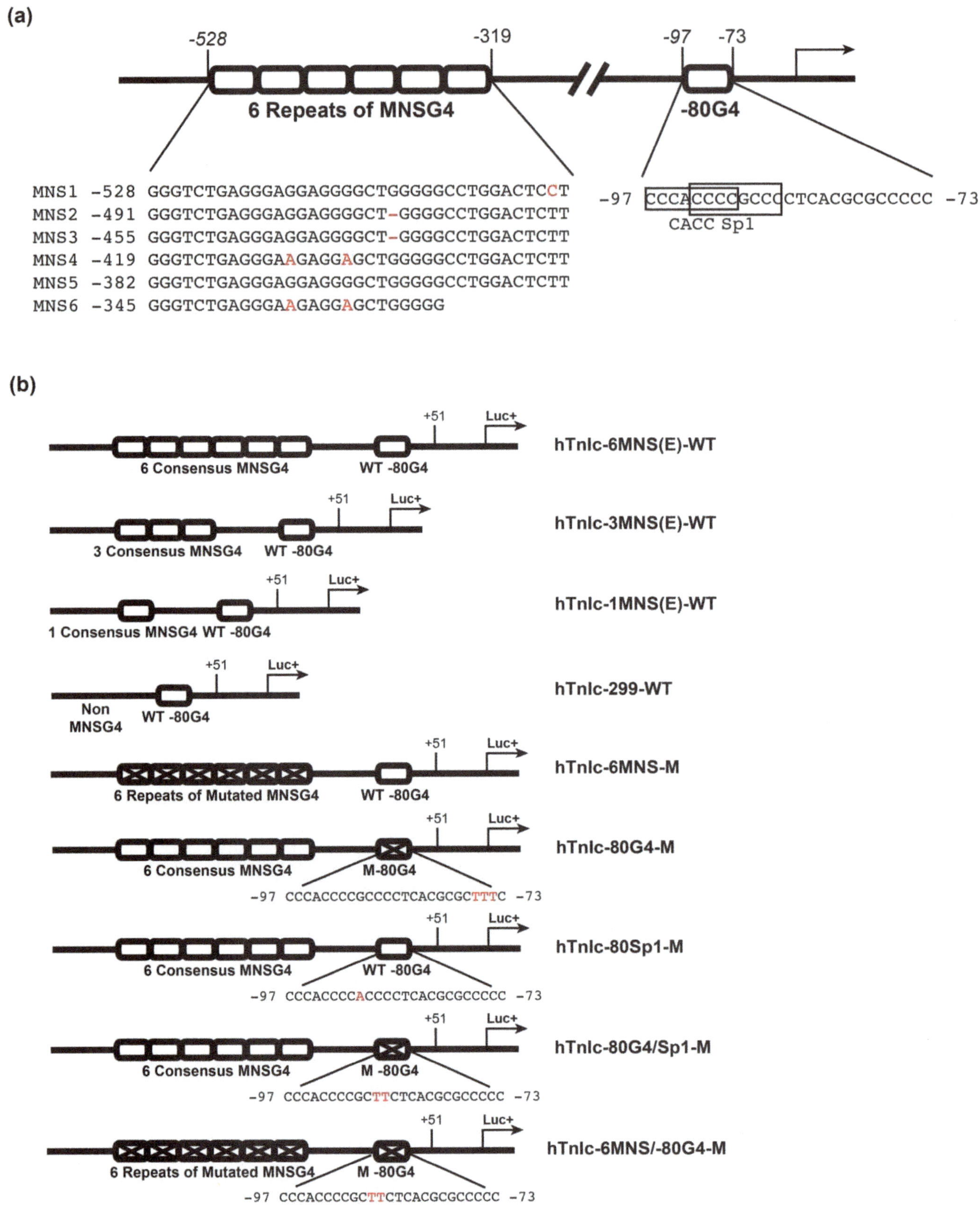

Figure 1. Structure of human TnIc gene promoter and constructed luciferase expression vectors. (**a**) Schematic illustration of the G4-forming motifs in the promoter of human *TnIc* gene. (**b**) the structure of luciferase expression vectors driven by engineered human *TnIc* promoters.

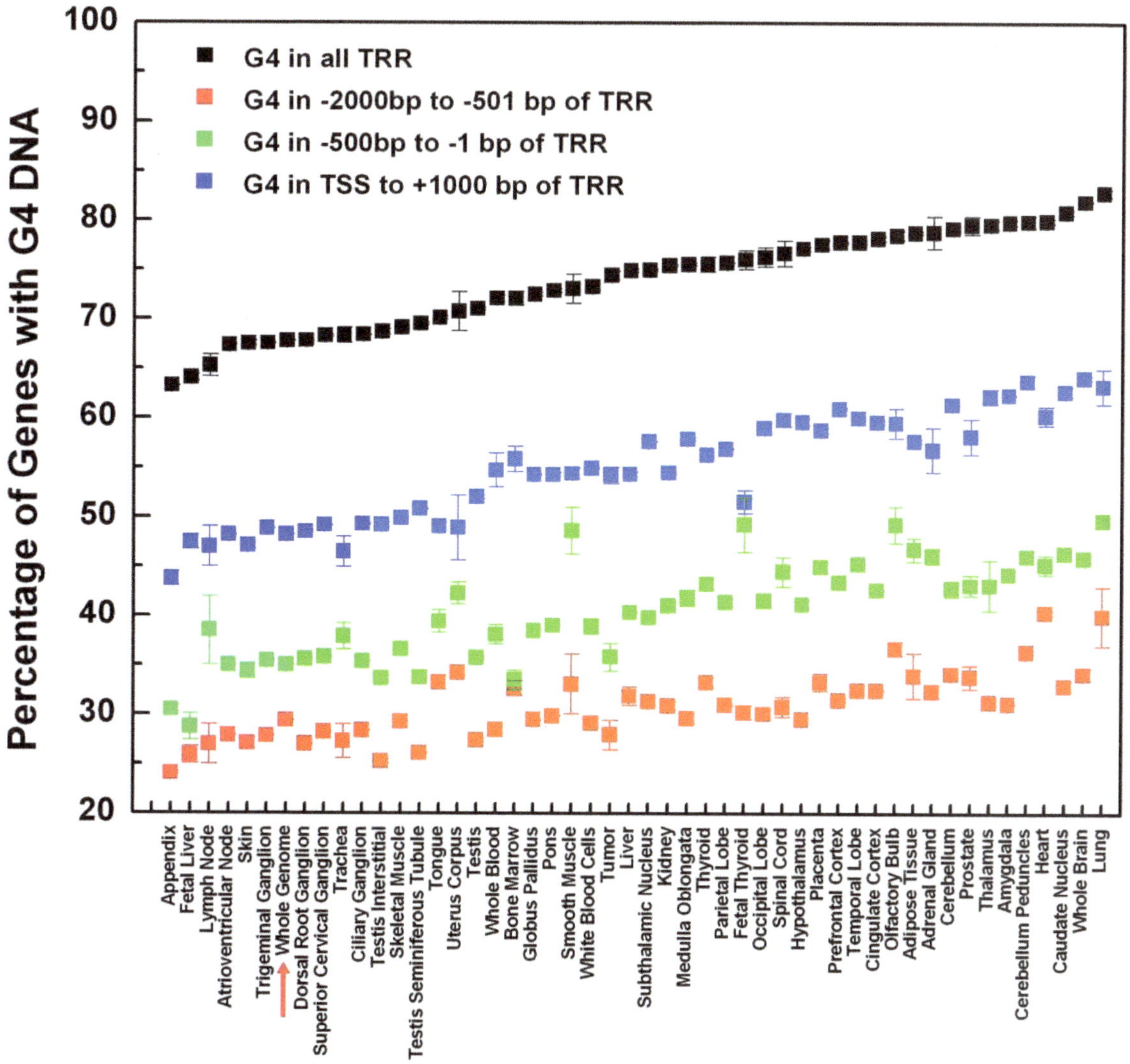

Figure 2. Enrichment of G4s in TRRs of genes active in different tissues. The percentage of genes active in different tissues containing at least one G4-forming sequence in the whole TRR (black), −2,000~−501 bp of TRR (red), −500~−1 bp of TRR (green), and TSS~+1,000 bp of TRR (blue). Enrichment of G4s in all protein-coding genes (G4 enrichment at genome level) is indicated by the red arrow. The calculation was repeated randomly for ten times (multiple transcripts of one gene were randomly chosen for ten times) to take into account the redundancy effects of multiple transcripts.

For example, in smooth muscle, although the overall G4 enrichment is only moderately higher than that found at the whole genome level, its G4 enrichment in the proximal promoter region is among the highest. Similar phenomena were found in tissues of adipose, olfactory bulb, and fetal thyroid (Table S1). This may indicate that in these tissues certain mechanism may exist to utilize G4s in the proximal promoter region, but not in other regions.

According to the above findings, several points became apparent. Firstly, enrichment of G4 in TRR of genes has a strong tissue-specificity. Secondly, in some tissues (smooth muscle, adipose tissue, olfactory bulb, and fetal thyroid), G4s are enriched specifically in certain regions of TRRs (location-specificity) reflecting possible tissue-related regulatory roles of G4s. These tissues, with either tissue- or location-specific enrichment of G4s in TRRs, are worthy of further investigation in the future.

From the studies described here, G4s appear to be highly abundant in TRRs of genes active in the cardiac function-related tissues, such as the heart and smooth muscle. To further investigate the biological significance of G4s in the heart, we analysed the abundance and location significance of G4s in TRRs of genes involved in different pathways described within the KEGG (Kyoto Encyclopedia of Genes and Genomes) PATHWAY database, especially those related to cardiac functions. The importance of G4s in each TRR was evaluated by two scores: G4 abundance score (*F* score, and its cumulative frequency: *CF*) and G4 location significance score (*Q* score, and its cumulative frequency *CQ*), which will be described separately in following sections.

F Score

As compared to duplex DNA, G4s are normally less stable (reviewed in [4,62,63]). Thus, high G4 content (abundance) in a particular TRR can enhance the competition of G4 DNA with adjacent duplex DNA, resulting in the increased probability of the involvement of G4s in regulating downstream gene expression. Based on this assumption, the *F* score representing the abundance of G4s in a particular TRR was used as the first indicator to evaluate the importance of G4 in the TRR of each gene (methods modified from Eddy *et al.* [37]). For each transcript, the *F* score was calculated based on the percentage of nucleotides involved in G4 formation in the TRR of that transcript (Equation 2). By including all transcripts with redundancy, the coding strand, template strand (non-coding strand), and both strands were analysed separately to give the F_{cd}, F_{ncd}, and F_o scores, respectively (Table 2). Figure S1 shows the distribution of these *F* scores obtained by including all transcripts with redundancy. A similar skewed distribution of G4 abundance was reported previously in an analysis showing G4 formation potential (G4P) in RefSeq of genes [37]. However, due to the higher resolution in this study (using the percentage of nucleotides involved rather than the percentage of a shifting searching window containing G4-forming motifs), two peaks were found in the distribution of G4 abundance scores. By comparing the peak position and corresponding values in the distribution histograms of G4 abundance on coding and template strand (Figure S1 a and b, lower graphs), G4-forming motifs are more enriched in the coding strand. All values were re-calculated with 10 iterations with randomly chosen transcripts, and compared with values from transcripts with redundancy, and similar results were obtained (Table 2). The CF_{cd}, CF_{ncd}, and CF_o values for each transcript were calculated as the cumulative frequency according to the F_{cd}, F_{ncd}, and F_o histograms (Figure S1, upper graphs).

Q Score

Besides the abundance, the locations of G4s in a particular TRR are also important in estimating the potential biological significance of G4. For example, Du *et al.* suggested that the colonization of G4-forming motifs with TF binding sites in the proximal promoters indicates potential regulatory roles of G4s [32]. Thus, we believe that the biased probability distribution of G4-forming motifs in TRRs at whole genome level, which was reported by Huppert *et al.* [58], may reflect the preference of G4 location in TRRs through evolution. Following Equation 3 (modified from Huppert *et al.* [58]), the probability distribution histogram was generated (Figure S2 a). Similar probability distribution of G4 around TSS has been reported in the genomes of other warm-blooded animals [59]. The location and strand polarity-dependent distribution of G4s appears to reflect an evolutionary pressure, and it is necessary to take into account the location of G4 motifs when considering their importance in the TRR of a particular gene. Therefore, a location significance score *Q* was introduced in order to characterise the location significance of G4s. For each TRR, its G4 location significance scores on the coding (Q_{cd}) and template (Q_{ncd}) strand were calculated according to Equations 3 and 4. The overall G4 location significance score (Q_o) was determined as the sum of Q_{cd} and Q_{ncd}. Histograms Q_{cd}, Q_{ncd}, and Q_o for TRRs of all transcripts with redundancy are shown in Figure S2 b~d (lower graphs). The larger maximum, median, and average *Q* scores on the coding strand also reflect the higher location significance of G4s on the coding strand as compared to the template strand. Transcripts without redundancy were also analysed with 10 iterations of randomly chosen transcripts, and similar results were obtained as compared to the results from transcripts with redundancy (Table 3). The *CQ* values for each transcript were calculated as its cumulative frequency in *Q* histogram and are shown in Figure S2 b~d (upper graphs).

Correlation between G4 Abundance and G4 Location Scores

The correlation between *CF* and *CQ* scores (abundance and location significance of G4s in TRRs) were evaluated. As shown in Figure S3, *CF* and *CQ* are plotted as correlation maps. According to the maps, *CF* values are positively correlated to *CQ* values. Particularly, in G4-important TRRs (TRRs with both values higher than 50%), their G4 abundance is more positively correlated to their G4 location significance as compared to G4-less-important TRRs (TRRs with both values lower than 50%). To investigate whether this positive correlation of G4-important TRRs is only caused by the increase of G4 content or not, G4s found in different regions of TRRs were subjected to further investigation. The total number of G4s identified in different regions in G4-important and G4-less-important transcripts were counted and normalised to the length of different regions. If the increase of *CQ*, the location significance score, was solely dependent on the increase of *CF*, the G4 abundance score, then the proportion of the normalised total number of G4 found in different regions should be in constant between G4-important and G4-less-important TRRs. We found that the proportion of G4s

Table 2. Comparison of G4 abundance in TRRs of transcripts (the *F* score).

	Median Value	Mean Value	Max Value
Transcripts with Redundancy (46,205 Transcripts)			
F_{cd}	1.6	2.1	33.2
F_{ncd}	1.4	1.9	29.6
F_o	2.2	2.9	33.2
*Transcripts without Redundancy (19,633 Transcripts), 10 iterations**			
F_{cd}	1.55±0.02	2.03±0.01	33.2
F_{ncd}	1.42±0.02	1.93±0.01	25.88±3.96
F_o	2.13±0.01	2.90±0.01	33.2

*The F_{cd}, F_{ncd}, and F_o values from 1st iteration were used as templates and compared with those from following iterations by one-way ANOVA test. No differences among iterations were observed ($P \geq 0.05$).

Table 3. Comparison of G4 location significance in TRRs of transcripts (the *Q* score).

	Median Value	Mean Value	Max Value
Transcripts with Redundancy (46,205 Transcripts)			
Q_{cd}	2.5	3.3	26.9
Q_{ncd}	1.7	2.3	19.8
Q_o	3.2	4.2	29.3
*Transcripts without Redundancy (19,633 Transcripts), 10 iterations**			
Q_{cd}	2.48±0.02	3.13±0.02	23.4±1.6
Q_{ncd}	1.85±0.02	2.40±0.02	19.7±0.2
Q_o	3.20±0.02	4.07±0.01	28.3±0.2

*The Q_{cd}, Q_{ncd}, and Q_o values from 1st iteration were used as templates and compared with those from following iterations by one-way ANOVA test. No differences among iterations were observed ($P > 0.23$).

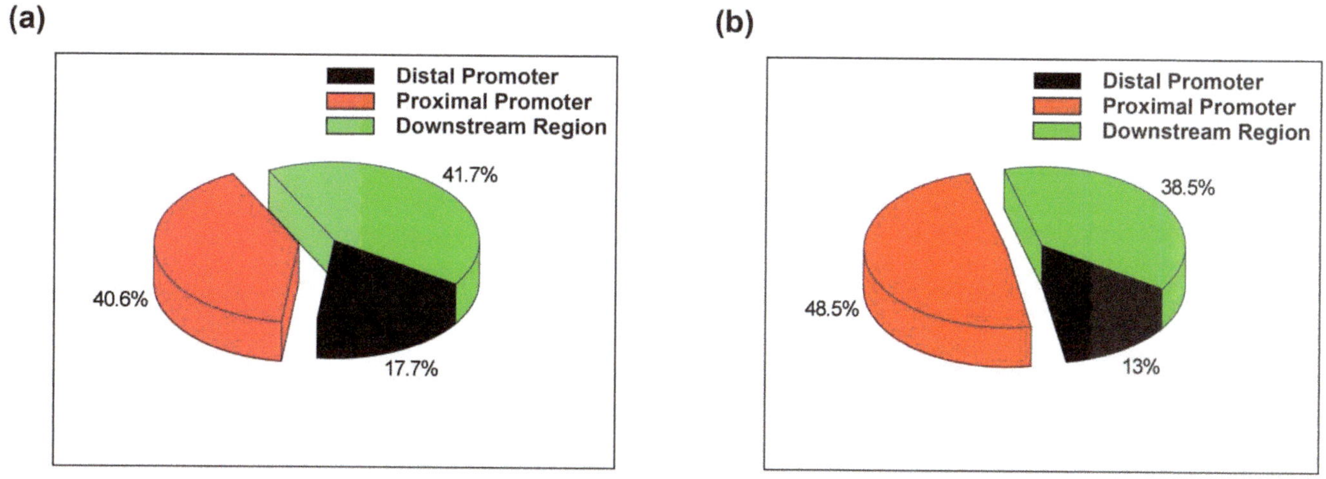

Figure 3. The normalised percentage of G4s in different regions of TRRs of G4-less-important and G4-importance transcripts. (**a**) By normalising to the total length of different regions, 17.7%, 40.6%, and 41.7% G4s were found in the Region I (distal promoter, −2,000~500 bp, black), II (proximal promoter, −500~−1 bp, red) and III (downstream region, TSS~1,000 bp, green) in TRRs of G4-less-important transcripts. (**b**) By normalising to the total length of different regions, 13.0%, 48.5%, and 38.5% G4s were found in the Region I (distal promoter, black), II (proximal promoter, red) and III (downstream region, green) in TRRs of G4-important transcripts. The proportion of G4s found in Region II, the proximal promoter region increases by 20% from G4-less-important to G4-important TRRs.

found in the proximal promoter region (−500 bp to −1) was increased by 20% from less G4-important TRRs to G4-important TRRs (Figure 3). This indicates that the larger *CQ* value in G4-important TRR reflects the combined effects of increased total number of G4s and particularly the increased number of G4 in key promoter regions (with the high W_k values). Most well-studied genes containing promoter G4s with potential regulatory functions are in the group of G4-important transcripts in the correlation map, including *MYC*, *VEGF-A*, *BCL-2*, *HIF1-α*, *MYB*, *PDGF-A*, *PDGFR-β*, *KRAS*, and *TERT* (Table S2). One interesting exception is *KIT*, with CF_o and CQ_o of 43.5% and 73.4%. This may suggest that G4 location might be more relevant than the abundance of G4 in evaluating potential biological significance of G4s in a TRR.

G4-important Pathways

All pathways composed of more than 50 genes available in KEGG PATHWAY database were extracted, and CF_o and CQ_o values of each gene in a particular pathway were calculated. For genes with multiple transcripts, one transcript was picked randomly, and genes with both values (CF_o and CQ_o) equaling to zero were excluded. Then, in each pathway, percentages of genes with TRRs having both CF_o and CQ_o values higher than 50% were counted. Pathways with more than 50% genes' transcripts identified as G4-important were regarded as G4-important pathways as listed in Table 4. Ten iterations were carried out to evaluate effects from transcripts redundancy. For all pathways, no significant difference was found in CF_o and CQ_o values among iterations of each individual pathway (one-way ANOVA test, $P \geq 0.05$). Based on this criterion, G4-important pathways could be grouped into six categories with distinct functions. As shown in Table 4, the first category is cancer-related pathways where G4s are enriched in proto-oncogenes, in accordance with previously reported results [37]. Given the fact that G4s are enriched in cancer-related pathways, it is not difficult to understand why the off-target effects of certain G4-binding ligands with anti-cancer potential could well bring some benefits in anti-cancer therapy, such as the G4-binding fluoroquinolone derivative Quarfloxin [44], and the naphthalene diimide derivative 1 [40]. The second category is signaling pathways such as those involving VEGF, Wnt, Hedgehog, and MAPK (mitogen-activated protein kinase), which have key component containing promoter G4s with possible regulatory functions. Pathways related to cell junction, neuron function, and bacterial infection are also found to be G-important. The last category, heart function-related pathways, is found to be highly G4-important. There are seven pathways in the KEGG PATHWAY database involved in cardiac functions or heart diseases, while six of them are identified as highly G4-important. These pathways are not only involved in normal heart function such as calcium signaling and muscle contraction, but also in various cardiomyopathies, making targeting promoter G4s at pathway level rather than in single gene an attractive strategy (the multi-targeting strategy) in the treatment of certain cardiovascular diseases.

Based on the high importance of G4s in cardiac function related pathways, we assume that G4s might function as a general regulatory element in cardiac related genes. To validate our bioinformatics results and our assumption, we have chosen human cardiac troponin I (*TnIc*) as a model system to investigate possible biological functions of cardiac promoter G4s. As one of the central components in cardiac muscle contraction pathway, TnIc is also involved in most heart function related pathways. Furthermore, TRR of *cTnI* is highly G4 important according to our bioinformatics analysis (CF_o 92.6%, and CQ_o 95.2%). In the promoter of human *TnIc*, two types of G4-forming motifs were identified, which are designated as "MNSG4" (between −528 and −319 bp upstream to the TSS), and "−80 G4" (centred at −80 bp to the TSS) (Figure 1 a). Both MNSG4 and −80 G4 are characterised in solution and in HEK293 cells and the results are shown below.

Evidence of G4s Formation in Solution

EMSA experiments were first used to probe the formation of G4s in the unlabeled MNS and −80 G4-forming sequences from the promoter of human *TnIc*. The fast migrating band of MNSG4-forming sequences (oligos $\mathbf{Tr_{MNS}}$**-I**) in native gels indicated that intramolecular G4(s) is the only conformation at 5 μM oligo concentration. However, at similar concentration, slow migrating bands of −80 G4-forming sequence (oligo $\mathbf{Tr_{-80}}$**-II**) were observed

Table 4. KEGG pathways with high G4-importance.

KEGG Pathway ID	Pathway Description	% of High CF_o and CQ_o	No. of Gene	% of Gene with G4 in TRR	P_{max} of CQ_o*	P_{max} of CF_o*
Cancer-related pathways						
hsa05210	Colorectal cancer	57.9±4.5	62	86.3±2.2	6E-4	0.001
hsa05211	Renal cell carcinoma	57.6±3.3	71	83.9±2.8	7E-4	0.003
hsa05213	Endometrial cancer	56.9±3.7	52	84.8±3.1	0.019	0.037
hsa05217	Basal cell carcinoma	53.7±2.9	55	93.8±1.3	4E-5	6E-6
hsa05200	Pathways in cancer	53.5±1.3	327	82.5±1.4	3E-11	3E-11
hsa05223	Non-small cell lung cancer	53.5±2.6	54	86.9±3.1	0.003	0.02
Signaling pathways with promoter G4s in key components						
hsa04340	Hedgehog signaling pathway	60.9±3.4	56	87.3±1.8	2E-4	1E-4
hsa04370	VEGF signaling pathway	60.7±3.0	80	87.9±2.1	8E-6	1E-4
hsa04010	MAPK signaling pathway	59.5±1.5	271	83.5±1.3	7E-16	1E-16
hsa04310	Wnt signaling pathway	58.9±2.5	150	89.9±0.9	2E-12	4E-12
Cell junction-related pathways						
hsa04520	Adherens junction	67.0±4.2	72	81.3±1.6	2E-5	6E-5
hsa04510	Focal adhesion	56.9±1.3	199	80.4±1.3	4E-8	8E-8
hsa04810	Regulation of actin cytoskeleton	55.7±2.4	213	79.1±2.3	3E-6	2E-5
hsa04530	Tight junction	53.9±2.1	129	77.0±1.8	8E-4	5E-4
hsa04540	Gap junction	53.8±2.0	88	86.1±1.9	8E-5	6E-5
Neural cell regulation-related pathways						
hsa04724	Glutamatergic synapse	61.8±1.6	129	87.1±1.6	5E-9	7E-9
hsa04360	Axon guidance	61.0±2.7	128	86.7±2.2	4E-9	4E-9
hsa04720	Long-term potentiation	60.2±2.0	70	90.6±1.2	1E-6	8E-6
hsa04722	Neurotrophin signaling pathway	55.6±2.1	125	88.9±1.9	4E-8	1E-7
hsa04912	GnRH signaling pathway	55.5±1.3	104	85.9±1.7	1E-6	1E-6
Bacterial infection-related pathways						
hsa05130	Pathogenic E.coli infection	58.6±4.9	55	82.7±3.8	0.004	0.008
hsa05131	Shigellosis	55.8±3.4	62	86.3±2.3	8E-5	2E-4
hsa05100	Bacterial invasion of epithelial cells	54.5±2.8	71	82.4±3.7	0.006	0.01
Heart function-related pathways						
hsa05412	Arrhythmogenic right ventricular cardiomyopathy	60.0±3.5	73	80.1±3.5	0.004	0.01
hsa05410	Hypertrophic cardiomyopathy	57.5±3.1	83	77.1±2.7	0.01	0.04
hsa05414	Dilated cardiomyopathy	57.2±3.1	90	78.2±2.8	0.009	0.02
hsa04020	Calcium signaling pathway	54.4±2.1	178	83.1±1.5	5E-6	5E-6
hsa04260	Cardiac muscle contraction	54.2±3.0	72	73.2±2.9	0.1	0.4
hsa04270	Vascular smooth muscle contraction	53.8±1.0	121	85.2±1.8	1E-7	9E-7

*P_{max} values are calculated by comparing the CQ_o and CF_o scores of genes from individual pathways with those from all transcripts without redundancy in each iteration by Wilcoxon rank sum test. $P \leq 0.05$ means the G4 abundance (CF_o) or location significance (CQ_o) of gene involved in the pathway is significantly different from those at genome level. For all pathways, no significant difference was found in CF_o and CQ_o values among iterations of each individual pathway (one-way ANOVA test, $P \geq 0.05$).

in a native gel, indicating it can adopt a mixture of intramolecular (the major form,~60% according to band intensity) and intermolecular G4s. EMSA results are shown in Figure S4.

Comparative CD spectra analysis was also used to provide primary evidence of the formation and general conformations of both MNSG4 and −80 G4 in solution. Briefly, a single repeat of MNSG4 (oligo **Tr$_{\mathbf{MNS}}$-I**) may adopt parallel structure(s) in LiCl solution. In KCl solution, it is very likely that the majority of single repeat of MNSG4 adopts a similar folding pattern as the TBA (thrombin binding aptamer) and forms an anti-parallel G4 (or G4s) [64,65] (Figure 4 a). For the −80 G4, the −80 G4-forming sequence (oligo **Tr$_{\mathbf{-80}}$-I**) may adopt a hybrid parallel/anti-parallel structure similar to the BCL-2 Pu39WT in LiCl solution [30], while in NaCl and KCl solutions, it adopts typical parallel conformation (Figure 4 b). However, as an empirical method, CD spectra are not conclusive to confirm the existence of G4s in solution [66]. Indeed, CD spectra with a peak at 260 nm and a trough at 240 nm, which is believed to represent the existence of parallel G4s, can also be observed in duplexes, hairpins and single-stranded DNA samples [67]. Thus, to further confirm the existence of both MNSG4 and −80 G4, fast migrating bands

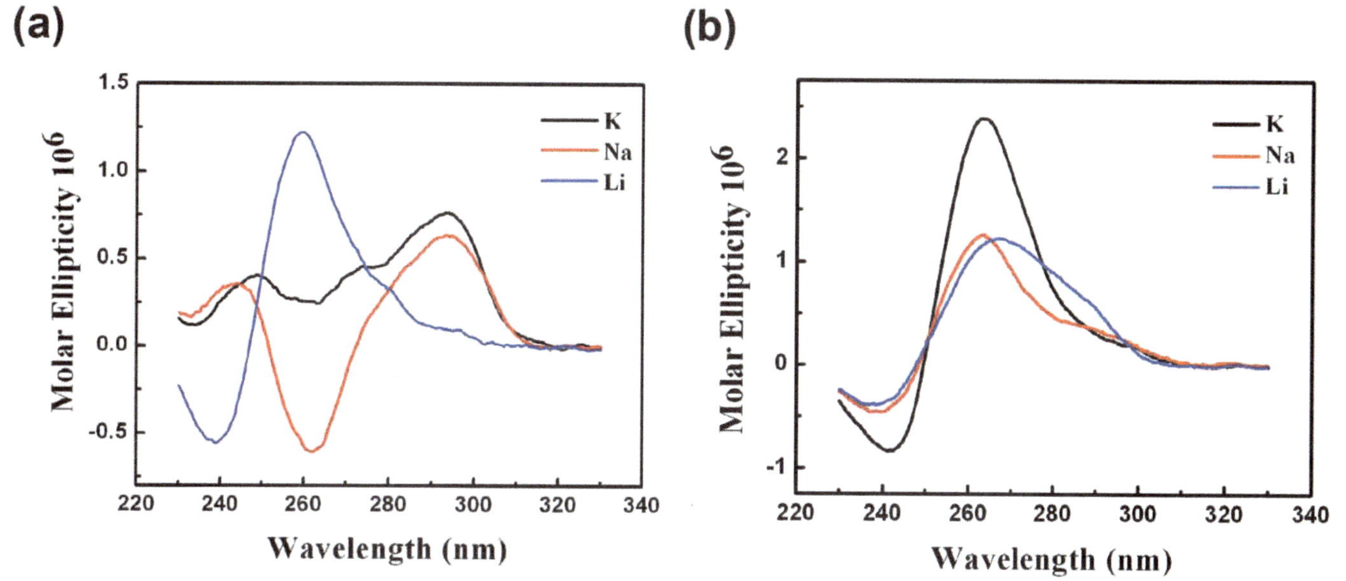

Figure 4. CD spectra of the two promoter G-quadruplex sequences in solutions containing different salts. (**a**) single-repeat MNSG4. (**b**) −80 G4.

from EMSA experiments (in the presence of 100 mM K^+) were extracted and subjected to DMS footprinting.

According to the different protection effects in DMS footprinting, a preferred folding pattern of MNSG4 is proposed as $G_3N_5G_3N_5G_3N_2G_3$, while for the −80 G4 multiple folding patterns may coexist with first loop ranging between 7~10 nucleotides, and second and third loops ranging between 1~2 nucleotides (Figure S5). Both of them are distinct from the general folding pattern of promoter G4s indentified in cancer-related genes. Although the exact G4 structures of MNSG4 and −80 G4 cannot be proposed based on DMS footprinting results alone, those fully protected guanines in both sequences clearly indicate the formation of G4s.

Thermal Denaturation of MNSG4 and −80 G4 Measured by CD Melting

CD melting experiments were used to investigate the thermal stability of these G4s in the presence of 100 mM K^+. For the MNSG4 (oligo **Tr_{MNS}-I**), by fitting the melting curve obtained at 295 nm, the melting temperature and unfolding enthalpy and entropy of the anti-parallel MNSG4 were derived to be 71.4±0.5°C, 170±11 kJ mol^{-1} and 494±31 J mol^{-1} K^{-1} respectively by a model as previously reported (Figure S6 a). For the −80 G4 (oligo **Tr_{-80}-I**), the parallel G4 with lower thermal stability, which accounts for 57.3% of CD signals at 263 nm, was almost fully denatured at 90°C, while no denaturing was observed for the other parallel species with high thermal stability. Since it has been shown by EMSA experiments that about 40% of the −80 G4 is able to adopt an intermolecular G4 conformation at 5 μM concentration, it is likely that the undenatured parallel species corresponds to intermolecular parallel G4s (Figure S4 b and S6 b).

Unfolding Kinetics of MNSG4 and −80 G4 Measured by FRET

Two subpopulations with different unfolding behaviors were revealed by fitting the unfolding curves of MNSG4, (Figure 5 a and Table S3). The fast hybridised subpopulation is believed to be partially folded or randomly coiled oligos with hybridisation rate constant around 0.4×10^4 $M^{-1}s^{-1}$. As compared to a typical hybridisation rate constant, which is around 10^5 $M^{-1}s^{-1}$, the much slower hybridisation rate constants observed here clearly indicated that the C-rich oligos still need to unfold certain unstable structures before fully hybridising the partially unfolding G-rich strand. The slow decaying component represents the unfolding process of MNSG4. As compared to the KIT1 G4 studies that used a similar system [68], the MNSG4 is more kinetically stable at high temperature (time constants of KIT1 G4 at 45 and 50°C are 2,300 and 1,000 s respectively), which echoes the higher thermal stability of MNSG4 in single-stranded form. The activation energy of the unfolding of MNSG4 was determined as 22.1±0.4 kJ mol^{-1} by Arrhenius equation (Figure S7 a and Table S3), and the activation enthalpy and entropy were derived by Eyring equation as 19.6±0.4 kJ mol^{-1}, and −249.2±1.2 J mol^{-1} respectively. This relatively low activation energy/enthalpy and very negative entropy indicates that the unfolding of MNSG4 is entropically driven, and this may suggest that the MNSG4 unfolds through a "pre-organized" transition state, which possibly relates to the long loops in the folded MNSG4 (two 5-nucleotide adjacent loops). In the process of MNSG4 opening, these fluctuating long loops may partially anneal to the C-rich strand, resulting in a series of relatively "ordered" intermediates with negative entropy, and the invasion of C-rich strand may further destabilise the folded G4 structures without transiting G4 into other partially unfolded structures (more disordered structures with positive entropies).

Similar experiments were carried out on the TnIc −80 G4, and two subpopulations (fast and slow unfolding components) were revealed with different kinetic behaviours (Figure 5 b, Table S3). The fast-unfolding −80 G4 (−80 G4-F) exhibits comparable kinetic stability to the MNSG4, while the slow-unfolding −80 G4 (−80 G4-S) is at least 30-fold more stable than the −80 G4-F, and of similar stability as the KIT2 G4 studied previously by using an identical method [68]. Both −80 G4-F and −80 G4-S were characterised by a large activation enthalpy and small unfavoured activation entropy, suggesting the existence of more significant enthalpic barriers (Figure S7 b and Table S3). As compared to the MNSG4, only one long loop was found in the −80 G4 (the first

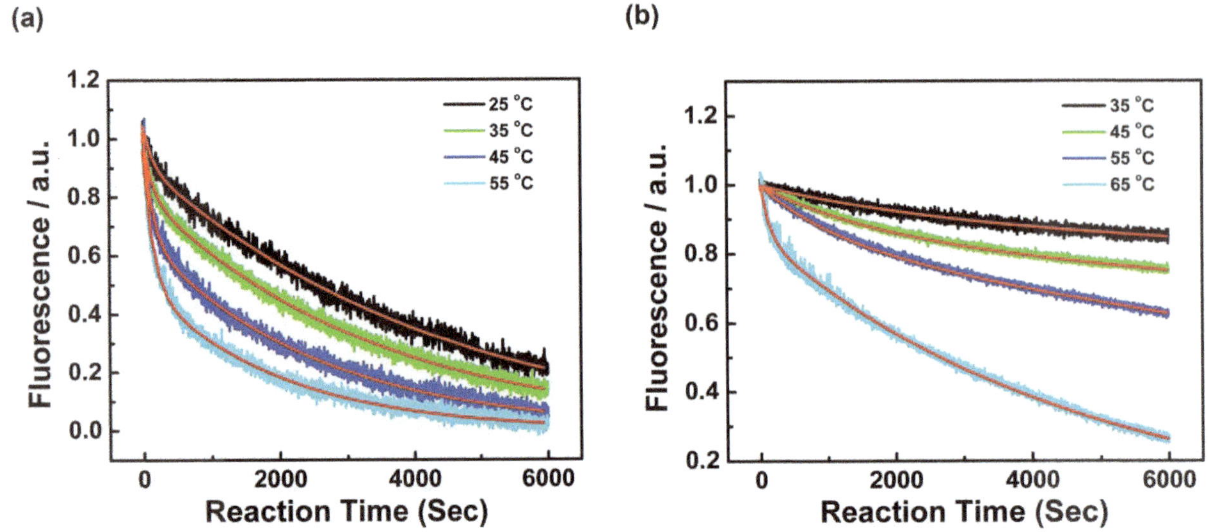

Figure 5. TnIc G4s unfolding curves and corresponding fitted curves. (**a**) Decrease of Cy5 fluorescence in the process of MNSG4 (formed between oligos **Tr$_{MNS}$-III** and **Comp-Cy5**) opened by its C-rich complimentary strand (oligo **Tr$_{MNS}$-II**) at different temperature. Black, green, blue and cyan curves are from experimental data obtained at 25, 35, 45, and 55°C, respectively, and red curves are fitted results by a double exponential decay model. (**b**) Decrease of Cy5 fluorescence in the process of −80 G4 (formed between oligos **Tr$_{-80}$-III** and **Comp-Cy5**) opened by its C-rich complimentary strand (oligo **Tr$_{-80}$-II**) at different temperature. Black, green, blue and cyan curves are from experimental data obtained at 35, 45, 55, and 65°C, respectively, and red curves are fitted results by a double exponential decay model. Hybridisation was performed between 100 nM G4 forming oligos and 2 μM complimentary C-rich oligo. The reaction was carried out in 100 mM KCl solution containing 10 mM Tris-HCl (pH 7.4).

loop, length between 10~7 nucleotides), and correspondingly less negative activation entropies were identified in unfolding processes of the −80 G4. As the −80 G4-F has a more negative $\Delta S^{\ddagger}$ than the −80 G4-S, the destabilising contribution from the loop (partially annealed more ordered states) of −80 G4-F is more prominent than that from −80 G4-S, and thus less energy is required for the −80 G4-F to reach the transition state for the full hybridisation to happen. Indeed, a smaller $\Delta H^{\ddagger}$ of −80 G4-F was also found.

Stabilisation of MNSG4 and −80 G4 by Synthetic G4-binding Ligands

Once it was established that these sequences can indeed form quadruplex structures, it was of interest to determine whether small molecules could facilitate the formation and stabilise the quadruplex structure. Over the past few years some of us have demonstrated that metal-containing compounds (where the metal is tightly bound to an organic ligand) can be excellent quadruplex DNA binder [55,69]. Therefore, we evaluated the ability of three of these metal compounds (compound **1–3**, Figure 6) to stabilise the quadruplex structures in the MNSG4 and −80 G4 sequences. The compounds were selected since they were found to be stronger quadruplex binders via a combination of Π–Π end-stacking interactions (with the terminal guanine tetrad) and electrostatic interactions (with the loops and grooves of quadruplexes) [55].

We first determined their relative binding affinity towards both quadruplexes via the well-established Fluorescent Intercalator Displacement (FID) assay. The results, which are summarized in Table 5, show that the two di-metallic compounds (**2** and **3**) bind strongly to MNSG4 with DC50 values under the 0.5 μM threshold for good binders. The mono-platinum compound **1**, also displays some interaction but its DC50 value is twice as big as those of **2** and **3**. Interestingly, there is a wider range of binding affinities towards the −80 G4 quadruplex with the di-copper compound (**3**) displaying a very low DC50 (i.e. strong binding), reasonably good affinity of **1** and poor affinity of di-platinum compound **2**. Furthermore, we had reported that these compounds are very selective binders to G4 DNA over duplex DNA (up to 100-fold selectivity [55]).

Thus, given the good binding affinity and selectivity of the di-copper compound to G4s, its G4-stabilising ability was further investigated by the CD melting experiment (by measuring the changes of G4 melting temperature, ΔT_m) in the presence of the ligand. In 10 μM compound **3**, the CD spectrum of MNSG4 dramatically changed from parallel to hybrid of parallel/anti-parallel. The melting temperatures measured at 263 nm and 295 nm are 90.4±0.7°C and 86.2±2.1°C, respectively (Figure S8 a and b). It is likely that the di-copper compound preferentially binds to parallel MNSG4 structures (possibly due to the exposed end G-tetrads in parallel conformations), and thus promotes the transition of MNSG4 from anti-parallel to parallel structures. For the −80 G4, the presence of the compound did not change its CD spectrum. Due to the strong stabilising ability of the compound, only a small fraction of the parallel −80 G4 was melted at 93°C, and the melting temperature is estimated to be over 100°C (Figure S8 c and d). The binding preference to parallel G4s observed here might be due to the long loops in both MNSG4 and −80 G4. Long lateral or diagonal loops may partially or fully cover the end G-tetrad, and impede the Π–Π stacking, while the long external loops could leave enough space for the π–π stacking at the end G-tetrad for the first metal moiety, and even provide ideal electrostatic interactions sites for the second metal moiety.

Ligand Screening in Cell Model

Since these ligands were able to bind to the MNSG4 and −80 G4 in solution with relatively high affinity and selectivity, we aim to investigate possible biological functions of these G4-binding ligands and their interactions with these G4 elements in living

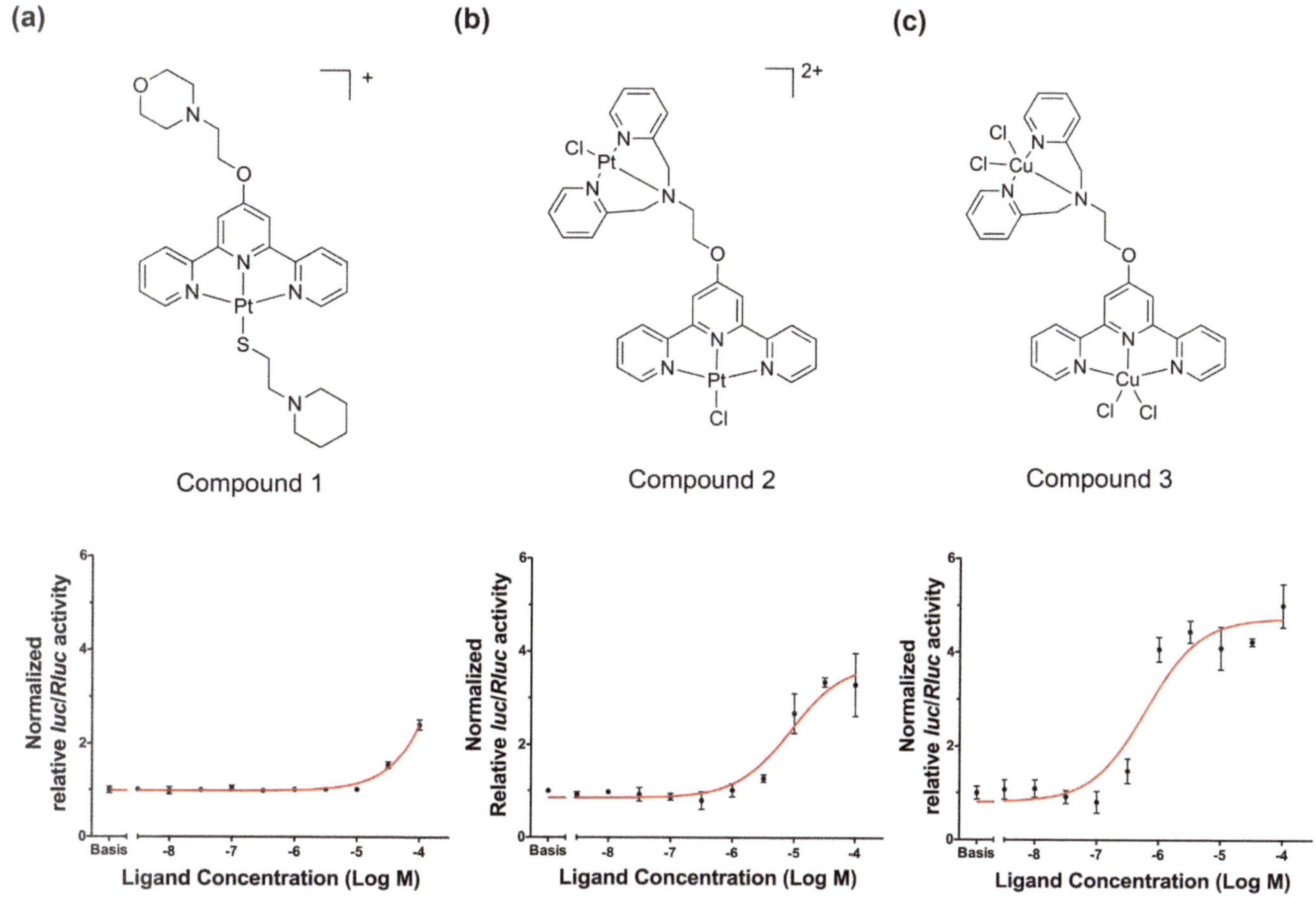

Figure 6. Structures and dose-response curves of compound 1, 2, and 3. (**a**) Compound **1** only slightly activated downstream luciferase expression (EC_{50} = 812.1 μM). (**b**) Compounds **2** moderately activated downstream luciferase expression (EC_{50} = 1.6 μM). (**c**) Compound **3** significantly activated downstream gene expression (EC_{50} = 0.6 μM). Experiments were carried out by measuring the firefly (hTnIc-6MNS(E)-WT) and *Renilla* luciferase ratio changes in the presence of ligand with different concentrations.

cells. Since both MNSG4 and −80 G4 are only conserved in *TnIc* promoters in the primate genomes, HEK293 cell line, rather than rat/mouse cardiomyocytes, was used. Dose-response curves of compound **1**–**3** were generated by measuring the firefly (vector hTnIc-6MNS(E)-WT) and *Renilla* luciferase ratios of lysates of transfected HEK293 cells incubated in the presence of ligands at different concentrations. Vector hTnIc-6MNS(E)-WT was used because it contains the greater number of G4 repeats, which was predicted to give the most significant response if the ligand-stabilisation of G4s is involved in determining the promoter activity. As shown in Figure 6 a, the mono-platinum compound **1**, showed a weak activating effect towards downstream gene transcription with the promoter activity being increased 2-fold at 100 μM concentration (EC_{50} = 812 μM). The di-platinum compound **2** exhibited modest activating effects towards downstream luciferase transcription with the promoter activity being increased by more than 3 times at 100 μM concentration (EC_{50} = 8.6 μM) (Figure 6 b). By replacing platinum with copper, the third compound (compound **3**) demonstrated most significant activation effect, which elevated the promoter activity by approximately 5-fold at 100 μM concentration (EC_{50} = 0.6 μM) (Figure 6 c). These results are in good agreement with the FID results that compound **3** is the best G4-binder among the three. Together, these observations suggest that the di-copper compound **3** has the best performance in living cells.

Table 5. $^{MSN1}DC_{50}$ and $^{-80\ G4}DC_{50}$ values (μM) determined using FID assay for complexes 1–3 (values are average of three independent measurements and error is estimated to be below 5%).

Compound	$^{MSN1}DC_{50}$	$^{-80\ G4}DC_{50}$
1	0.76	0.55
2	0.39	2.22
3	0.31	0.19

G4-binding Ligand is Critical in Promoting Positive Regulatory Functions of *TnIc* Promoter G4s

To investigate the regulatory function of *TnIc* promoter G4s, firefly and *Renilla* luciferase ratios of different vectors were measured following incubation with and without the G4-binding ligand compound **3**. We found that compared to the engineered wild type construct containing six repeats of consensus MNSG4, human *TnIc* constructs containing three, one repeat, or no repeat, exhibited similar activities in driving downstream luciferase transcription ($P \geq 0.05$, Figure 7 a and Figure S9). This indicates

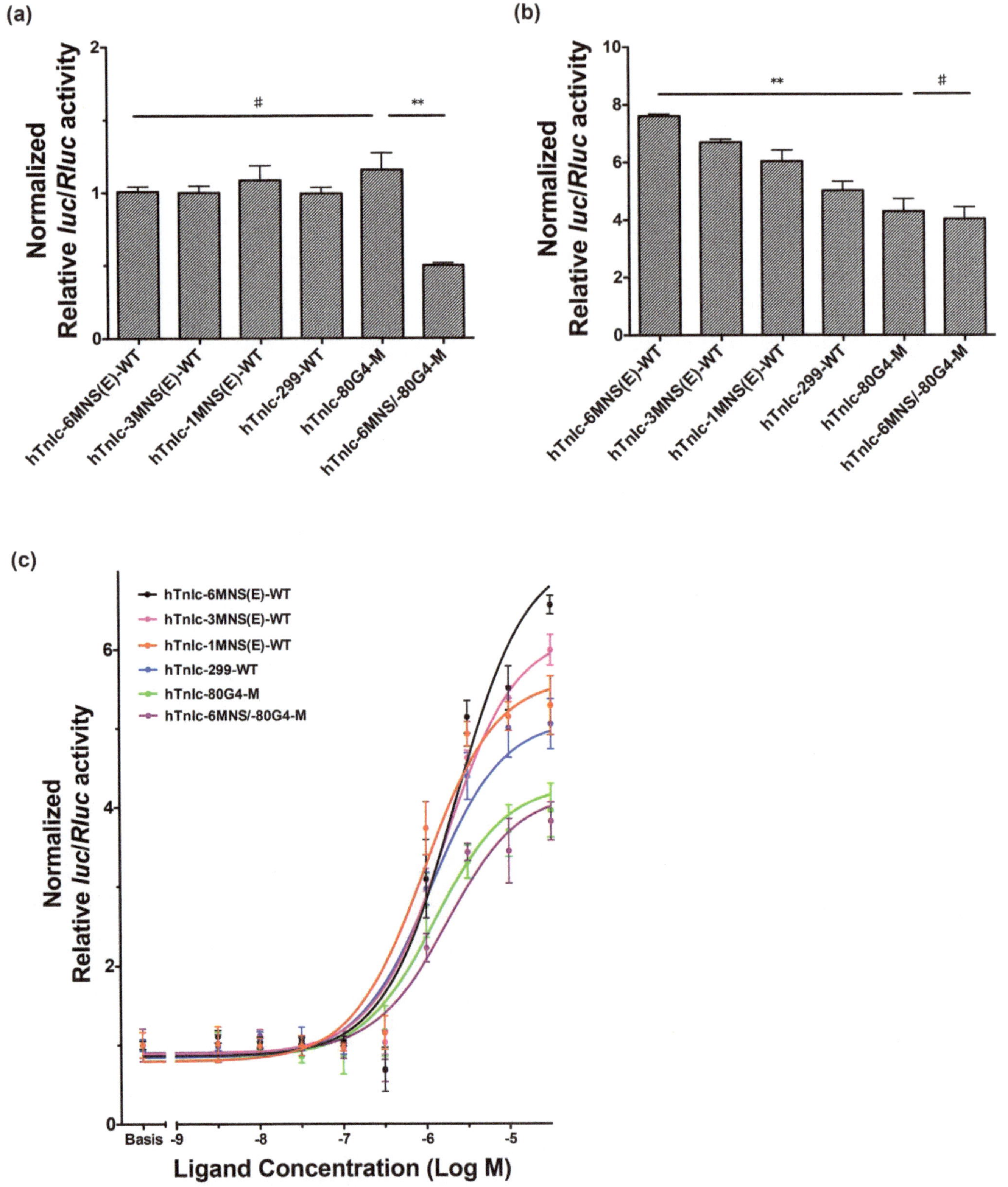

Figure 7. Transcription activity changes of human *TnIc* promoters in the presence of compound 3. (**a**) Transcription activity of constructed human *TnIc* promoters in the absence of compound **3**. (**b**) Transcription activity of constructed human *TnIc* promoters in the presence of 100 μM compound **3**. Columns in a and b are normalised to the luciferase transcription level of the wild type control 6MNS(E)-WT in the absence of ligand. (**c**) Dose-response curves of compound **3** with different luciferase vectors. (**$P<0.01$, *$P<0.05$ and ‡$P\geq0.05$, no significant differences compared to WT construct).

that in the absence of G4-binding ligands, neither the MNSG4 nor Sp1 binding sites in the MNSG4 element appear to be involved in regulating downstream gene transcription, at least in HEK293 cells.

Regulatory roles of the −80 G4 together with a conserved Sp1 binding site in the promoter of human *TnIc* were also investigated by dual luciferase reporter assays in the HEK293 cell line in the absence of G4-binding ligand. As shown in Figure 7 a and Figure S10, the human *TnIc* construct containing −80 G4 with disrupted G4-forming sequence but intact Sp1 binding site (hTnIc-80 G4-M) has similar transcription activity to the engineered wild type promoter (hTnIc-6MNS(E)-WT). However, when the Sp1 binding site is mutated (hTnIc-80Sp1-M), transcriptional activity decreases by~25%. Furthermore, when both the −80 G4-forming sequence and Sp1 binding site are mutated (hTnIc-80 G4/Sp1-M, Figure S10), transcriptional activity further decreases to 50% compared to the engineered wild type promoter. This may relate to the further mutated Sp1 binding site in vector hTnIc-80 G4/Sp1-M as compared to that in vectors hTnIc-80Sp1-M, rather than the disruption of the −80 G4 formation. Although the −80 G4 is not directly involved in regulating downstream gene transcription, the Sp1 binding site seems to be critical in determining downstream gene expression. This is similar to the situation where an Sp1 binding site found in the G4-forming sequence (KIT1) in the promoter of human *KIT* gene appears to be critical in determining the maximal activity of the human *KIT* promoter [70].

The −80 G4 mutation (with an intact Sp1 binding site, hTnIc-80 G4-M) which disrupts G4 formation, slightly elevated the transcriptional activity of *TnIc* promoter. The overall G4 mutant, hTnIc-6MNS/−80 G4-M, exhibited lowest transcriptional activity (Figure 7 a). The decreased luciferase expression from this vector may result from a combination of multiple mutation sites in MNSG4 and −80 G4, which have disrupted the critical Sp1 binding site within the −80 G4, and may have introduced certain inhibitory factor binding sites.

We conclude that in the absence of G4-binding ligand, both the MNSG4 and −80 G4 are not involved in regulating downstream gene transcription in HEK293 cells, and the depressed transcription levels mainly relate to the disruption of the critical Sp1 binding site within −80 G4. As *TnIc* is exclusively expressed in cardiomyocytes, we suspect that the formation and stabilisation of these promoter G4s may be associated with certain cardiac-specific proteins or TFs, which are absent in HEK293 cells. Since it has been reported that short proximal promoter of human *TnIc* is sufficient to confer cardiac-specific expression [71], it is possible that the luciferase expression level of these vectors without ligands is just the basal expression level in HEK293 cells. To test this hypothesis, luciferase assays were repeated in the presence of 100 μM compound **3**. As shown in Figure 7 b, when incubated with compound **3** for 24 hours before cell lysis, the expression activities of luciferase from all vectors increase differently. Generally, the transcription activity of human *TnIc* promoter gradually decreases when the number of MNSG4 repeat decreases, while the lowest transcription activities were observed in vectors with mutated −80 G4. It's worth noting that in the presence of ligand, the overall mutant, hTnIc-6MNS/−80 G4-M, was also activated for about 4 times as compared to the wild type construct hTnIc-6MNS(E)-WT in the absence of ligand. This may relate to the interaction between the ligand and a G-rich region in the *TnIc* promoter (between −260 and −150 bp), which may adopt multiple two-tetrad G4s.

To further discriminate activation effects of MNSG4 and −80 G4, dose-response curves of different luciferase vectors treated with compound **3** were generated. As shown in Figure 7 c, the strongest activation effect was observed in the vector with wild type MNSG4 and −80 G4 (hTnIc-6MNS(E)-WT), in which downstream luciferase expression increases 6.5 times in the presence of 3.2 μM compound **3**. When the −80 G4 is intact, the activation effects of downstream luciferase decreases proportionally with the decrease of repeat number of MNSG4 (hTnIc-6MNS(E)-WT>hTnIc-3MNS(E)-WT>hTnIc-1MNS(E)-WT>hTnIc-299-WT). However, when the −80 G4 is mutated (hTnIc-80 G4-M, hTnIc-6MNS/−80 G4-M), lowest activation effects were observed again. These results suggest that the −80 G4, which is located in a region with high location significance (according to our hypothesis in bioinformatics studies), may operate as a gatekeeper in regulating downstream luciferase expression in HEK293 cells; the MNSG4, on the other hand, acts as to a fine tune regulation when the −80 G4 is intact. It is likely that both types of G4s orchestrate the gene expression of cardiac troponin I in the heart.

Discussion

The correlations between G4s and gene functions, and the regulatory roles of promoter G4s in proto-oncogenes have been thoroughly investigated previously, which has led to the application of several G4-binding ligands as potential anticancer drugs (as reviewed in [38,72]). There is now compelling evidence to support the existence of G4 *in vivo*, such as G4-induced guanine-protection revealed by *in vivo* DMS footprinting [23] (reviewed in [73]). However, although it is clear that G4 complexes function in living cells and there has been some good successes in designing specific G4 binding small molecule ligands (reviewed in [38]), discriminating and targeting specific G4s via small synthetic G4-binding ligands remains very challenging. Such efforts are hampered by many factors including: (i) structural polymorphism and dynamic properties of G4s, (ii) similar major recognition sites (G-tetrad) in different G4s, (iii) limited precise structural data of G4s available, (iv) low abundance and instability of G4s in the human genome as compared to duplex DNA, (v) obstacles in synthesis of natural G4-binding ligands and (vi) poor pharmaceutical properties of some synthetic compounds. Among all these issues, the selectivity of G4-binding ligands is the most notable one that warrants further improvement. Although there are some ligands reported to interact with G4s much more selectively than with duplex DNA, such as Mn^{III} porphyrin (10^4-fold) [74], and dimetallic terpyridine-based ligand (over 100-fold) [55], the selectivity of most ligands are in the range between 10- and 100-fold (reviewed in [38,39]). Furthermore, very few ligands have been reported to exert satisfactory selectivity between different G4s. Thus, we reason that G4-binding ligands with ideal selectivity and affinity to a particular promoter G4 structure would be difficult to achieve at the moment, and off-target effects would be inevitable, limiting the therapeutic potential of these ligands *in vivo*. However, off-target effects or multi-targeting effects of G4-binding ligands may in turn become advantageous and enhance their therapeutic potential in some cases as mentioned in the Introduction section. This implies that targeting multiple similar G4s in a particular pathway might be feasible, and furthermore, this could be an alternative to overcome selectivity problem of G4 ligands. Therefore, G4-pathway correlation is particularly important in applying G4s as potential targets for the treatment of certain diseases.

In this study, we first investigated the abundance of G4s in the TRRs of genes active in different tissues. The tissue-specific and location-specific enrichment of G4s in TRRs of genes were clearly shown, particularly for the genes active in the heart. Enrichment of G4s in different regions around TSS is known to associate with diverse biological processes. For example, G4s in the distal or proximal promoters of proto-oncogenes normally act as repressors in regulating downstream gene transcription (*MYC* promoter G4 [10,11]). G4s in the downstream regions relative to TSS can modulate mRNA translation efficiency differently depending on the strand polarity [35,36]. Thus, this tissue-specific, location-specific, and asymmetric G4 enrichment may also imply that tissues with high G4 abundance in TRRs or certain regions of TRRs have evolved certain tissue-specific mechanisms to tolerate or even to utilize G4s in the processes of transcription regulation.

Previous genome-wide studies have found that promoter G4s have a strong positional bias towards TSS in different species [6,35,59,61,75,76]. This positional bias of promoter G4 is believed to be a result from evolution selection [6,37]. Thus, in addition to G4 abundance (*F* score, and *CF*), the location significance of G4 (*Q* score, and *CQ*) is also considered when evaluating the biological importance of G4s in each TRR. Generally, correlation between *F* and *Q* scores in G4-less-important TRRs is not as strong as that in G4-important TRRs. Furthermore, we notice that when G4 importance increases, the probability of finding G4s in the proximal (−500 bp~−1 bp) promoter regions of TRR increases. Because the majority of promoter G4s with proven regulatory functions locate in the proximal promoter regions, it is highly likely that the accumulation of proximal promoter G4s in pathways with high G4 importance is biologically relevant. In regions downstream the TSS, however, the probability is relatively constant, suggesting that the transcription induced G4 formation might be ubiquitous in different genes.

G4-important pathways show significant connection with their functions (Table 4). Particularly, in accordance to previous findings, most cancer-related pathways are also G4-important, which may suggest that off-target effects of certain anticancer G4-binding ligands can provide extra benefits for cancer treatment. Heart function-related pathways, especially those involved in various cardiomyopathies, are also found to be highly G4-important, indicating the therapeutic potential of G4s in cardiovascular diseases.

Because of its high G4 importance and unique expression pattern, the proximal promoter of human *TnIc* was chosen as a model to investigate possible regulatory functions of G4s in cardiac function-related genes. Two G4 elements, the MNSG4 and −80 G4 were analysed separately. Results from CD, EMSA and DMS footprinting experiments have confirmed formation of these G4 elements in solution. Different folding patterns between these cardiac promoter G4s and proto-oncogene promoter G4s may relate to their distinct biological functions and therefore warrant further investigation by NMR spectroscopy.

The stability of MNSG4 and −80 G4 were both evaluated by their unfolding processes either thermodynamically and kinetically. Both have similar thermal stability (Table 6) as compared to the human telomeric G4 [77,78], and other G4s with similar loop length, such as MYC, KIT, WNT1 and VEGF G4s [4]. Kinetic stability studies (Table S3) revealed that the unfolding of MNSG4 is entropically driven and that it can unfold relatively easily [79,80]. On the other hand, the unfolding of the −80 G4s is enthalpy-driven. According to limited kinetic stability data of other G4s, the MNSG4 and the fast unfolding −80 G4 are slightly less kinetically stable than the others, while the slow unfolding −80 G4 is much more stable than the others (*Oxytricha* telomeric G4 [81], KIT1 G4 [68], KIT2 G4 [68], and human telomeric G4 [60]). As the biological roles of human telomeric, *MYC*, *VEGF*, and *KIT* G4s have already been established, it is reasonable to assume that both TnIc MNSG4 and −80 G4 are stable enough to possess certain biological roles in the promoter of human *TnIc* in the context of adjacent duplex.

Dose response curves from different constructs demonstrated that the stabilisation of −80 G4 is critical in activating downstream luciferase expression, while the individual MNSG4 acts as a fine-tuner. Since in HEK293 cells, most cardiac specific proteins and transcription factors are absent, this activation effect may be due to the negative twist from the formation of G4s in the promoter region, which can facilitate the unwinding of supercoiled plasmid DNA. The regulatory roles of G4 elements in human *TnIc* promoter were mainly measured by comparing the transcription activities of engineered human *TnIc* promoter in HEK293 cells with and without the G4-binding ligand by dual luciferase reporter assays. This is based on the assumption that the G4-binding ligand, compound **3**, is able to directly interact with these TnIc G4 elements in living cells, and subsequently to induce transcription activity changes. However, similar to most studies published so far, direct evidence indicating the interaction between G4-binding ligands and G4s in living cells is extremely difficult to obtain and could not be obtained in this study. However, several pieces of indirect evidence support the conclusion that the positive regulatory roles of these *TnIc* promoter G4s relate to the stabilisation of these elements by the ligand. The first evidence is the selectivity of the ligand to the TnIc G4 elements as compared to the dsDNA. Previous studies showed that compound **3** exhibits a relatively high selectivity to G4s as compared to dsDNA (~ 100-fold [55]), although its selectivity among different G4s is relatively low. These features of compound **3** eliminate the doubts that this ligand may interact with dsDNA, and change the transcription activity of *TnIc* promoter in HEK293 cells indirectly (such as the inhibitory effects of Quarfloxin towards *MYC* [82,83]). The second evidence is the preference of the ligand between the TnIc MNSG4 and −80 G4. Biophysical studies of G4-ligand interactions indicated that the Cu-Cu compound has better performance when interacting with the −80 G4. Similarly, in cell studies, activation effects from the ligand-stabilised −80 G4 are more prominent. Finally, the prevalent activating effect from −80 G4 in regulating downstream gene expression is in good agreement with our hypothesis in the bioinformatics studies about location significance of promoter G4s (G4s found in the region with high distribution probability are more important). Thus, although additional experiments may be necessary to fully clarify the interaction between compound 3 and TnIc G4 elements in the context of a living cell, it is very likely that the activation effects of these G4 elements are directly induced by interacting with the G4-binding ligand.

In contrast to the widely observed G4-mediated phenomena in promoters of oncogenes that formation and stabilisation of these promoter G4s could impede downstream gene transcription, we found that the formation and stabilisation of the MNSG4 and −80 G4 could activate downstream gene transcription. This is similar to the G4s found in the ILPR [8] and skeletal muscle gene promoters [84]. The formation and functions of G4s around gene promoters are dependent on local environment (transcription-induced supercoiling, existence of transcription bubbles, etc.), and the availability of G4-interacting proteins in a particular cell type. Thus, we believe that the actual function of individual promoter G4 needs to be studied case by case. Here we suggest two possible mechanisms for this G4-mediated transactivation.

The first mechanism is related to the superhelicity of the testing plasmid. It has been known that the formation of G4 structures within the DNase I or S1 nuclease hypersensitive pPu/pPy tracts in gene promoters are often associated with negative supercoiling stress, which facilitates the local unwinding or melting of duplex DNA [21,85,86], such as in cases of ILPR [87–89], VEGF [23,24] and MYC [90]. Furthermore, the RNA polymerase (RNApol) can

Table 6. Energetics of TnIc MNSG4 and −80 G4 unfolding in solutions determined by CD melting.

Oligo Names	Wavelength/nm	T_m/°C	ΔH/kJ mol^{-1}	ΔS/J mol^{-1} K^{-1}	$\Delta G_{(310\ K)}$/kJ mol^{-1}
Tr_{MNS}-I (TnIc MNSG4)	263	N/A	N/A	N/A	N/A
	295	71.4±0.5	170±11	494±31	17.2±1.3
Tr_{-80}-I (TnIc −80 G4)	263	75.2±1.0	187±22	538±62	20.5±1.7
	295	N/A	N/A	N/A	N/A

Experiments were carried out in solutions containing 10 mM Tris-HCl (pH 7.4) and 100 mM K^+.

generate negative supercoiling during transcription [91]. It is worth noting that this transcription-induced supercoiling force was found to enable DNA transitions to conformations other than B-DNA in an *in vivo* study of the MYC promoter [92]. Therefore, it is possible that in the case of human *TnIc* promoter, the stabilisation of multiple G4s by the ligand facilitates the local unwinding of the negatively supercoiled plasmid, and resolves the transcription-induced supercoiling forces, hence increasing transcription rate. Since the −80 G4-forming motif locates very close to the TSS, the formation of this G4 may exert stronger forces to release the transcription-induced supercoiling stresses, as compared to the MNSG4. In addition, the formation of the −80 G4 may result in the exposure of multiple TF binding sites around it and facilitate the initiation of the transcription. This may explain the "gatekeeper" role of the −80 G4 and "fine-tuner" role of the MNSG4. We are aware that *in vitro* and *in vivo* DMS footprinting [23,24] to probe the formation of these G4 elements would be necessary to fully characterise this mechanism.

The G4/E-box-mediated mechanism is another possibility. Recently, Shklover et al. proposed that the promoter G4s formed in muscle-specific genes may function as binding sites for homodimeric MyoD and increase its concentration in the micro-environment around the promoters [84]. This accumulated homodimeric MyoD then associate with constitutive E-proteins to form heterodimers, which bind to adjacent E-box (E47) and activate downstream gene expression [84]. Interestingly, in the promoter of human *TnIc*, potential E-boxes are also identified around both MNSG4 and −80 G4 [49]. In HEK293 cells, these co-localized G4-forming motifs and adjacent E-boxes may provide binding sites for constitutively expressed bHLH factors and E-proteins, and hence transactivate the downstream luciferase gene. This mechanism may also exist in the human cardiomyocyte, since several cardiac-restricted bHLH proteins have been found to be important in heart development, such as HAND1 and HAND2 [93–95].

In summary, this study shows that potential G4-forming motifs are highly enriched in genes active in the heart as well as those involved in cardiac function-related pathways. As shown in the model of *TnIc* promoter, it is very likely that multiple G4s act cooperatively in regulating downstream gene transcription. Thus, multi-targeting strategy via promoter G4s could be utilised in manipulating certain gene expression or even modulating pathway performance. On the other hand, it is necessary that all the G4 forming motifs in a given promoter are carefully investigated to understand their potential regulatory functions. It should be noted that HEK293 cell line used in this study is a limited cell model, but we have shown that it is still valuable to understand the ligand-induced promoter activity changes of human *TnIc*. Cardiomyocytes could be used in future studies to fully resolve the G4-mediated regulatory mechanism.

Supporting Information

Figure S1 Distribution of G4 abundance scores (the *F* scores) and corresponding cumulative frequencies (the *CF* scores) on coding strand (a), template strand (b), and both strands (c). In the analysis, TRRs from 46,205 transcripts exported from ENSEMBL with redundancy were included. TRRs of transcripts without G4-forming motifs were excluded in calculating the distribution and corresponding cumulative frequency.

Figure S2 Probability distribution and location significance scores (the *Q* scores) distribution of G4s in TRRs of transcripts. (**a**) Frequency of PQS (potential quadruplex sequence) at each position (W_k as described in Equation 3) on coding strand (black), template strand (red), and both strands (green) in TRRs of all transcripts. TRRs from 46,205 transcripts exported from ENSEMBL with redundancy were included. (**b~d**) Distribution of G4 location significance (lower graphs) and corresponding cumulative frequency (upper graphs) of all transcripts available in ENSEMBL database on the coding strands, template strand, and both strands, respectively. TRRs of transcripts without G4-forming motifs were excluded in calculating the distribution and corresponding cumulative frequency.

Figure S3 Correlation between the location significance (the *CQ* score) and abundance (the *CF* socre) of G4s identified in TRRs of all transcripts with redundancy. (a~c) Correlation between G4 location significance and G4 abundance on the coding strand (CQ_{cd} vs. CF_{cd}), template strand (CQ_{ncd} vs. CF_{ncd}), and both strands (CQ_o vs. CF_o), respectively. In each map, transcripts with the TRRs *CQ* and *CF* scores larger than 50% are identified as G4-important transcripts, and those with both values smaller than 50% are G4-less-important (or less G4-important) transcripts. Correlation coefficients between *CF* and *CQ* on coding strand (CF_{cd} vs. CQ_{cd}), template strand (CF_{ncd} vs. CQ_{ncd}), and both strands (CF_o vs. CQ_o) are 0.57, 0.50, and 0.61, respectively. However, corresponding correlation coefficients for G4-important TRRs are 0.61 (coding strand), 0.58 (template strand), and 0.66 (both strands), respectively, while those for G4-less important TRRs are 0.10, 0.09, and 0.34 respectively. Apparently, in G4-important TRRs, their G4 abundance is more positively correlated to their G4 location significance.

Figure S4 Native and denaturing gel electrophoresis indicating the formation of G4s. (**a**) TnIc MNSG4, 1 repeat: Lane1 and 2 are oligos **Tr_{MNS}-II** (C-rich) and **Tr_{MNS}-I** (G-rich) run in a native gel. Lane 3 and 4 are oligos run in a denaturing gel as controls. (**b**) TnIc −80 G4: Lane1 and 2 are oligos **Tr_{-80}-II** (C-rich) and **Tr_{-80}-I** (G-rich) run in a native gel. Lane 3 and 4 are oligos run in a denaturing gel as controls. Formation of intramolecular G4s is proved by species with fast mobility under native conditions (indicated by black arrows) but not under denaturing conditions. Intermolecular G4s formed by the −80 G4 forming sequence are indicated by gray arrows in (**b**).

Figure S5 DMS footprinting results of TnIc MNSG4 and −80 G4. (**a**) DMS footprinting of MNSG4 (oligo **Tr_{MNS}-I**). Strong DMS protection is observed at G1~G3, G5~G7, G11~G13, and G14~G15. Two guanines at the 3′ of MNSG4 (G17~18, oligo **Tr_{MNS}-I**) are not involved in G4 formation. Compared to other guanines involved in loops (G4, G8, and G9), G10 is partially protected from DMS. Because G13 is fully protected from DMS, this partial protection of G10 may be due to the gyration hindrance from the adjacent G4 structure, which blocks its access by DMS. (**b**) DMS footprinting of −80 G4 (oligo **Tr_{MNS}-I**). In the first 5′ run of guanines (G1~G5), partial protection of guanines is apparent. Meanwhile, in the second and third runs of guanines (G9~G12 and G13~G16), partially protected flanking guanines in each G-run are also found (G9/G12 and G12/G16) with two fully protected guanines in the middle (G10~G11 and G14~G15). This poorly defined footprinting pattern of −80 G4 element is possibly due to the long first loop (at least 7 nucleotides) which induces flexibility of the G4 conformations. Together with CD experiments, it's highly likely that this sequence can fold into a mixture of exchanging parallel intramolecular G4s with different loop arrangements in KCl solution. Furthermore, in the second and third G-runs, it is clear that G12 and G13 are better protected from DMS than G9 and G16, which may reflect the preference of different intramolecular G4s.

Figure S6 CD melting of the two promoter G-quadruplex sequences. (a) single-repeat MNSG4. **(b)** −80 G4.

Figure S7 Arrhenius plot for unfolding of TnIc G4s. (**a**) Arrhenius plot for the slow decaying component identified in the unfolding process of TnIc MNSG4. Empty circles represent opening rates measured at different temperatures, and red line is the result fitted by Arrhenius equation. The activation energy is determined to be 22.1±0.4 kJ mol^{-1}. (**b**) Arrhenius plot for the slow and fast decaying components identified in the unfolding process of TnIc −80 G4. Empty triangles and squares represent opening rates of the fast and slow decaying components, respectively, and red lines are results fitted by Arrhenius equation. The activation energies for the fast and slow decaying components are 86.1±23.7 and 100.6±6.6 kJ mol^{-1}, respectively, as determined by the fits.

Figure S8 CD melting of TnIc MNSG4 and −80 G4 in the presence of the G4-binding ligand (complex 3). (**a**) CD melting spectra of TnIc MNSG4 (oligo **Tr_{MNS}-I**, 5 μM) in the presence of 10 μM complex **3**. (**b**) CD signal changes of TnIc MNSG4 (oligo **Tr_{MNS}-I**) at 295 nm (black dots) and 263 (red dots) nm wavelengths in response to temperature changes, both of which were fitted by a sigmoidal model (indicated by red line and black line respectively). The melting temperature measured at 265 nm and 295 nm are 90.4±0.7°C and 86.2±2.1°C respectively (**c**) CD melting spectra of TnIc −80 G4 (oligo **Tr_{-80}-I**, 5 μM) in the presence of 10 μM complex **3**. (**d**) CD signal changes of TnIc −80 G4 (oligo **Tr_{-80}-I**) at 295 nm (black dots) and 263 (red dots) nm wavelengths in response to temperature changes, both of which were fitted by a sigmoidal model (indicated by red line and black line respectively). All experiments were carried out in 10 mM Tris-HCl buffer (pH 7.4) containing 100 mM K^+.

Figure S9 Comparison of the transcription activities of human *TnIc* promoters containing variable number of MNSG4. Transcription activities of human *TnIc* promoters containing 6 repeats of consensus MNSG4 (hTnIc-6MNS(E)-WT), 3 repeats of consensus MNSG4 (hTnIc-3MNS(E)-WT), 1 repeat of consensus MNSG4 (hTnIc-1MNS(E)-WT), and no repeat of MNSG4 (hTnIc-299-WT). No transcription activity difference was found among these human *TnIc* promoters ($P \geq 0.05$, no significant differences).

Figure S10 Transcription activities of human *TnIc* promoters with wild type and mutated −80 G4. Mutation of −80 G4 alone (hTnIc-80 G4M) did not change the transcription activity, while the mutation of Sp1 binding site (hTnIc-80sp1M-1/2) led to~25% depression in the transcription activity. When both G4 formation and Sp1 binding sites were mutated (hTnIc-80 G4/sp1M), 50% decrease of the transcription activity was found (* $P<0.05$ and ♯ $P \geq 0.05$ no significant differences).

Table S1 The rankings and scores of G4 enrichment in TRRs of genes active in different tissues. G4s in the full-length TRR (Whole TRR), distal promoter region (−2,000~−501 bp), proximal promoter region (−500~−1 bp), and downstream region (TSS~+1,000 bp) were searched and the procedure was repeated randomly for five times.

Table S2 CF_o and CQ_o values of TRRs of genes containing well-studied promoter G4s.

Table S3 Parameters obtained from curve fittings of TnIc MNSG4 and −80 G4 unfolding in solution.

Protocols S1 Supporting protocols including G4-binding ligand synthesis, Electrophoretic mobility shift assay, CD characterization, DMS footprinting, unfolding kinetics and the human genome sequence data bases used in the bioinformatic analysis.

Acknowledgments

The authors thank Julian Huppert and Thomas Branch for critical reading of the manuscript.

Author Contributions

Conceived and designed the experiments: WZ NJB LY. Performed the experiments: WZ KS. Analyzed the data: WZ KS NJB PJB RV LY. Contributed reagents/materials/analysis tools: WZ KS NJB PJB RV LY. Wrote the paper: WZ NJB RV LY.

References

1. Watson JD, Crick FH (1953) Molecular structure of nucleic acids; a structure for deoxyribose nucleic acid. Nature 171: 737–738.
2. Phan AT, Mergny JL (2002) Human telomeric DNA: G-quadruplex, i-motif and Watson-Crick double helix. Nucleic Acids Res 30: 4618–4625.
3. Qin Y, Hurley LH (2008) Structures, folding patterns, and functions of intramolecular DNA G-quadruplexes found in eukaryotic promoter regions. Biochimie 90: 1149–1171.
4. Kumar N, Maiti S (2008) A thermodynamic overview of naturally occurring intramolecular DNA quadruplexes. Nucleic Acids Res 36: 5610–5622.
5. Lane AN, Chaires JB, Gray RD, Trent JO (2008) Stability and kinetics of G-quadruplex structures. Nucleic Acids Res 36: 5482–5515.
6. Huppert JL, Balasubramanian S (2007) G-quadruplexes in promoters throughout the human genome. Nucleic Acids Res 35: 406–413.
7. Henderson E, Hardin CC, Walk SK, Tinoco I, Jr, Blackburn EH (1987) Telomeric DNA oligonucleotides form novel intramolecular structures containing guanine-guanine base pairs. Cell 51: 899–908.
8. Kennedy GC, German MS, Rutter WJ (1995) The minisatellite in the diabetes susceptibility locus IDDM2 regulates insulin transcription. Nat Genet 9: 293–298.
9. Lew A, Rutter WJ, Kennedy GC (2000) Unusual DNA structure of the diabetes susceptibility locus IDDM2 and its effect on transcription by the insulin promoter factor Pur-1/MAZ. Proc Natl Acad Sci U S A 97: 12508–12512.
10. Siddiqui-Jain A, Grand CL, Bearss DJ, Hurley LH (2002) Direct evidence for a G-quadruplex in a promoter region and its targeting with a small molecule to repress c-MYC transcription. Proc Natl Acad Sci U S A 99: 11593–11598.
11. Grand CL, Han H, Munoz RM, Weitman S, Von Hoff DD, et al. (2002) The cationic porphyrin TMPyP4 down-regulates c-MYC and human telomerase reverse transcriptase expression and inhibits tumor growth in vivo. Mol Cancer Ther 1: 565–573.
12. Cogoi S, Xodo LE (2006) G-quadruplex formation within the promoter of the KRAS proto-oncogene and its effect on transcription. Nucleic Acids Res 34: 2536–2549.
13. Cogoi S, Paramasivam M, Spolaore B, Xodo LE (2008) Structural polymorphism within a regulatory element of the human KRAS promoter: formation of G4-DNA recognized by nuclear proteins. Nucleic Acids Res 36: 3765–3780.
14. Qin Y, Rezler EM, Gokhale V, Sun D, Hurley LH (2007) Characterization of the G-quadruplexes in the duplex nuclease hypersensitive element of the PDGF-A promoter and modulation of PDGF-A promoter activity by TMPyP4. Nucleic Acids Res 35: 7698–7713.
15. Phan AT, Kuryavyi V, Burge S, Neidle S, Patel DJ (2007) Structure of an unprecedented G-quadruplex scaffold in the human c-kit promoter. J Am Chem Soc 129: 4386–4392.
16. Fernando H, Reszka AP, Huppert J, Ladame S, Rankin S, et al. (2006) A conserved quadruplex motif located in a transcription activation site of the human c-kit oncogene. Biochemistry 45: 7854–7860.
17. Todd AK, Haider SM, Parkinson GN, Neidle S (2007) Sequence occurrence and structural uniqueness of a G-quadruplex in the human c-kit promoter. Nucleic Acids Res 35: 5799–5808.
18. Hsu ST, Varnai P, Bugaut A, Reszka AP, Neidle S, et al. (2009) A G-rich sequence within the c-kit oncogene promoter forms a parallel G-quadruplex having asymmetric G-tetrad dynamics. J Am Chem Soc 131: 13399–13409.
19. Kuryavyi V, Phan AT, Patel DJ (2010) Solution structures of all parallel-stranded monomeric and dimeric G-quadruplex scaffolds of the human c-kit2 promoter. Nucleic Acids Res 38: 6757–6773.
20. Guo K, Pourpak A, Beetz-Rogers K, Gokhale V, Sun D, et al. (2007) Formation of pseudosymmetrical G-quadruplex and i-motif structures in the proximal promoter region of the RET oncogene. J Am Chem Soc 129: 10220–10228.
21. Sun D, Guo K, Rusche JJ, Hurley LH (2005) Facilitation of a structural transition in the polypurine/polypyrimidine tract within the proximal promoter region of the human VEGF gene by the presence of potassium and G-quadruplex-interactive agents. Nucleic Acids Res 33: 6070–6080.
22. Sun D, Liu WJ, Guo K, Rusche JJ, Ebbinghaus S, et al. (2008) The proximal promoter region of the human vascular endothelial growth factor gene has a G-quadruplex structure that can be targeted by G-quadruplex-interactive agents. Mol Cancer Ther 7: 880–889.
23. Sun D, Guo K, Shin YJ (2011) Evidence of the formation of G-quadruplex structures in the promoter region of the human vascular endothelial growth factor gene. Nucleic Acids Res 39: 1256–1265.
24. Sun D (2010) In vitro footprinting of promoter regions within supercoiled plasmid DNA. Methods Mol Biol 613: 223–233.
25. Murchie AI, Lilley DM (1992) Retinoblastoma susceptibility genes contain 5′ sequences with a high propensity to form guanine-tetrad structures. Nucleic Acids Res 20: 49–53.
26. Xu Y, Sugiyama H (2006) Formation of the G-quadruplex and i-motif structures in retinoblastoma susceptibility genes (Rb). Nucleic Acids Res 34: 949–954.
27. Xu Y, Sugiyama H (2005) Structural and functional characterizations of the G-quartet and i-motif elements in retinoblastoma susceptibility genes (Rb). Nucleic Acids Symp Ser (Oxf): 177–178.
28. Dexheimer TS, Sun D, Hurley LH (2006) Deconvoluting the structural and drug-recognition complexity of the G-quadruplex-forming region upstream of the bcl-2 P1 promoter. J Am Chem Soc 128: 5404–5415.
29. Dai J, Chen D, Jones RA, Hurley LH, Yang D (2006) NMR solution structure of the major G-quadruplex structure formed in the human BCL2 promoter region. Nucleic Acids Res 34: 5133–5144.
30. Dai J, Dexheimer TS, Chen D, Carver M, Ambrus A, et al. (2006) An intramolecular G-quadruplex structure with mixed parallel/antiparallel G-strands formed in the human BCL-2 promoter region in solution. J Am Chem Soc 128: 1096–1098.
31. De Armond R, Wood S, Sun D, Hurley LH, Ebbinghaus SW (2005) Evidence for the presence of a guanine quadruplex forming region within a polypurine tract of the hypoxia inducible factor 1alpha promoter. Biochemistry 44: 16341–16350.
32. Du Z, Zhao Y, Li N (2009) Genome-wide colonization of gene regulatory elements by G4 DNA motifs. Nucleic Acids Res 37: 6784–6798.
33. Todd AK, Neidle S (2008) The relationship of potential G-quadruplex sequences in cis-upstream regions of the human genome to SP1-binding elements. Nucleic Acids Res 36: 2700–2704.
34. Kumar P, Yadav VK, Baral A, Saha D, Chowdhury S (2011) Zinc-finger transcription factors are associated with guanine quadruplex motifs in human, chimpanzee, mouse and rat promoters genome-wide. Nucleic Acids Res 39: 8005–8016.
35. Verma A, Halder K, Halder R, Yadav VK, Rawal P, et al. (2008) Genome-wide computational and expression analyses reveal G-quadruplex DNA motifs as conserved cis-regulatory elements in human and related species. J Med Chem 51: 5641–5649.
36. Du Z, Zhao Y, Li N (2008) Genome-wide analysis reveals regulatory role of G4 DNA in gene transcription. Genome Res 18: 233–241.
37. Eddy J, Maizels N (2006) Gene function correlates with potential for G4 DNA formation in the human genome. Nucleic Acids Res 34: 3887–3896.
38. Monchaud D, Teulade-Fichou MP (2008) A hitchhiker's guide to G-quadruplex ligands. Org Biomol Chem 6: 627–636.
39. Georgiades SN, Abd Karim NH, Suntharalingam K, Vilar R (2010) Interaction of metal complexes with G-quadruplex DNA. Angew Chem Int Ed Engl 49: 4020–4034.
40. Gunaratnam M, Swank S, Haider SM, Galesa K, Reszka AP, et al. (2009) Targeting human gastrointestinal stromal tumor cells with a quadruplex-binding small molecule. J Med Chem 52: 3774–3783.
41. Riou JF, Guittat L, Mailliet P, Laoui A, Renou E, et al. (2002) Cell senescence and telomere shortening induced by a new series of specific G-quadruplex DNA ligands. Proc Natl Acad Sci U S A 99: 2672–2677.
42. Gomez D, Mergny JL, Riou JF (2002) Detection of telomerase inhibitors based on g-quadruplex ligands by a modified telomeric repeat amplification protocol assay. Cancer Res 62: 3365–3368.
43. Gomez D, Lemarteleur T, Lacroix L, Mailliet P, Mergny JL, et al. (2004) Telomerase downregulation induced by the G-quadruplex ligand 12459 in A549 cells is mediated by hTERT RNA alternative splicing. Nucleic Acids Res 32: 371–379.
44. Drygin D, Siddiqui-Jain A, O'Brien S, Schwaebe M, Lin A, et al. (2009) Anticancer activity of CX-3543: a direct inhibitor of rRNA biogenesis. Cancer Res 69: 7653–7661.
45. Alpert JS, Thygesen K, Antman E, Bassand JP (2000) Myocardial infarction redefined–a consensus document of The Joint European Society of Cardiology/American College of Cardiology Committee for the redefinition of myocardial infarction. J Am Coll Cardiol 36: 959–969.
46. Thygesen K, Alpert JS, White HD (2007) Universal definition of myocardial infarction. Eur Heart J 28: 2525–2538.
47. Bhavsar PK, Dhoot GK, Cumming DV, Butler-Browne GS, Yacoub MH, et al. (1991) Developmental expression of troponin I isoforms in fetal human heart. FEBS Lett 292: 5–8.
48. Bhavsar PK, Brand NJ, Yacoub MH, Barton PJ (1996) Isolation and characterization of the human cardiac troponin I gene (TNNI3). Genomics 35: 11–23.
49. Bhavsar PK, Dellow KA, Yacoub MH, Brand NJ, Barton PJ (2000) Identification of cis-acting DNA elements required for expression of the human cardiac troponin I gene promoter. J Mol Cell Cardiol 32: 95–108.
50. Kojima T, Yoshikawa Y, Takada S, Sato M, Nakamura T, et al. (2001) Genomic organization of the Shc-related phosphotyrosine adapters and characterization of the full-length Sck/ShcB: specific association of p68-Sck/ShcB with pp135. Biochem Biophys Res Commun 284: 1039–1047.
51. Leem SH, Kouprina N, Grimwood J, Kim JH, Mullokandov M, et al. (2004) Closing the gaps on human chromosome 19 revealed genes with a high density of repetitive tandemly arrayed elements. Genome Res 14: 239–246.
52. Das HK, Jackson CL, Miller DA, Leff T, Breslow JL (1987) The human apolipoprotein C-II gene sequence contains a novel chromosome 19-specific minisatellite in its third intron. J Biol Chem 262: 4787–4793.
53. Barton PJ, Cullen ME, Townsend PJ, Brand NJ, Mullen AJ, et al. (1999) Close physical linkage of human troponin genes: organization, sequence, and expression of the locus encoding cardiac troponin I and slow skeletal troponin T. Genomics 57: 102–109.
54. Ying L, Green JJ, Li H, Klenerman D, Balasubramanian S (2003) Studies on the structure and dynamics of the human telomeric G quadruplex by single-

molecule fluorescence resonance energy transfer. Proc Natl Acad Sci U S A 100: 14629–14634.
55. Suntharalingam K, White AJ, Vilar R (2010) Two metals are better than one: investigations on the interactions between dinuclear metal complexes and quadruplex DNA. Inorg Chem 49: 8371–8380.
56. Zhou W, Brand NJ, Ying L (2011) G-quadruplexes-novel mediators of gene function. J Cardiovasc Transl Res 4: 256–270.
57. Baumann P (2009) G-Quadruplex DNA: Methods and Protocols. New York: Humana Press.
58. Huppert JL, Bugaut A, Kumari S, Balasubramanian S (2008) G-quadruplexes: the beginning and end of UTRs. Nucleic Acids Res 36: 6260–6268.
59. Zhao Y, Du Z, Li N (2007) Extensive selection for the enrichment of G4 DNA motifs in transcriptional regulatory regions of warm blooded animals. FEBS Lett 581: 1951–1956.
60. Green JJ, Ying L, Klenerman D, Balasubramanian S (2003) Kinetics of unfolding the human telomeric DNA quadruplex using a PNA trap. J Am Chem Soc 125: 3763–3767.
61. Du Z, Kong P, Gao Y, Li N (2007) Enrichment of G4 DNA motif in transcriptional regulatory region of chicken genome. Biochem Biophys Res Commun 354: 1067–1070.
62. Li W, Wu P, Ohmichi T, Sugimoto N (2002) Characterization and thermodynamic properties of quadruplex/duplex competition. FEBS Lett 526: 77–81.
63. Chaires JB (2010) Human telomeric G-quadruplex: thermodynamic and kinetic studies of telomeric quadruplex stability. FEBS J 277: 1098–1106.
64. Dapic V, Abdomerovic V, Marrington R, Peberdy J, Rodger A, et al. (2003) Biophysical and biological properties of quadruplex oligodeoxyribonucleotides. Nucleic Acids Res 31: 2097–2107.
65. Padmanabhan K, Padmanabhan KP, Ferrara JD, Sadler JE, Tulinsky A (1993) The structure of alpha-thrombin inhibited by a 15-mer single-stranded DNA aptamer. J Biol Chem 268: 17651–17654.
66. Bryan TM, Baumann P (2011) G-quadruplexes: from guanine gels to chemotherapeutics. Mol Biotechnol 49: 198–208.
67. Kypr J, Kejnovska I, Renciuk D, Vorlickova M (2009) Circular dichroism and conformational polymorphism of DNA. Nucleic Acids Res 37: 1713–1725.
68. Fegan A, Shirude PS, Ying L, Balasubramanian S (2010) Ensemble and single molecule FRET analysis of the structure and unfolding kinetics of the c-kit promoter quadruplexes. Chem Commun (Camb) 46: 946–948.
69. Suntharalingam K, White AJ, Vilar R (2009) Synthesis, structural characterization, and quadruplex DNA binding studies of platinum(II)-terpyridine complexes. Inorg Chem 48: 9427–9435.
70. Park GH, Plummer HK, 3rd, Krystal GW (1998) Selective Sp1 binding is critical for maximal activity of the human c-kit promoter. Blood 92: 4138–4149.
71. Gallo P, Grimaldi S, Latronico MV, Bonci D, Pagliuca A, et al. (2008) A lentiviral vector with a short troponin-I promoter for tracking cardiomyocyte differentiation of human embryonic stem cells. Gene Ther 15: 161–170.
72. Neidle S (2010) Human telomeric G-quadruplex: the current status of telomeric G-quadruplexes as therapeutic targets in human cancer. FEBS J 277: 1118–1125.
73. Lipps HJ, Rhodes D (2009) G-quadruplex structures: in vivo evidence and function. Trends Cell Biol 19: 414–422.
74. Dixon IM, Lopez F, Tejera AM, Esteve JP, Blasco MA, et al. (2007) A G-quadruplex ligand with 10000-fold selectivity over duplex DNA. J Am Chem Soc 129: 1502–1503.
75. Rawal P, Kummarasetti VB, Ravindran J, Kumar N, Halder K, et al. (2006) Genome-wide prediction of G4 DNA as regulatory motifs: role in Escherichia coli global regulation. Genome Res 16: 644–655.
76. Hershman SG, Chen Q, Lee JY, Kozak ML, Yue P, et al. (2008) Genomic distribution and functional analyses of potential G-quadruplex-forming sequences in Saccharomyces cerevisiae. Nucleic Acids Res 36: 144–156.
77. Balagurumoorthy P, Brahmachari SK (1994) Structure and stability of human telomeric sequence. J Biol Chem 269: 21858–21869.
78. Mergny JL, Phan AT, Lacroix L (1998) Following G-quartet formation by UV-spectroscopy. FEBS Lett 435: 74–78.
79. Lee JY, Okumus B, Kim DS, Ha T (2005) Extreme conformational diversity in human telomeric DNA. Proc Natl Acad Sci U S A 102: 18938–18943.
80. Kumar N, Maiti S (2004) Quadruplex to Watson-Crick duplex transition of the thrombin binding aptamer: a fluorescence resonance energy transfer study. Biochem Biophys Res Commun 319: 759–767.
81. Raghuraman MK, Cech TR (1990) Effect of monovalent cation-induced telomeric DNA structure on the binding of Oxytricha telomeric protein. Nucleic Acids Res 18: 4543–4552.
82. Gonzalez V, Guo K, Hurley L, Sun D (2009) Identification and characterization of nucleolin as a c-myc G-quadruplex-binding protein. J Biol Chem 284: 23622–23635.
83. Brooks TA, Hurley LH (2010) Targeting MYC Expression through G-Quadruplexes. Genes Cancer 1: 641–649.
84. Shklover J, Weisman-Shomer P, Yafe A, Fry M (2010) Quadruplex structures of muscle gene promoter sequences enhance in vivo MyoD-dependent gene expression. Nucleic Acids Res 38: 2369–2377.
85. McCarthy JG, Heywood SM (1987) A long polypyrimidine/polypurine tract induces an altered DNA conformation on the 3′ coding region of the adjacent myosin heavy chain gene. Nucleic Acids Res 15: 8069–8085.
86. Michelotti GA, Michelotti EF, Pullner A, Duncan RC, Eick D, et al. (1996) Multiple single-stranded cis elements are associated with activated chromatin of the human c-myc gene in vivo. Mol Cell Biol 16: 2656–2669.
87. Hammond-Kosack MC, Docherty K (1992) A consensus repeat sequence from the human insulin gene linked polymorphic region adopts multiple quadriplex DNA structures in vitro. FEBS Lett 301: 79–82.
88. Hammond-Kosack MC, Dobrinski B, Lurz R, Docherty K, Kilpatrick MW (1992) The human insulin gene linked polymorphic region exhibits an altered DNA structure. Nucleic Acids Res 20: 231–236.
89. Hammond-Kosack MC, Kilpatrick MW, Docherty K (1993) The human insulin gene-linked polymorphic region adopts a G-quartet structure in chromatin assembled in vitro. J Mol Endocrinol 10: 121–126.
90. Sun D, Hurley LH (2009) The importance of negative superhelicity in inducing the formation of G-quadruplex and i-motif structures in the c-Myc promoter: implications for drug targeting and control of gene expression. J Med Chem 52: 2863–2874.
91. Droge P (1994) Protein tracking-induced supercoiling of DNA: a tool to regulate DNA transactions in vivo? Bioessays 16: 91–99.
92. Maizels N (2006) Dynamic roles for G4 DNA in the biology of eukaryotic cells. Nat Struct Mol Biol 13: 1055–1059.
93. Srivastava D (2006) Making or breaking the heart: from lineage determination to morphogenesis. Cell 126: 1037–1048.
94. Risebro CA, Smart N, Dupays L, Breckenridge R, Mohun TJ, et al. (2006) Hand1 regulates cardiomyocyte proliferation versus differentiation in the developing heart. Development 133: 4595–4606.
95. Scott IC, Anson-Cartwright L, Riley P, Reda D, Cross JC (2000) The HAND1 basic helix-loop-helix transcription factor regulates trophoblast differentiation via multiple mechanisms. Mol Cell Biol 20: 530–541.

2

Genome-Wide Genetic Diversity and Differentially Selected Regions among Suffolk, Rambouillet, Columbia, Polypay, and Targhee Sheep

Lifan Zhang[1], Michelle R. Mousel[2], Xiaolin Wu[3], Jennifer J. Michal[1], Xiang Zhou[1], Bo Ding[1], Michael V. Dodson[1], Nermin K. El-Halawany[4], Gregory S. Lewis[2], Zhihua Jiang[1]*

1 Department of Animal Sciences, Washington State University, Pullman, Washington, United States of America, **2** USDA/ARS US Sheep Experiment Station, Dubois, Idaho, United States of America, **3** Department of Dairy Science, University of Wisconsin-Madison, Madison, Wisconsin, United States of America, **4** Cell Biology Department, Division of Genetic Engineering and Biotechnology, National Research Center, Dokki, Gueza, Egypt

Abstract

Sheep are among the major economically important livestock species worldwide because the animals produce milk, wool, skin, and meat. In the present study, the Illumina OvineSNP50 BeadChip was used to investigate genetic diversity and genome selection among Suffolk, Rambouillet, Columbia, Polypay, and Targhee sheep breeds from the United States. After quality-control filtering of SNPs (single nucleotide polymorphisms), we used 48,026 SNPs, including 46,850 SNPs on autosomes that were in Hardy-Weinberg equilibrium and 1,176 SNPs on chromosome × for analysis. Phylogenetic analysis based on all 46,850 SNPs clearly separated Suffolk from Rambouillet, Columbia, Polypay, and Targhee, which was not surprising as Rambouillet contributed to the synthesis of the later three breeds. Based on pair-wise estimates of F_{ST}, significant genetic differentiation appeared between Suffolk and Rambouillet ($F_{ST}=0.1621$), while Rambouillet and Targhee had the closest relationship ($F_{ST}=0.0681$). A scan of the genome revealed 45 and 41 differentially selected regions (DSRs) between Suffolk and Rambouillet and among Rambouillet-related breed populations, respectively. Our data indicated that regions 13 and 24 between Suffolk and Rambouillet might be good candidates for evaluating breed differences. Furthermore, ovine genome v3.1 assembly was used as reference to link functionally known homologous genes to economically important traits covered by these differentially selected regions. In brief, our present study provides a comprehensive genome-wide view on within- and between-breed genetic differentiation, biodiversity, and evolution among Suffolk, Rambouillet, Columbia, Polypay, and Targhee sheep breeds. These results may provide new guidance for the synthesis of new breeds with different breeding objectives.

Editor: Zhanjiang Liu, Auburn University, United States of America

Funding: This work was supported by USDA/FAS grant (58-3148-0-138) to ZJ. The funders had no role in study design, data collection and analysis, decision to publish, or preparation of the manuscript.

Competing Interests: The authors have declared that no competing interests exist.

* E-mail: jiangz@wsu.edu

Introduction

During the last five years, the animal genome community has made significant progress in mapping, sequencing, assembly, and annotation of the ovine genome. Based on BAC (bacterial artificial chromosome) end sequences, Dalrymple and colleagues [1] first reported a virtual sheep genome by painting a total of 84,624 sheep BACs (about 5.4-fold genome coverage) to orthologous regions in the human genome, which were assembled into 1,172 sheep BAC comparative genome contigs that covered 91.2% of the human genome. In 2009, Goldammer and coworkers [2] constructed a cytogenetic map of the sheep genome with 566 loci, which helped link and order genome regions, such as sequence contigs, genes, and polymorphic DNA markers to ovine chromosomes. Approximately two years ago, the International Sheep Genomics Consortium (ISGC) began assembly of a draft reference genome of sheep (*Ovis aries*) using both Sanger sequencing and the next-generation sequencing platforms [3]. This large scale sequencing of the ovine genome led to discovery of more than 2.8 million ovine single nucleotide polymorphisms (SNPs; http://www.ncbi.nlm.nih.gov/SNP/). In collaboration with the ISGC, Illumina developed the OvineSNP50 Genotyping BeadChip that contains a total of 54,241 SNPs with a marker placed approximately every 46 Kb along the sheep genome (www.illumina.com).

The Illumina OvineSNP50 Genotyping BeadChip has been successfully used in sheep and goat genome research. For example, BeadChip analysis revealed that the *PITX3* gene is responsible for microphthalmia [4]. A similar approach also helped identify the dentin matrix protein 1 gene (*DMP1*) as responsible for inherited rickets in Corriedale sheep [5] and the solute carrier family 13 (sodium/sulphate symporters), member 1 (*SLC13A1*) gene for chondrodysplasia in Texel sheep [6]. Both OvineSNP50 BeadChips and microsatellite markers were used to refine two quantitative trait loci (QTL) mapped on OAR5 and 13 for resistance to *Haemonchus contortus* in sheep [7]. Other applications of OvineSNP50 BeadChip include investigating gene drivers of pigmentation in Merino sheep [8], long range linkage disequilibrium analysis in wild sheep [9], inbreeding coefficient and pairwise relatedness in Finnsheep [10], and genomic selection in different sheep breeds from around the world by the ISGC [11].

It is well known that Suffolk and Rambouillet were developed in England and France, respectively, but the breeding history of American synthetic breeds may be unfamiliar to readers. In brief, Columbia was one of the first breeds of sheep developed in the United States. In 1912, rams of the long wool breeds were crossed with high quality Rambouillet ewes to produce large ewes yielding more pounds of wool and more pounds of lamb. The original cross was made at Laramie, Wyoming, and then moved to the Sheep Experiment Station, Dubois, Idaho, in 1918. Subsequently, Columbia sheep were released to the public [12]. Polypay sheep were developed at the U.S. Sheep Experiment Station starting in 1968. The objective was to develop a breed with a reproductive capacity markedly superior to that of domestic Western range breeds. The final composition of the Polypay is 1/4 Dorset × 1/4 Finnsheep × 1/4 Targhee × 1/4 Rambouillet. The first "Polypay" ewes and rams were sold 1975–1977 [13]. Targhee sheep were developed at the U.S. Sheep Experiment Station, Dubois, Idaho in 1926. A group of cross-bred ewes, consisting of Rambouillet, Lincoln, and Corriedale blood, was bred to USSES Rambouillet rams. After three years, first generation ewes were carefully selected and bred intensely. The U.S. Targhee Sheep Association was founded in 1951 (http://www.ustargheesheep.org/).

Generally speaking, sheep breeds can be classified into three groups: meat, wool, or dual-purpose breeds based on their breeding objectives. For example, Suffolk is a typical meat breed as the animals possess large body size, rapid growth rate, and high cutability carcasses (http://u-s-s-a.org/). On the other hand, Rambouillet sheep represent a fine wool breed with a well-developed flocking instinct, an extended breeding season, and high-quality fleece (http://www.countrylovin.com/ARSBA/facts.htm). Columbia, Targhee, and Polypay are, however, considered as dual-purpose breeds, because they are fast-growing, high-quality market lambs that also yield heavy, medium-wool fleeces with good staple length [12–13] (http://www.sheepusa.org/).

Previously, microsatellite markers were the main source of markers used to investigate genetic diversity of sheep breeds. For instance, Bayesian cluster analysis on microsatellite genotypes of 666 animals for 28 U.S. sheep breeds derived from 222 producers located in 38 states was able to distinguish meat vs. wool producers due to physiological differences rather than geographic origin [14]. In the present study, our goal was to test the power of the Illumina OvineSNP50 Genotyping BeadChip in evaluating genetic diversity, genome selection, and breed differentiation among Suffolk, Rambouillet, Columbia, Polypay, and Targhee sheep breeds. These results may provide new guidance for the synthesis of new breeds with different breeding objectives.

Results

Illumina OvineSNP50 BeadChip Genotyping Basics

Among the 54,241 SNPs on the Illumina OvineSNP50 BeadChip that were genotyped on the 94 sheep DNA samples, we observed that 695 SNPs had no calls, 1,019 SNPs were not genotyped for at least 95% of all the individuals, 1,235 SNPs were monomorphic in all breeds, 350 SNPs could not be assigned to chromosome locations and 2,057 SNPs had MAF $\leq$0.05 for the whole dataset. By excluding the SNPs described above, the remaining 48,885 SNPs, including 47,597 autosomal SNPs and 1,288 SNPs on chromosome X, were used for further analysis. Of the 1,288 SNPs on chromosome X, 1,176 SNPs identified non-heterozygous males, while 112 SNPs were heterozygous in some rams. This might be related to a homologous region between chromosomes X and Y because our data do not show that the heterozygous regions are random. As suggested by Gautier et al [15], we further excluded a total of 747 autosomal SNPs, which showed significant ($P<0.01$) deviations from the Hardy-Weinberg equilibrium (HWE) test due to the small number of samples. We did not test chromosome X because the SNPs on chromosome X in males carry only one copy. As a consequence, 46,850 autosomal SNPs were included in linkage disequilibrium, genetic diversity and DSR analyses, while the 1,176 SNPs with non-heterozygous males on chromosome X were used for DSRs analysis only. As shown in Figure S1, the final 46,850 and 1,176 SNPs were uniformly distributed on different autosomes (1–26) and the X chromosome, and were comparable to the initial distribution of the 54,241 SNPs on these chromosomes, although the numbers of SNPs in each chromosome were different between them.

r^2 Measurements by Chromosomes

The r^2 values for pairs of loci were measured along with the physical distance separating the loci and averaged within each breed. As the sheep genome is currently estimated to be 2.86 Gb in size (http://genome.ucsc.edu/), the 46,850 SNPs used in linkage disequilibrium (LD) analysis would have an average inter-marker distance of approximately 60 Kb. As shown in Figure S2, the average within-population pairwise r^2 dropped quickly toward its asymptotic value when physical distances reached 200 Kb. More interestingly, the decreasing trends of the average r^2 values remained similar among these five breeds, but the Suffolk breed had the highest r^2, followed by Columbia, Rambouillet, Targhee, and Polypay, respectively.

The Genetic Structure at the Individual Level

Figure 1 demonstrates a neighbor-joining (NJ) tree based on allele sharing distances (ASD) among 94 rams derived from Columbia, Polypay, Rambouillet, Suffolk, and Targhee breeds. The results clearly showed that there were no conflicts about the origin of individuals assigned to each breed. Also, the individuals from different sheep breeds were clearly clustered with closer genetic distances observed among Targhee, Columbia, Rambouillet, and Polypay breeds as compared to the Suffolk population.

Accessing the Genetic Structure at the Population Level

As shown in Figure S3, the gene diversity, heterozygosity, and polymorphism information content (PIC) among five sheep populations were 0.3291–0.3576, 0.3496–0.3722 and 0.2619–0.2837 respectively. Polypay and Targhee populations had the highest gene diversity, heterozygosity, and PIC while the Suffolk population had the lowest values in these indexes. Classical F-statistics showed that most variation originated from individuals within a breed, while only 11% of the variation resulted from different breeds (Table S1). In particular, Targhee sheep had the highest within-breed variation.

Furthermore, a multidimensional scaling plot clearly showed the genetic origin of breed between Suffolk and Rambouillet and among Rambouillet-related sheep breeds (Figure 2). Based on pair-wise estimates of F_{ST}, significant genetic differentiation appeared between Suffolk (meat breed) and Rambouillet (fine wool breed) ($F_{ST} = 0.1621$). In comparison, Rambouillet-related breeds were not significantly separated ($F_{ST} = 0.0681$–0.0952). In particular, Rambouillet and Targhee had the closest relationship ($F_{ST} = 0.0681$) (Figure 2).

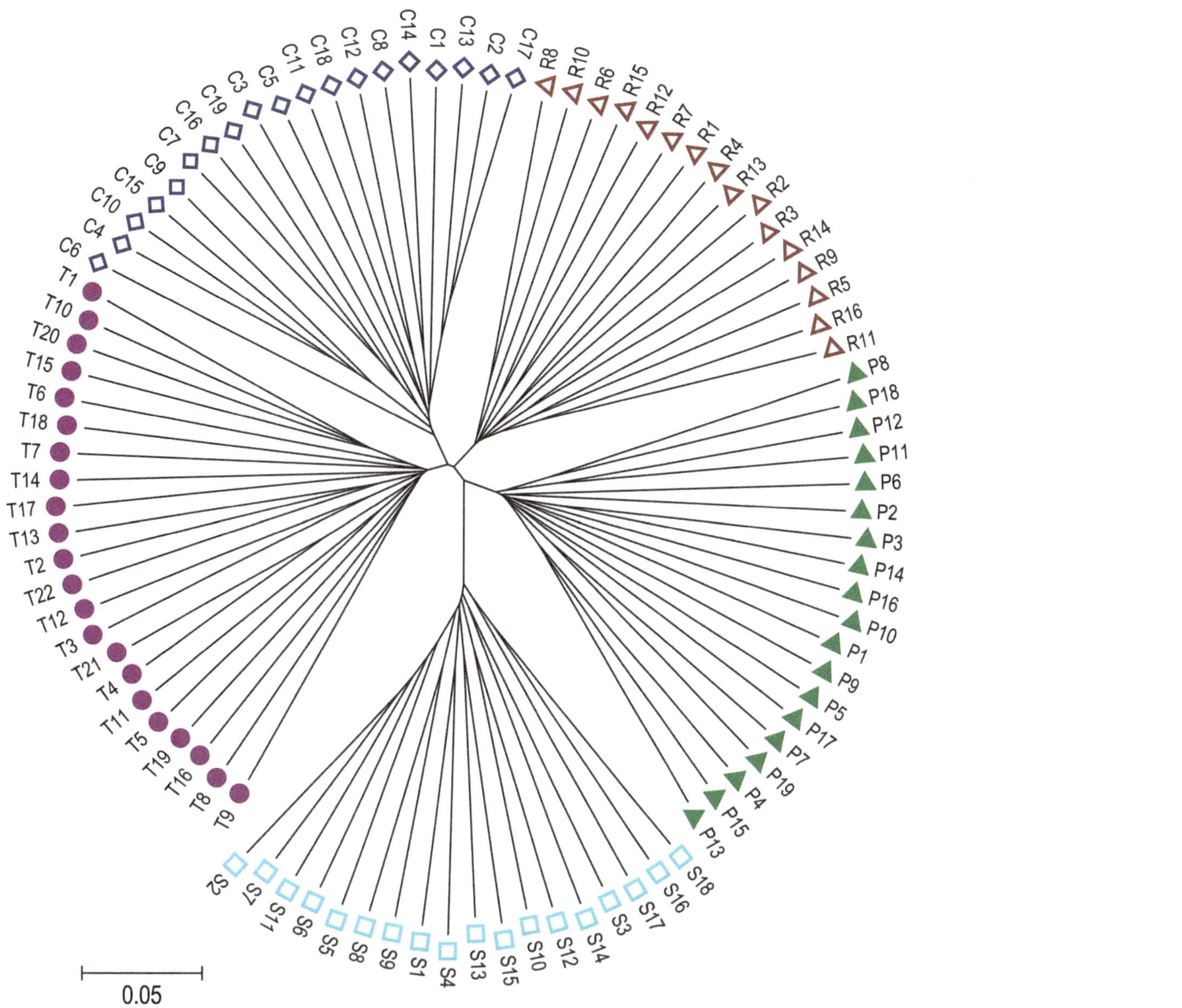

Figure 1. Neighbor-Joining tree relating the 94 individuals. The tree was constructed using allele sharing distances averaged over 46,850 SNPs. Different colors in labels represent the origin of breed individuals. S, R, C, P, and T represent Suffolk, Rambouillet, Columbia, Polypay, and Targhee sheep breeds, respectively. The meanings of S, R, C, P, and T are same in the following figures.

Population Differences in Minor Allele Frequencies

Based on minor allele frequency of SNPs, different levels of variation across breeds were observed. As shown in Figure S4, over 80% of the SNPs with one allele in all breeds had a MAF > 0.10. In all MAF ranges, the proportion of loci were significantly different among all sheep breeds ($\chi^2 = 204.1084$–1510.9140, $P = 0.0000$), indicating that each sheep breed had different numbers of SNPs in each MAF range.

Characterization of DSRs between Suffolk and Rambouillet

Based on the SNP F_{ST} estimates, a total of 45 DSRs were identified in genomes between Suffolk and Rambouillet, which contained the top 0.1% of markers (48 SNPs in autosomal chromosomes and 6 SNPs in chromosome X; Table 1). Further examination of these DSRs identified 608 unique known genes, including 507 from autosomal DGRs and 101 from X chromosome DSRs (Table S2). The GO analysis revealed pathways enriched for a wide range of biological processes, such as regulation of organelle/cytoskeleton organization, translational elongation, protein catabolic processes, and cilium morphogenesis (Table S3). Among these 45 DSRs between Suffolk and Rambouillet, 13 also appeared as DSRs in cattle (Table S4). Among autosomal DSRs, the highest F_{ST} signal (OAR3_163921101, $F_{ST} = 0.95$, region 13) (Table 1 and Figure 3) was located at 153.26 Mb on ovine chromosome 3 (Ovine v3.1 genome), where glutamate receptor interacting protein 1 (GRIP1) resides. On the other hand, the DSR (named region 24, spans 28.58 to 29.84 Mb) with the highest number of top 0.1% SNPs was located on chromosome 10 (Table 1 and Figure 3). Furry homolog (Drosophila) (*FRY*), which included four of the top 0.1% SNPs

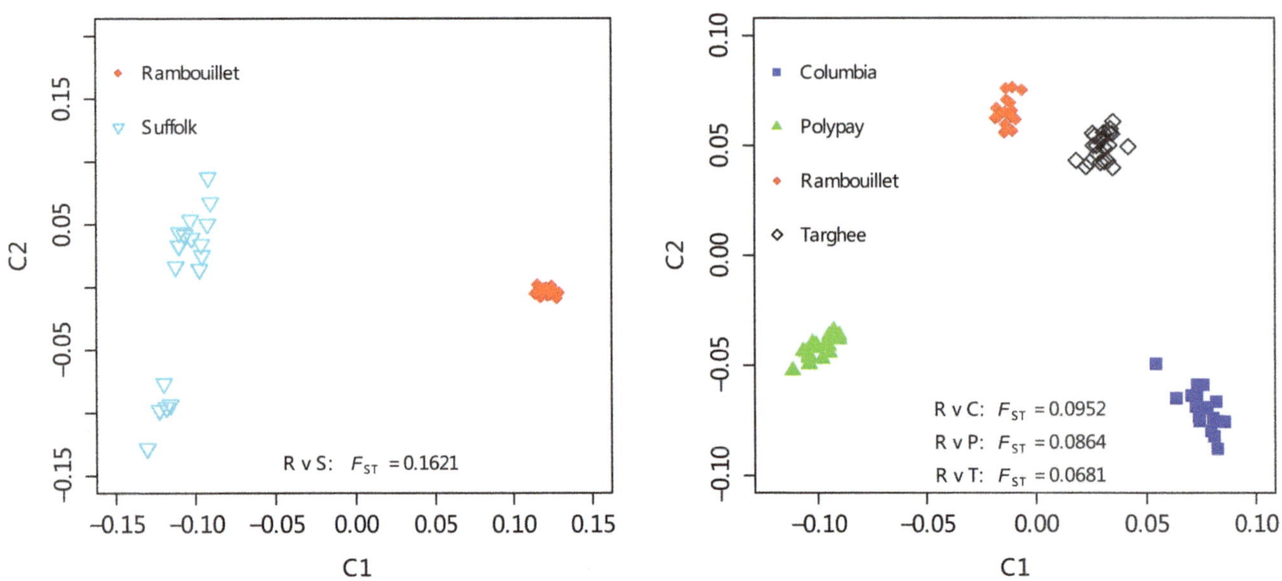

Figure 2. Multidimensional scaling plots for the genetic differentiations between Suffolk and Rambouillet (left) and among four Rambouillet-related breeds (right). F_{ST} represent the pair-wise F_{ST} between any two sheep breeds.

of this region, is an evolutionarily conserved protein implicated in cell division and morphology [16]. Additionally, selection signals were detected for genes associated with economically important traits, i.e., *MITF* and *GHR*.

Characterization of DSRs among Rambouillet-related Breeds

Genome-wide distribution of F_{ST} among Rambouillet, Columbia, Polypay, and Targhee are shown in Figure 4. Among these four Rambouillet-related breeds, 41 DSRs were identified with the top 0.1% of markers ranked by SNP F_{ST} (46 in autosomal and 1 in chromosome X, Table 2). These DSRs harbor a total of 526 unique genes, including 524 from autosomal DSRs and 2 from chromosome X DSRs (Table S5). Interestingly, GO analysis revealed that the enriched pathways were mainly related to cell adhesion processes (Table S6). Among these four sheep breeds, the OAR21_19719146 SNP (F_{ST} = 0.65, region 37) (Table 2) that belongs to potassium channel tetramerisation domain containing 14 (*KCTD14*) gene ranked highest, but unfortunately, little is known about this gene. Our data also show that both sheep and cattle may share eight DSRs identified among Rambouillet, Columbia, Polypay, and Targhee (Table S7).

Discussion

In the present study, we used the Illumina OvineSNP50 Genotyping BeadChip to analyze the genetic diversity and genome selection among Suffolk, Rambouillet, Columbia, Polypay, and Targhee sheep breeds from the USDA, ARS, U.S. Sheep Experiment Station. Our present study determined that close genetic relationships exist within Rambouillet-related breeds: Rambouillet, Columbia, Polypay, and Targhee, while Suffolk sheep are well separated from the Rambouillet-related breeds. The F_{ST} results showed significant genetic difference between Suffolk and Rambouillet (F_{ST} = 0.1621). Between these two distinct breeds, 45 DSRs and 608 candidate genes were identified using the Ovine Genome v3.1 Assembly as a reference. On the other hand, 41 DSRs and 526 genes were also determined among the four Rambouillet-related breeds.

Polypay (Finn × Targhee × Rambouillet × Dorset) [13], Targhee (Rambouillet × Lincoln × Corriedale), and Columbia (Lincoln × Rambouillet) are three breeds that were originally developed at the U.S. Sheep Experiment Station decades ago. Rambouillet and Suffolk were developed in France and England, respectively. In the present study, the highest gene diversity, heterozygosity and PIC were shown in Polypay and Targhee. This is not surprising because these two sheep breeds are the most recently developed breeds and we expect them to retain greater heterozygosity than the three other sheep breeds. Also, we found the F_{ST} averaged 0.1140 but Rambouillet-related breeds were not significantly separated, suggesting that the genetic differentiation is mainly between Suffolk and Rambouillet-related breeds. Not unexpectedly, cluster analysis also clearly showed that Suffolk is genetically distant from the other four sheep breeds (Figure 1). These results are rational because Columbia, Targhee, and Polypay are Rambouillet-related breeds. In particular, pair-wise F_{ST} estimation suggested the Targhee should be considered genetically most similar to Rambouillet (Figure 2). These results are reasonable because they are supported by our records and breed selection history of Columbia, Targhee, and Rambouillet. Recently, the same SNP chip was used to assign population of origin between wild sheep breeds, including bighorn and thinhorn sheep [9,] and to determine the historic selection of 74 sheep breeds [11]. And in cattle, the bovine SNP chip had also been used to reveal genetic history or population diversity [17–18]. Now, our results further confirm that the SNP chip is a powerful tool to discover the population genetic diversity in livestock and these data provided strong evidence of the genetic structure in these five sheep breeds.

Meat, wool, and dual-purpose breeds of sheep were developed because these are highly valued traits in sheep production. Based on genetic distance, our results indicated that the meat breed (Suffolk) is very distant from the fine wool breed (Rambouillet), which is very much in line with the functional purpose of the different breeds. For example, sheep selected for meat production

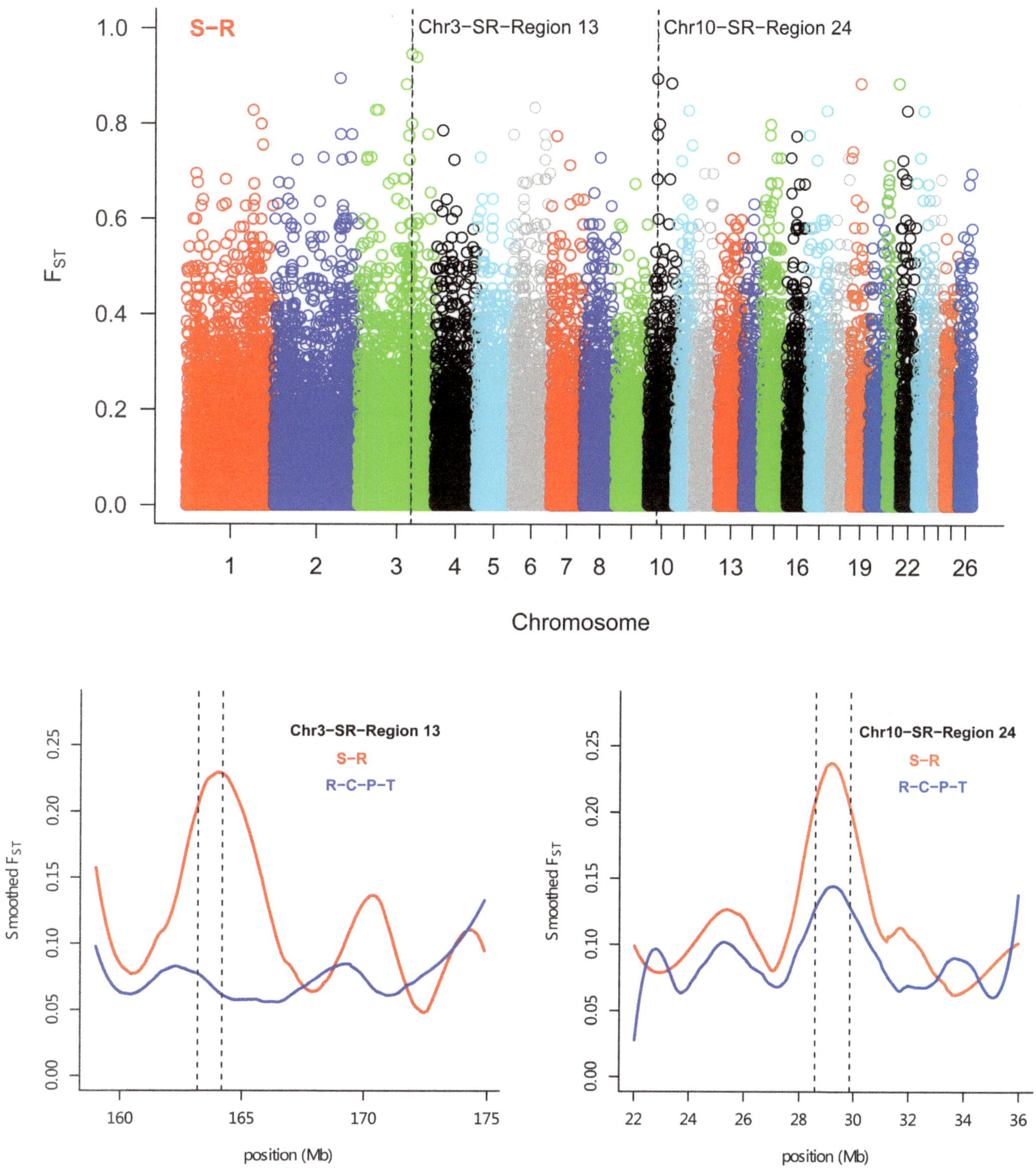

Figure 3. Genome-wide distribution of F_{ST} between Suffolk and Rambouillet. Based on OvineSNP50 BeadChip position, smoothed F_{ST} show that strong selection signals are observed in regions 13 and 24. S-R represents Suffolk-Rambouillet, while R-C-P-T means Rambouillet-Columbia-Polypay-Targhee.

generally have greater body weights. Mature weights of Suffolk rams, which have been historically selected for meat production, range from 113 to 159 kg and the fleece is considered a medium wool type with a staple length of 5 to 8.75 cm (http://u-s-s-a.org/). In comparison, mature Rambouillet rams, that have been bred to produce high quality wool, are smaller and weigh between 113 to 135 kg while the fleece staple length varies from 5 to10 cm and fiber diameter ranges from 18.5 to 24.5 microns (http://www.countrylovin.com/ARSBA/facts.htm).

Table 1. Differential Genomic Regions between Suffolk and Rambouillet.

Region	Chr	OvineSNP50 BeadChip Position (Mb)	Ovine version 3.1 Position (Mb)	Peak SNP (*F*st)	Top 0.1%	Top 5%	Gene Candidates	
1	1	223,692,105.224,089,918	207,151,194. 207,545,071	OAR1_224053727 (0.83)	1	3	0	NA
2	1	250,742,595.252,518,595	232,582,008. 233,998,801	OAR1_251993749 (0.80)	1	7	3	NA
3	1	257,381,496.260,183,641	238,188,431. 240,583,116	OAR1_258333225 (0.76)	1	4	6	NA
4	2	202,599,712.203,260,925	191,179,157. 191,834,956	OAR2_202880426 (0.89)	1	5	3	NA
5	2	204,712,172.204,901,274	193,176,635. 193,374,050	OAR2_204724536 (0.78)	1	3	1	NA
6	2	238,542,709.240,388,073	225,842,690. 227,694,352	OAR2_23999476 7(0.78)	1	1	3	NA
7	3	26,522,262.27,337,380	24,478,117. 25,283,037	s53615 (0.73)	1	3	1	NA
8	3	36,636,298.39,008,465	34,218,962. 36,296,085	s07478 (0.73)	1	1	29	*PLB1*
9	3	51,184,039.53,141,887	48,017,454. 50,434,716	OAR3_51654982 (0.83)	1	1	10	*HSPA4*
10	3	59,599,168.61,257,663	56,387,458. 57,906,999	OAR3_59779674 (0.83)	1	2	24	NA
11	3	146,832,060.147,380,630	137,368,282. 137,726,913	OAR3_147229298 (0.88)	1	2	2	NA
12	3	154,771,125.155,555,840	144,946,795. 145,685,656	OAR3_154940822 (0.78)	1	4	3	NA
13	3	163,186,840.164,185,125	152,644,200. 153,519,437	OAR3_163921101 (0.95)	2	7	4	*GRIP1,HELB*
14	3	179,997,804.181,039,892	167,562,727. 168,548,617	OAR3_180711543 (0.94)	1	4	10	*ANKS1B*
15	3	216,437,540.217,530,305	201,198,665. 202,134,428	OAR3_217179573 (0.78)	1	3	11	*HEBP1*
16	4	22,765,057.24,779,248	21,632,745. 23,648,876	OAR4_23178649 (0.79) OAR4_23188917 (0.79)	2	3	3	NA
17	5	15,497,764.16,376,239	13,268,366. 14,154,057	s48780 (0.73)	1	4	31	*ARHGEF18*
18	6	7,001,684.8,507,164	5,094,411. 6,363,162	OAR6_7822475 (0.78)	1	2	6	NA
19	6	74,570,690.77,508,495	68,044,077. 71,068,886	s32922 (0.83)	1	0	18	NA
20	6	109,205,577.109,456,098	99,193,263. 99,428,191	OAR6_109386725 (0.78)	1	3	4	NA
21	6	111,022,637.112,892,917	100,900,651. 102,765,374	s61454 (0.76)	1	4	15	*PPP2R2C*
22	7	18,503,023.20,430,110	17,777,251. 19,581,841	OAR7_18600852 (0.78)	1	2	12	*THSD4*
23	8	47,992,613.49,442,593	44,527,752. 45,986,304	OAR8_48260342 (0.73)	1	2	0	NA
24	10	28,598,904.29,867,192	28,584,727. 29,842,383	OAR10_29223007 (0.89) OAR10_29341212 (0.89)	4	8	9	*FRY*
25	10	34,081,682.36,999,163	33,723,894. 36,238,012	DU468275_284 (0.80)	1	0	22	*SGCG*
26	10	80,478,253.82,187,881	73,523,887. 75,018,763	OAR10_80709863 (0.89)	1	1	7	NA
27	11	49,315,455.51,153,763	46,334,232. 48,164,176	s53904 (0.83)	1	5	32	NA
28	11	62,887,032.64,089,979	58,151,039. 59,339,383	OAR11_63882013 (0.76)	1	3	0	NA
29	15	28,713,885.31,235,158	27,396,963. 29,749,473	s10361 (0.80)	3	6	54	*TMPRSS4,* *PVRL1*
30	15	55,184,101.56,938,649	50,486,417. 51,939,057	s34973 (0.73)	1	2	18	NA
31	16	15,458,830.17,547,585	14,215,465. 15,879,198	OAR16_16902918 (0.73)	1	1	11	*RNF180*
32	16	32,916,237.34,666,518	30,290,785. 31,926,643	OAR16_34620156 (0.78)	1	4	12	*GHR*
33	17	172,595.2,807,057	47,789. 2,338,716	s47560 (0.78)	1	2	9	NA
34	17	64,389,126.64,669,860	58,960,116. 59,261,632	OAR17_64627979 (0.83)	1	2	0	NA
35	19	1,721,354.4,144,231	1,700,469. 3,900,485	OAR19_2610848 (0.73)	1	2	7	NA
36	19	5,578,148.5,861,149	5,354,385. 5,635,464	OAR19_5820545 (0.74)	1	2	0	NA
37	19	31,917,811.33,355,170	30,269,881. 31,674,797	OAR19_33278780 (0.89)	1	1	2	*MITF*
38	21	47,142,810.49,259,843	42,654,067. 44,580,177	OAR21_47788299 (0.89)	1	1	73	*OVOL1*
39	22	25,234,645.28,159,147	21,379,260. 23,858,085	OAR22_26729825 (0.83)	1	0	46	*CNNM2*
40	23	14,719,599.17,399,614	13,512,270.16,274,959	OAR23_15999547 (0.73)	1	1	3	NA
41	23	25,829,771.28,768,759	24,678,908. 27,575,686	OAR23_27272887 (0.83)	1	0	16	NA
42	X	51,771,664.53,703,822	51,193,144. 53,231,464	OARX_52914005_X (1.00)	2	2	35	*SHROOM4*
43	X	60,238,540. 64,779,887	56,827,620. 61,295,149	OARX_63571789 (1.00)	2	6	42	*MED12* *FAM155B*
44	X	97,956,675.99,324,958	77,903,961.79,702,349	S47111 (0.78)	1	2	18	NA
45	X	103,934,729.106,069,046	83,501,497.85,670,129	OARX_105162278 (0.80)	1	1	6	NA

Note: A total of 45 genomic regions contained the top 0.1% of SNPs ranked using F_{ST} (48 SNPs in autosomal and 6 SNPs in chromosome X). The candidate genes are given within the SNPs of top SNP (0.1%) for each region.

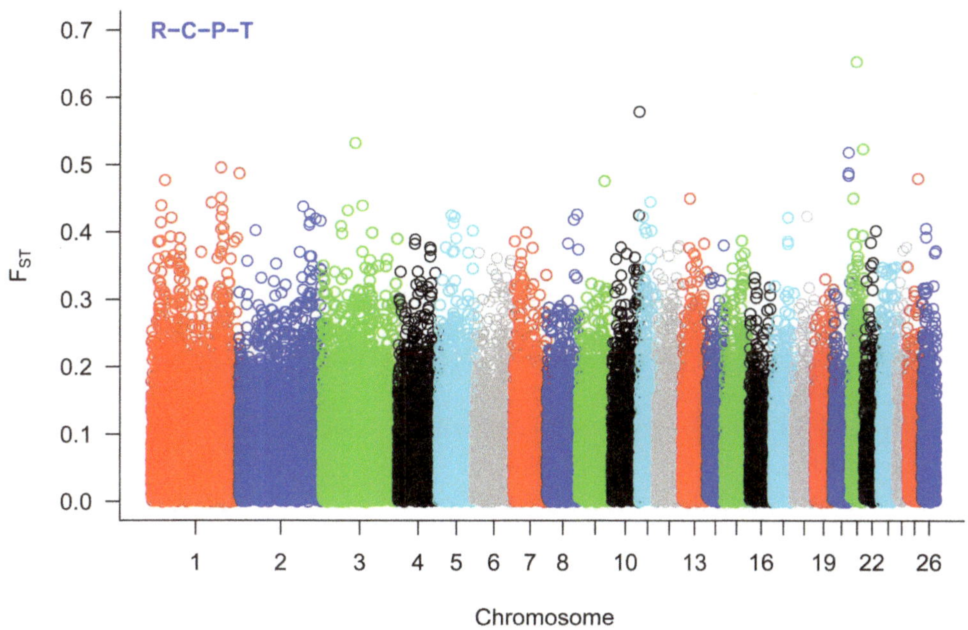

Figure 4. Genome-wide distribution of F_{ST} among Rambouillet, Columbia, Polypay and Targhee. R-C-P-T means Rambouillet-Columbia-Polypay-Targhee.

In the present study, a genome-wide scan or differentiation analysis using F_{ST} revealed 45 chromosomal regions with evidence for selection. Interestingly, three regions, 19, 24, and 37, are almost identical to the regions identified in the 74 sheep breeds examined in [11], implying that important genomic selections might appear in these regions. Interestingly, region 24, which includes the *RXFP2* gene that is involved in horn morphology had a selection signal that was reconstituted only when comparing horned with polled populations [11,19], was discovered in this study. Kijas et al (2012) [11] indicated this gene had the strongest selection signal due to the long-standing nature of selection. But in our study, only had two Rambouillet rams had horns., Therefore, the sample size was most likely too small to detect a difference in our study. Not unexpectedly, *GHR*, an important growth-related gene, was identified (region 32 on OAR 16). It is well-known that this gene affects body growth and decreases fatness [20], and its genetic variations are associated with growth traits in sheep or cattle [20–21]. Therefore, our study provides additional information for interpreting the difference in growth ability between Suffolk and Rambouillet. The highest ranked SNPs (F_{ST}>0.90) were located in glutamate receptor interacting protein 1 (GRIP1) and ankyrin repeat and sterile alpha motif domain containing 1B (ANKS1B). Many studies have found GRIP1 plays an important role in receptor trafficking, synaptic organization, transmission in glutamatergic and GABAergic synapses and modulating autistic phenotype [22–24]. But unfortunately, little is known about the function of *GRIP1* in livestock. Here our results might provide a new clue for its role in sheep production. Recently, *ANKS1B* gene had been shown to be associated with body weight index and waist circumference in human GWAS studies [25], and Parker et al [26] indicated this gene may underlie the QTL associated with body weight in mice. Our studies also suggested *ANKS1B* gene might be a good growth trait candidate. However, additional studies are required to confirm this speculation. Interestingly, we discovered *MITF* gene in DSRs (region 37 on OAR 19). This gene accounts for pigmentation phenotypes in cattle [27]. However, *ASIP*, which controls a series of alleles of black and white coat color [28], was not included in our DSRs. In sheep, gene duplications might also cause black fleece [28]. In this study, the Suffolk has black head and legs, while the Rambouillet does have recessive black. It appears as though the key gene of pigmentation may provide evidence for selection between the two sheep breeds. In this study we also identified *FRY*, a gene involved in growing wing hairs [29] and bristles [30] in Drosophila. Mutations in *FRY* resulted in the formation of a strong multiple hair cell phenotypes that consisted of clusters of epidermal hairs and branched hairs [31]. But there is a little known about the role of *FRY* in livestock. In the present study, Rambouillet is often considered a fine wool breed while Suffolk has rather poor quality wool. Therefore, our results provide strong evidence for the role of *FRY* in sheep wool development. Additionally, 13 of the 45 DSRs identified in sheep represent those in cattle, suggesting that these genes are targets for selection across multiple species.

As described above, Columbia, Polypay, and Targhee are related to Rambouillet sheep. Among these four sheep breeds, a total of 41 DSRs were identified (Table 2). Interestingly, GO terms analyses of functionally known genes in these regions discovered pathways related to hemophilic/cell adhesion, translational elongation, germ cell development, sexual reproduction, and macromolecule biosynthetic processes. Some signature genes suggested strong selection given their roles, such as *CNTNAP5*, *ADAM23*, and *PCDHB4* in cell adhesion, *ME3*, *RIMS2*, and *TDRD7* in cellular respiration, cellular macromolecule/protein localization, and multicellular organism reproduction, respectively. These might result from long-term selection for improved reproduction and wool traits in these four sheep breeds [32–33].

In summary, we revealed the genetic diversity among Suffolk, Rambouillet, Columbia, Polypay, and Targhee sheep breeds of the United States using the Illumina OvineSNP50 BeadChip. Meanwhile, DSRs between Suffolk and Rambouillet and among Rambouillet-related sheep breeds were also identified with production of a list of candidate genes in these regions based on Ovine Genome v3.1 Assembly. Estimation of genome-wide diversity and identification of DSRs regions provide a powerful

Table 2. Differential Genomic Regions among Rambouillet, Columbia, Polypay and Targhee.

Region	Chr	OvineSNP50 BeadChip Position (Mb)	Ovine version 3.1 Position (Mb)	Peak SNP (*Fst*)	Top 0.1%	Top 5%	Gene	Candidates
1	1	30,388,117.30,798,950	29,732,489. 30,143,388	OAR1_30552484 (0.41)	1	3	0	NA
2	1	31,470,501.32,656,978	30,814,041. 31,918,917	OAR1_31791056 (0.44)	1	2	6	*C1ORF168*
3	1	43,609,775.44,153,575	42,128,684. 42,656,182	s54617 (0.48)	1	2	5	*IL23R*
4	1	65,172,267.67,951,620	61,589,657. 64,314,437	OAR1_66495276 (0.42)	1	1	24	*CLCA1*
5	1	202,259,180.203,821,806	187,627,690. 189,001,556	s48685 (0.44)	1	2	16	NA
6	1	234,341,600.234,884,834	217,261,740. 217,734,182	OAR1_234474464 (0.50)	3	4	3	*GOLIM4*
7	1	235,109,323.236,540,864	217,827,095. 219,332,381	OAR1_235893252 (0.42)	1	9	4	NA
8	1	238,519,182.238,893,868	221,199,906. 221,519,414	OAR1_238532462 (0.43)	1	2	0	NA
9	1	243,197,038.243,803,482	225,691,787. 226,307,550	OAR1_243366256 (0.41)	1	8	3	NA
10	2	52,036,310.54,944,048	48,501,546. 51,384,849	s67253 (0.40)	1	1	33	*TDRD7*
11	2	199,204,762.200,621,949	187,885,054. 189,186,475	OAR2_200574953 (0.44)	1	3	1	*CNTNAP5*
12	2	218,777,904.220,621,478	206,643,029. 208,373,192	OAR2_219736519 (0.41)	1	3	13	*ADAM23*
13	2	221,160,555.223,548,439	208,910,057. 211,274,257	s71750 (0.43)	2	6	16	NA
14	2	240,718,168.241,220,099	228,019,818. 228,451,875	OAR2_241048971 (0.42)	1	3	4	*COL4A4*
15	2	258,064,261.258,169,201	244,129,054. 244,237,213	s36316 (0.42)	1	2	0	NA
16	3	57,803,870.59,844,188	54,668,289. 56,584,259	OAR3_59312021 (0.41)	1	1	4	NA
17	3	79,812,234.81,533,150	75,538,807. 77,148,543	OAR3_79936157 (0.43)	1	2	17	NA
18	3	101,572,092.102,519,143	95,521,406. 96,371,620	s65088 (0.53)	1	2	32	*LOXL3*
19	3	124,569,037.125,411,739	116,863,473. 117,540,039	OAR3_125387325 (0.44)	1	3	3	NA
20	5	41,108,751.43,897,574	37,314,853. 40,020,855	OAR5_42607507 (0.43)	1	0	20	NA
21	5	50,423,296.51,450,144	46,337,373. 47,282,405	OAR5_50467325 (0.41)	1	2	18	NA
22	5	52,451,051.55,344,981	48,225,599. 51,058,240	OAR5_53870555 (0.42)	1	1	66	*PCDHB4*
23	5	110,241,515.110,742,508	101,265,722. 101,776,198	OAR5_110440932 (0.40)	1	2	0	NA
24	8	82,084,037.82,635,994	76,042,356. 76,469,230	s02095 (0.42)	1	2	4	NA
25	8	91,759,596.92,094,265	85,080,887. 85,410,438	s19702 (0.43)	1	3	1	NA
26	9	76,165,857.78,371,606	71,768,469. 73,808,972	OAR9_77597438 (0.48)	1	3	9	*RIMS2*
27	10	92,069,072.93,254,917	84,440,241. 85,596,458	DU434120_194 (0.58)	2	6	9	*ARHGEF7, TUBGCP3*
28	11	13,571,422.16,047,304	13,680,542. 15,833,930	OAR11_13845361 (0.42)	1	4	36	NA
29	11	21,539,612.22,907,346	20,832,059. 22,055,035	s50268 (0.41)	1	3	19	NA
30	11	42,268,122.42,641,540	39,778,473. 40,158,382	OAR11_42371614 (0.45)	1	3	16	NA
31	11	46,846,457.49,449,235	44,059,749. 46,470,862	s17065 (0.40)	1	4	33	NA
32	13	25,313,379.26,260,934	22,757,476. 23,728,650	OAR13_25941147 (0.45)	1	2	5	NA
33	17	53,094,617.53,266,879	48,796,359. 48,949,569	OAR17_53158609 (0.42)	1	3	1	NA
34	18	44,151,635.46,578,380	41,525,466. 43,758,521	OAR18_44175536 (0.42)	1	2	5	NA
35	20	45,499,537.49,145,527	41,875,844. 45,195,888	OAR20_47285319 (0.52)	3	0	22	NA
36	21	9,818,736.10,702,527	8,402,965. 9,308,676	s42617 (0.45)	1	3	8	*ME3*
37	21	18,653,273.20,194,612	16,451,437. 17,888,494	OAR21_19719146 (0.65)	1	5	16	*KCTD14*
38	21	46,857,187.48,167,168	42,551,316. 43,585,536	s67834 (0.52)	1	4	51	*BATF2*
39	25	31,163,834.32,984,881	29,829,111. 31,596,285	OAR25_32856981 (0.48)	1	2	12	NA
40	26	13,102,143.13,432,722	10,556,336. 10,883,077	OAR26_13165931 (0.41)	1	3	0	NA
41	X	112,364,319.112,895,373	107,382,577. 107,865,244	OARX_112730737 (0.44)	1	5	2	NA

Note: A total of 41 genomic regions contained the top 0.1% of SNPs ranked using F_{ST} (46 in autosomal and 1 in chromosome X). The candidate genes are given within the SNPs of top SNP (0.1%) for each region.

method to identify economically important trait-related genes that have been enriched during a long-term selection for different breeding objectives. Furthermore, our results also provide a foundation to further investigate sheep evolution and gene functions in the near future.

Materials and Methods

Ethics Statement

The U.S. Sheep Experiment Station Animal Institutional Care and Use Committee specifically approved this study (Protocol number: 11–01). All efforts were made to minimize any discomfort during blood collection.

Sheep, DNA Preparation, and Genotyping on Illumina OvineSNP50 BeadChips

In the present study, blood samples were collected from 19 Columbia, 19 Polypay, 16 Rambouillet, 18 Suffolk, and 22 Targhee rams at the U.S. Sheep Experiment Station in Dubois, Idaho. Rams were produced from unique dams and 12, 12, 9, 10, and 17 unique sires of the Columbia, Polypay, Rambouillet, Suffolk, and Targhee breeds, respectively. The number of sheep per breed in this study is similar to the average number of sheep per breed Kijas and coworkers [11] used to quantify breed mixture and selection using the OvineSNP50 BeadChip. Blood was collected via jugular venipuncture into EDTA coated vacutainer tubes. Thereafter, DNA was extracted from 200 μL of whole blood with the GenElute Blood Genomic DNA extraction kit (Sigma, St. Louis, MO) according to the manufacturer's instructions. All DNA samples were genotyped with standard procedures at GeneSeek (Lincoln, NE, US) on the OvineSNP50 genotyping BeadChip. Basic information on the 54,241 SNPs on the BeadChip, including SNP name, chromosome, and map location was provided by the service provider. The genotype quality control process was as previously described [34].

Population Genetic Basics Analysis

Analysis of minor gene allele frequencies (MAF) was conducted with the chi-squared test using SAS Software for Windows v9.2 (SAS Institute Inc., Cary, NC). An exact test for Hardy-Weinberg Equilibrium (HWE) [35] of polymorphic SNPs was further carried out within each breed separately. We also computed the r^2 measure between each marker pair within each breed separately using Haploview 4.1 [36]. Allele sharing distances of the neighbor-joining tree relating sheep individuals were computed by PowerMarker V3.25 software [37], and then the neighbor-joining tree was constructed by MEGA 5 [38].

Gene diversity, heterozygosity, polymorphism information content (PIC), and classical F-statistics [39] were calculated in the present study using PowerMarker V3.25. FSTAT 2.9.3.2 [40] was used to evaluate population relatedness using pair-wise estimates of F_{ST}. An IBS matrix of distance (D) was constructed by Plink v1.07 [41], and then multidimensional scaling (MDS) analysis of 46,850 autosomal SNPs was determined using R 2.14.0 (www.r-project.org).

Detection of Differentially Selected Regions (DSRs)

Fisher's exact test was performed by R 2.14.0 to compare the allele frequencies between Suffolk and Rambouillet and among Rambouillet-related breed populations first. A SNP with a P value <0.05 was considered to be a statistically significant SNP after Bonferroni correction. Then, estimation of SNP and population-specific F_{ST} were based on the model proposed by Nicholson et al [42] and Flori et al [43]. The DSR algorithm was described previously [11], but with slight modifications: 1) raw values were ranked and used to identify regions; 2) the significant SNPs with 0.1% or 5% highest F_{ST} values were selected as the top significant SNPs; 3) centered on the top significant SNP (0.1%), neighboring markers were included until markers were encountered more than three consecutive SNPs ranking outside of the top significant 5%. We only considered the range between the upstream and downstream 1.5 Mb of the top SNP (0.1%) if the length of candidate regions were more than 3 Mb and combined any two regions as one region if they overlapped. SNP-specific Fst values were smoothed over each chromosome with a local variable bandwidth kernel estimator [44].

Genes in these DSRs were examined for potential involvement in phenotypes using the Ovine Genome v3.1 Assembly (http://www.livestockgenomics.csiro.au/cgi-bin/gbrowse/oarv3.1/). The functional annotation of target genes for the gene ontology was performed using DAVID bioinformatics resources [45]. Allele frequency per breed for all DSR can be found at www.animalgenome.org/repository/pub/USDA2013.0411/.

Supporting Information

Figure S1 Distributions of SNPs on different chromosomes.

Figure S2 Decay of average pairwise r^2 with intermarker distance for the different sheep breeds.

Figure S3 Genetic diversity analysis in different sheep breeds.

Figure S4 Minor allele frequencies (MAF) with 46,850 SNPs for different sheep breeds.

Table S1 Classical F-statistics in different sheep breeds.

Table S2 List of ovine positional candidate genes based on the predicted protein coding genes from DSRs between Suffolk and Rambouillet.

Table S3 Gene ontology analysis related to the ovine positional candidate genes from DSRs between Suffolk and Rambouillet.

Table S4 Selection signals identified in both sheep and cattle using the ovine positional candidate genes from DSRs between Suffolk and Rambouillet.

Table S5 List of ovine positional candidate genes based on the predicted protein coding genes from DSRs among Rambouillet, Columbia, Polypay, and Targee.

Table S6 Gene ontology analysis related to the ovine positional candidate genes from DSRs among Rambouillet, Columbia, Polypay, and Targhee.

Table S7 Selection signals identified in both sheep and cattle using the ovine positional candidate genes from DSRs among Rambouillet, Columbia, Polypay, and Targhee.

Author Contributions

Conceived and designed the experiments: ZJ MRM GSL. Performed the experiments: JJM MRM. Analyzed the data: LFZ XLW. Contributed reagents/materials/analysis tools: MRM GSL JJM XZ BD MVD NKE. Wrote the paper: LFZ ZJ MRM GSL.

References

1. Dalrymple BP, Kirkness EF, Nefedov M, McWilliam S, Ratnakumar A, et al. (2007) Using comparative genomics to reorder the human genome sequence into a virtual sheep genome. Genome Biol 8: R152.
2. Goldammer T, Di Meo GP, Lühken G, Drögemüller C, Wu CH, et al. (2009) Molecular cytogenetics and gene mapping in sheep (Ovis aries, 2n = 54). Cytogenet Genome Res126: 63–76.
3. International Sheep Genomics Consortium, Archibald AL, Cockett NE, Dalrymple BP, Faraut T, et al. (2010) The sheep genome reference sequence: a work in progress. Anim Genet 41: 449–453.
4. Becker D, Tetens J, Brunner A, Bürstel D, Ganter M, et al. (2010) Microphthalmia in Texel sheep is associated with a missense mutation in the paired-like homeodomain 3 (PITX3) gene. PLoS One 5: e8689.
5. Zhao X, Dittmer KE, Blair HT, Thompson KG, Rothschild MF, et al. (2011) A novel nonsense mutation in the DMP1 gene identified by a genome-wide association study is responsible for inherited rickets in Corriedale sheep. PLoS One 6: e21739.
6. Zhao X, Onteru SK, Piripi S, Thompson KG, Blair HT, et al. (2012) In a shake of a lamb's tail: using genomics to unravel a cause of chondrodysplasia in Texel sheep. Anim Genet 43 Suppl 1: 9–18.
7. Sallé G, Jacquiet P, Gruner L, Cortet J, Sauvé C, et al. (2012) A genome scan for QTL affecting resistance to Haemonchus contortus in sheep. J Anim Sci 90: 4690–4705.
8. García-Gámez E, Reverter A, Whan V, McWilliam SM, Arranz JJ, et al. (2011) Using regulatory and epistatic networks to extend the findings of a genome scan: identifying the gene drivers of pigmentation in merino sheep. PLoS One 6: e21158.
9. Miller JM, Poissant J, Kijas JW, Coltman DW (2011) A genome-wide set of SNPs detects population substructure and long range linkage disequilibrium in wild sheep. Mol Ecol Resour 11: 314–322.
10. Li MH, Strandén I, Tiirikka T, Sevón-Aimonen ML, Kantanen J (2011) A comparison of approaches to estimate the inbreeding coefficient and pairwise relatedness using genomic and pedigree data in a sheep population. PLoS One 6: e26256.
11. Kijas JW, Lenstra JA, Hayes B, Boitard S, Porto Neto LR, et al. (2012) Genome-wide analysis of the world's sheep breeds reveals high levels of historic mixture and strong recent selection. PLoS Biol 10: e1001258.
12. Lambert WV, McPhee HC, Reed OE, Speh CF, Bishopp FC, et al. (1947) "Developing New Breeds". In Science in Farming: Year book of Agriculture 1943–1947 Part 1. Washington, DC: USDA.
13. Hulet CV, Ercanbrack SK, Knight AD (1984) Development of the Polypay breed of sheep. J Anim Sci 58: 15–24.
14. Blackburn HD, Paiva SR, Wildeus S, Getz W, Waldron D, et al. (2011) Genetic structure and diversity among sheep breeds in the United States: identification of the major gene pools. J Anim Sci 89: 2336–2348.
15. Gautier M, Faraut T, Moazami-Goudarzi K, Navratil V, Foglio M, et al. (2007) Genetic and haplotypic structure in 14 European and African cattle breeds. Genetics 177: 1059–1070.
16. Ikeda M, Chiba S, Ohashi K, Mizuno K (2012) Furry protein promotes aurora A-mediated Polo-like kinase 1 activation. J Biol Chem 287: 27670–27681.
17. Gautier M, Naves M (2011) Footprints of selection in the ancestral admixture of a New World Creole cattle breed. Mol Ecol 20: 3128–3143.
18. Engelsma KA, Veerkamp RF, Calus MP, Bijma P, Windig JJ (2012) Pedigree- and marker-based methods in the estimation of genetic diversity in small groups of Holstein cattle. J Anim Breed Genet 129: 195–205.
19. Johnston SE, McEwan JC, Pickering NK, Kijas JW, Beraldi D, et al. (2011) Genome-wide association mapping identifies the genetic basis of discrete and quantitative variation in sexual weaponry in a wild sheep population. Mol Ecol 20: 2555–2566.
20. Bai WL, Zhou CY, Ren Y, Yin RH, Jiang WQ, et al. (2011) Characterization of the GHR gene genetic variation in Chinese indigenous goat breeds. Mol Biol Rep 38: 471–479.
21. Sherman EL, Nkrumah JD, Murdoch BM, Li C, Wang Z, et al. (2008) Polymorphisms and haplotypes in the bovine neuropeptide Y, growth hormone receptor, ghrelin, insulin-like growth factor 2, and uncoupling proteins 2 and 3 genes and their associations with measures of growth, performance, feed efficiency, and carcass merit in beef cattle. J Anim Sci 86: 1–16.
22. Li RW, Serwanski DR, Miralles CP, Li X, Charych E, et al. (2005) GRIP1 in GABAergic synapses. J Comp Neurol 488: 11–27.
23. Takamiya K, Mao L, Huganir RL, Linden DJ (2008) The glutamate receptor-interacting protein family of GluR2-binding proteins is required for long-term synaptic depression expression in cerebellar Purkinje cells. J Neurosci 28: 5752–5755.
24. Mejias R, Adamczyk A, Anggono V, Niranjan T, Thomas GM, et al. (2011) Gain-of-function glutamate receptor interacting protein 1 variants alter GluA2 recycling and surface distribution in patients with autism. Proc Natl Acad Sci 108: 4920–4925.
25. Croteau-Chonka DC, Marvelle AF, Lange EM, Lee NR, Adair LS, et al. (2011) Genome-wide association study of anthropometric traits and evidence of interactions with age and study year in Filipino women. Obesity (Silver Spring), 19: 1019–1027.
26. Parker CC, Cheng R, Sokoloff G, Lim JE, Skol AD, et al. (2011) Fine-mapping alleles for body weight in LG/J × SM/J F_2 and F(34) advanced intercross lines. Mamm Genome 22: 563–571.
27. Hayes BJ, Pryce J, Chamberlain AJ, Bowman PJ, Goddard ME (2010) Genetic architecture of complex traits and accuracy of genomic prediction: coat colour, milk-fat percentage, and type in Holstein cattle as contrasting model traits. PLoS Genet 6: e1001139.
28. Norris BJ, Whan VA (2008) A gene duplication affecting expression of the ovine ASIP gene is responsible for white and black sheep. Genome Res 18: 1282–1293.
29. He Y, Fang X, Emoto K, Jan YN, Adler PN (2005) The tricornered Ser/Thr protein kinase is regulated by phosphorylation and interacts with furry during Drosophila wing hair development. Mol Biol Cell 16: 689–700.
30. Fang X, Lu Q, Emoto K, Adler PN (2010) The Drosophila Fry protein interacts with Trc and is highly mobile in vivo. BMC Dev Biol 10: 40.
31. Cong J, Geng W, He B, Liu J, Charlton J, et al. (2001) The furry gene of Drosophila is important for maintaining the integrity of cellular extensions during morphogenesis. Development 128: 2793–2802.
32. Bromley CM, Snowder GD, Van Vleck LD (2000) Genetic parameters among growth, prolificacy, and wool traits of Columbia, Polypay, Rambouillet, and Targhee sheep. J Anim Sci 78: 846–858.
33. Sawalha RM, Snowder GD, Keown JF, Van Vleck LD (2005) Genetic relationship between milk score and litter weight for Targhee, Columbia, Rambouillet, and Polypay sheep. J Anim Sci 83: 786–793.
34. Michelizzi VN, Wu X, Dodson MV, Michal JJ, Zambrano-Varon J, et al. (2010) A global view of 54,001 single nucleotide polymorphisms (SNPs) on the Illumina BovineSNP50 BeadChip and their transferability to water buffalo. Int J Biol Sci 7: 18–27.
35. Wigginton JE, Cutler DJ, Abecasis GR (2005) A note on exact tests of Hardy-Weinberg equilibrium. Am J Hum Genet 76: 887–893.
36. Barrett JC, Fry B, Maller J, Daly MJ (2005) Haploview: analysis and visualization of LD and haplotype maps. Bioinformatics 21: 263–265.
37. Liu K, Muse SV (2005) PowerMarker: an integrated analysis environment for genetic marker analysis. Bioinformatics 21: 2128–2129.
38. Tamura K, Peterson D, Peterson N, Stecher G, Nei M, et al. (2011) MEGA5: Molecular Evolutionary Genetics Analysis using Maximum Likelihood, Evolutionary Distance, and Maximum Parsimony Methods. Mol Biol Evol 28: 2731–2739.
39. Wright S (1965) The interpretation of population structure by F-statistics with special regard to systems of mating. Evolution 19: 395–420.
40. Goudet J (2002) FSTAT, a program to estimate and test gene diversities and fixation indices, version 2.9.3.2. Department of Ecology and Evolution, Université de Lausanne, http://www2.unil.ch/popgen/softwares/fstat.htm.
41. Purcell S, Neale B, Todd-Brown K, Thomas L, Ferreira MA, et al. (2007) PLINK: a toolset for whole-genome association and population-based linkage analysis. Am J Hum Genet 81: 559–575.
42. Nicholson G, Smith AV, Jonsson F, Gustafsson O, Stefansson K, et al. (2002) Assessing population differentiation and isolation from single-nucleotide polymorphism data. J R Stat Soc Ser B Stat Methodol 64: 695–715.
43. Flori L, Fritz S, Jaffrézic F, Boussaha M, Gut I, et al. (2009) The genome response to artificial selection: a case study in dairy cattle. PLoS One 4: e6595.
44. Herrmann E (1997) Local bandwidth choice in kernel regression estimation. J Graphic Comput Statist 6: 35–54.
45. Huang DW, Sherman BT, Lempicki RA (2009) Systematic and integrative analysis of large gene lists using DAVID Bioinformatics Resources. Nature Protoc 4: 44–57.

3

Biological Basis of miRNA Action when Their Targets Are Located in Human Protein Coding Region

Wanjun Gu[1]*, Xiaofei Wang[1], Chuanying Zhai[1], Tong Zhou[2,3]*, Xueying Xie[1]*

1 Research Center of Learning Sciences, Southeast University, Nanjing, Jiangsu, China, **2** Institute for Personalized Respiratory Medicine, The University of Illinois at Chicago, Chicago, Illinois, United States of America, **3** Section of Pulmonary, Critical Care, Sleep & Allergy, Department of Medicine, The University of Illinois at Chicago, Chicago, Illinois, United States of America

Abstract

Recent analyses have revealed many functional microRNA (miRNA) targets in mammalian protein coding regions. But, the mechanisms that ensure miRNA function when their target sites are located in protein coding regions of mammalian mRNA transcripts are largely unknown. In this paper, we investigate some potential biological factors, such as target site accessibility and local translation efficiency. We computationally analyze these two factors using experimentally identified miRNA targets in human protein coding region. We find site accessibility is significantly increased in miRNA target region to facilitate miRNA binding. At the mean time, local translation efficiency is also selectively decreased near miRNA target region. GC-poor codons are preferred in the flank region of miRNA target sites to ease the access of miRNA targets. Within-genome analysis shows substantial variations of site accessibility and local translation efficiency among different miRNA targets in the genome. Further analyses suggest target gene's GC content and conservation level could explain some of the differences in site accessibility. On the other hand, target gene's functional importance and conservation level can affect local translation efficiency near miRNA target region. We hence propose both site accessibility and local translation efficiency are important in miRNA action when miRNA target sites are located in mammalian protein coding regions.

Editor: Marc Tjwa, University of Frankfurt - University Hospital Frankfurt, Germany

Funding: This work was supported by grants from the National Basic Research Program of China (2012CB316501), the National High Technology Research and Development Program of China (2012AA020401), the National Natural Science Foundation of China (30900836, 61171143 and 61001175), and a Scientific Research Foundation for the Returned Overseas Chinese Scholars, State Education Ministry, China. The funders had no role in study design, data collection and analysis, decision to publish, or preparation of the manuscript.

Competing Interests: The authors have declared that no competing interests exist.

* E-mail: wanjungu@gmail.com (WG); tongzhou@uic.edu (TZ); xxying.rcls@seu.edu.cn (XX)

Introduction

miRNAs are a class of small non-coding RNAs that regulate gene expression in post-transcriptional stage [1]. In the process of miRNA action, miRNAs are first specifically bound to their target sites in mRNA transcripts [1]. After binding, targeted mRNA transcripts can be silenced by several mechanisms, such as gene silencing, translational repression and mRNA degradation [2]. In plants, most miRNA target sites are located in protein coding region of target gene. Unlike plants, animal miRNAs have their target sites mostly in 3′ untranslated region (3′-UTR) of mRNA transcripts [1]. Two recent studies have investigated the reason why mammalian miRNA target sites are restricted to the 3′-UTR of mRNA transcripts [3,4]. They have suggested active mRNA translation may impede miRNA association with target mRNAs in mammalian genomes [3,4]. But, increasing evidences have confirmed many functional miRNA target sites are located in protein coding region of mammalian mRNA transcripts as well [5–12]. Notably, genome-wide analysis of Argonaute (AGO) HITS-CLIP data [13] and PAR-CLIP data [14] have revealed almost half AGO binding sites are located in human protein coding region. Given the prevalence of miRNA target sites occurred in mammalian protein coding sequences, it is important to investigate biological factors that may affect miRNA action when their targets are located in mammalian protein coding regions.

Comparative genomic analyses have been successfully used in identifying important factors of miRNA action [1,15]. For example, miRNA genes are evolutionary conserved between species [1,16] and within species [17]. At the same time, nucleotides in miRNA target region are under negative selection as well [18,19]. These have implicated the importance of sequence conservation for proper miRNA function [1]. Comparing with miRNA target sites that are located in 3′-UTR of mammalian mRNA transcripts, those in mammalian protein coding region still need code amino acid information of their translated proteins. Previous studies have suggested the degeneracy of genetic code enables DNA sequences to code extra regulatory information as well as amino acid sequences [20]. Biased usage of synonymous codons have been related to many biological processes, such as DNA stability [21], nucleosome positioning [22], mRNA stability [23–25], mRNA splicing [26,27], nonsense mediated mRNA decay [28], translation initiation [29–31], translation elongation [32–36] and co-translational protein folding [37,38]. Additionally, many studies have suggested synonymous codon choices near miRNA target sites are related to miRNA function in mammalian genomes. For example, synonymous substitution rate in miRNA target region is reduced in some mammalian genomes [39–42]. Tay *et al.* [12] have found silent mutations occurred in miRNA

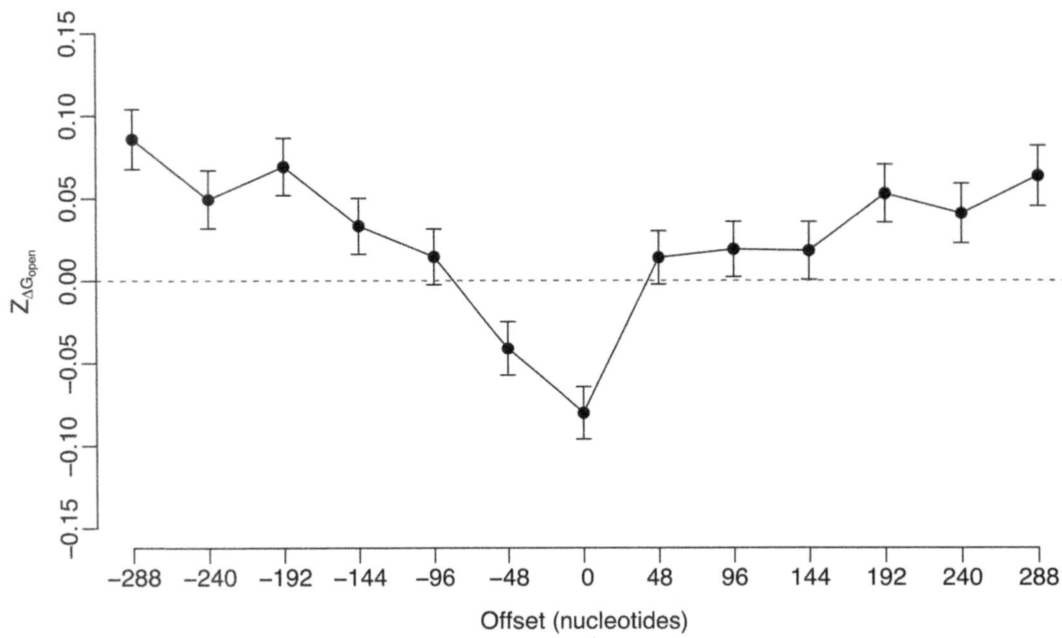

Figure 1. The mean and standard error of $Z_{\Delta G_{open}}$ of each sliding window near miRNA target region in the human genome.

target sites can eliminate miRNA activity in mouse. Brest *et al.* [43] have suggested that human Crohn's disease is caused by a synonymous mutation at the binding site of miR-196 in *IRGM*. Hence, analysis of synonymous codon usage in the flank region of miRNA target sites should be able to gain some insights of miRNA action.

In this paper, we analyzed the usage of synonymous codons in the flank region of miRNA target sites that are located in human protein coding sequences. We chose human as the example of mammalian species, since many functional miRNA targets have been experimentally identified in human protein coding region [14]. We considered site accessibility and local translation efficiency as two possible biological determinants of miRNA action when their target sites are located in human protein coding region. Site accessibility is one of the most important factors that affect miRNA binding when miRNA target sites are located in 3′-UTR of mammalian mRNA transcripts [44,45]. On the other hand, local translation efficiency around miRNA target region is particularly crucial for those targets located in mammalian protein coding region [3,4]. We hypothesized that if site accessibility and local translation efficiency are important in miRNA action, they should be selectively varied at synonymous codon sites that are in the flank region of miRNA target sites. We computed site accessibility and local translation efficiency for each miRNA target in human protein coding region. To estimate the selection pressure and its significance, we permuted mRNA sequences and assessed the deviation of local translation efficiency and site accessibility from random expectation. We addressed the following several problems: 1) is site accessibility important in miRNA action when their target sites are located in protein coding region of mammalian transcripts? 2) Is local translation efficiency near miRNA target region also important for proper miRNA function? 3) If site accessibility and/or local translation efficiency are selectively varied near miRNA target region, what are the factors that can affect the selection pressure?

Results

Site accessibility is selectively varied in the flank region of miRNA target sites

$Z_{\Delta G_{open}}$ measures the extent to which site accessibility deviates from random expectation. A negative $Z_{\Delta G_{open}}$ means that site accessibility is increased, and a positive $Z_{\Delta G_{open}}$ means it is decreased. We calculated $Z_{\Delta G_{open}}$ along mRNA sequences in sliding windows of 48 nucleotides in length. We started from the miRNA target region, which contains 21 nucleotides bound to miRNAs, 17 flank upstream nucleotides and 10 flank downstream nucleotides. We moved the sliding window both upward and downward along the mRNA sequences in a step of 48 nucleotides. We calculated $Z_{\Delta G_{open}}$ values in 13 consecutive windows for each miRNA target. For each window, we calculated a genomic mean value of $Z_{\Delta G_{open}}$ by averaging $Z_{\Delta G_{open}}$ values over all miRNA targets in the human genome.

Figure 1 shows the genomic mean value of $Z_{\Delta G_{open}}$ for all 13 consecutive windows in human. We observed a significant negative deviation of $Z_{\Delta G_{open}}$ from zero (*t*-test, $P << 10^{-6}$) in the central window. The negative values of $Z_{\Delta G_{open}}$ in the central window suggest selection for increased site accessibility in miRNA target region. When the sliding window moved upward or downward along the mRNA sequence, $Z_{\Delta G_{open}}$ values increase quickly and most windows have significant positive mean $Z_{\Delta G_{open}}$ values. The positive $Z_{\Delta G_{open}}$ values in these windows suggest decreased site accessibility is generally preferred in mRNA segments other than the miRNA target region. In following analysis, we used $Z_{\Delta G_{open}}$ in the central window to represent the selection signal of site accessibility in miRNA target region.

Translation efficiency is also selectively varied near miRNA target region

Z_{tAI} measures the deviation of local translation efficiency from random expectation. A negative Z_{tAI} value means local translation efficiency is reduced, and a positive Z_{tAI} value means it is increased. We calculated Z_{tAI} along mRNA sequences using a sliding window of nine codons (27 nucleotides) in length. We started from a window that is right downstream of miRNA target sites, and moved the window upward and downward along the mRNA sequence at a step of nine codons. We calculated Z_{tAI} in 29 consecutive windows for each miRNA target region. For each window, we calculated a genomic mean Z_{tAI} by averaging Z_{tAI} values over all miRNA target regions.

Figure 2 shows mean Z_{tAI} values in all 29 consecutive windows. We observed a significant negative deviation of Z_{tAI} value from zero (*t*-test, $P<<10^{-4}$) in a window that is nine codons downstream from the start point. The negative Z_{tAI} values in this window suggest decreased translation efficiency is selectively preferred near miRNA target sites. We did not observe any significant deviation of Z_{tAI} values from zero in other windows. This suggests reduced translation efficiency is only selectively preferred in the flank region of miRNA target sites. We used Z_{tAI} values in the window nine codons downstream of the start point to represent selection signal of local translation efficiency for following analysis.

GC-poor codons are preferred in miRNA target region

We observed site accessibility tended to increase in miRNA target region (Figure 1). Site accessibility is mainly determined by RNA secondary structure near miRNA target region. We calculated *Z*-score of local RNA secondary structure, $Z_{\Delta G_{local}}$, in all 13 sliding windows as we did in site accessibility analysis. We also observed decreased RNA stability in miRNA target region (Figure S1). $Z_{\Delta G_{local}}$ in miRNA target region is correlated well to $Z_{\Delta G_{open}}$ in that region (Pearson's product-moment correlation = −0.3604834, $P<<10^{-15}$; Figure S2). When looking into GC content in miRNA target region, we also observed a significant negative deviation of Z_{GC} values from zero ($Z_{GC} = -0.106+/-0.034$; *t*-test, $P = 1.3*10^{-9}$). A negative Z_{GC} value in miRNA target region suggests GC-poor codons are selectively preferred in that region. When comparing Z_{GC} and $Z_{\Delta G_{open}}$ of each miRNA target, we observed a significant positive correlation (Pearson's product-moment correlation = 0.3, $P<<10^{-15}$; Figure 3). We also observed a significant negative correlation (Pearson's product-moment orrelation = 0.25, $P<<10^{-15}$; Figure S3) between Z_{GC} and $Z_{\Delta G_{local}}$ in miRNA target region. This suggests GC-poor codons are selectively preferred in miRNA target region for increased site accessibility by loosing RNA structure in that region.

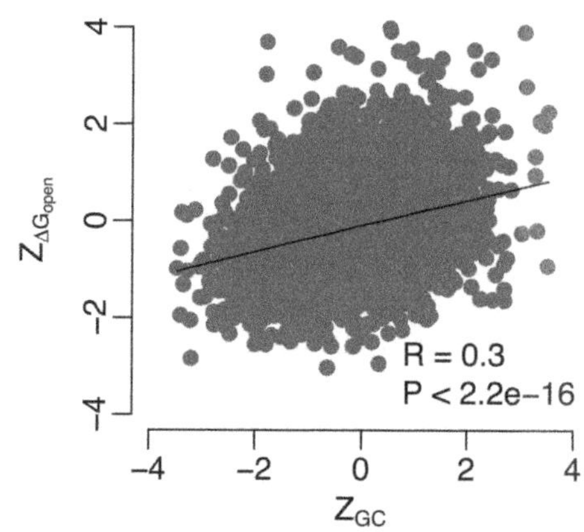

Figure 3. $Z_{\Delta G_{open}}$ in miRNA target region as a function of Z_{GC} in that region. Each point represents a miRNA target in human protein coding sequences.

Factors that influence the selection pressure among miRNA target regions

In previous sections, we considered the mean $Z_{\Delta G_{open}}$ and Z_{tAI} over all miRNA targets in protein coding region of the human genome. But, we observed substantial variations of site accessibility and translation efficiency among different miRNA targets in the genome (Figures 1 and 2). We next investigated the factors that may affect site accessibility and local translation efficiency of miRNA targets.

We first considered target gene's GC content. We selected miRNA targets in genes with the highest 5% and the lowest 5% GC content, and compared the mean $Z_{\Delta G_{open}}$ and Z_{tAI} of miRNA targets in these two groups. We observed $Z_{\Delta G_{open}}$ values of miRNA targets in GC-rich genes were significantly smaller than those in GC-poor genes (Welch Two Sample *t*-test, P = 0.04; Figure 4). This suggests miRNA targets in GC-rich genes are under higher selection pressures to increase site accessibility. But, Z_{tAI} values of miRNA targets in genes with the highest 5% GC content and those with the lowest 5% GC content are not statistically different (Welch Two Sample *t*-test, $P = 0.42$; Figure S4).

Next we considered the conservation level of miRNA target gene. We separated miRNA targets into three groups based on gene conservation, including primate-specific targets, mammal-specific targets and vertebrate-conserved targets. We compared

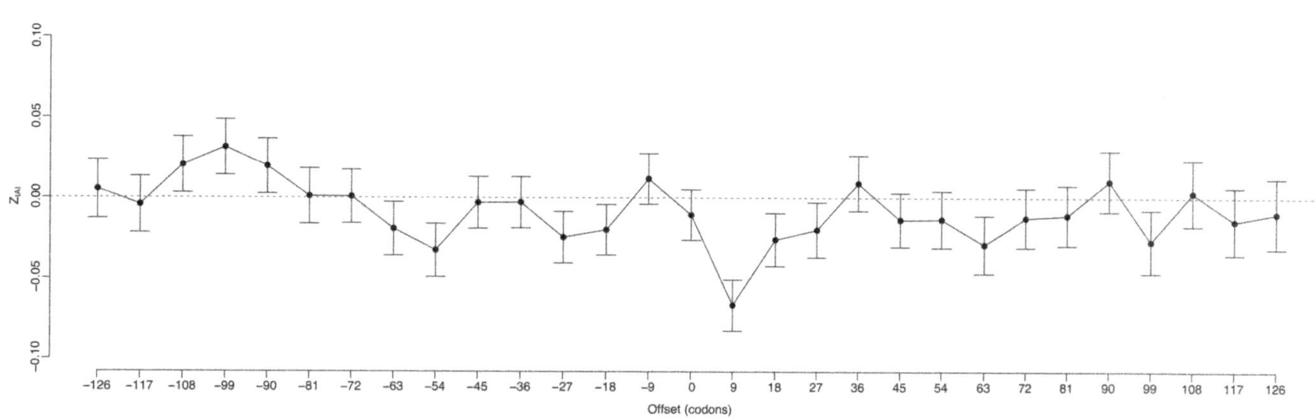

Figure 2. The mean and standard error of Z_{tAI} of each sliding window near miRNA target region in the human genome.

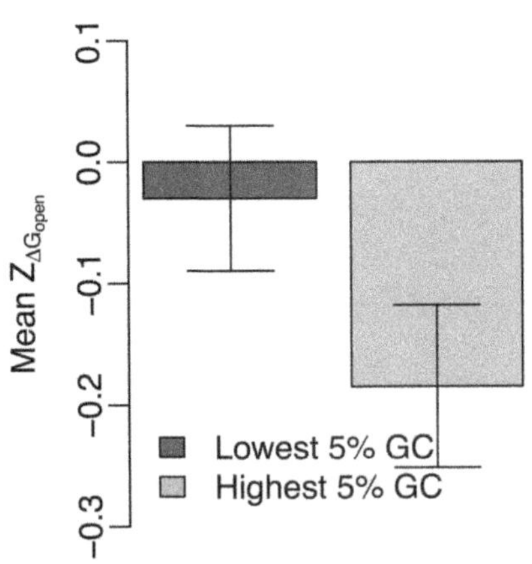

Figure 4. Comparison of the mean $Z_{\Delta G_{open}}$ between miRNA targets in genes with the highest 5% and the lowest 5% GC content.

$Z_{\Delta G_{open}}$ and Z_{tAI} values of miRNA targets in these three groups. Both $Z_{\Delta G_{open}}$ and Z_{tAI} mean of mammal-specific and vertebrate-conserved targets are smaller than those in primate-specific targets (Figures 5 and 6). While $Z_{\Delta G_{open}}$ values of mammal-specific targets (t-test, $P<<10^{-4}$) and vertebrate-conserved targets (t-test, $P=0.003$) are significantly deviated to zero, Z_{tAI} values of mammal-specific targets (t-test, $P<<10^{-4}$) and vertebrate-conserved targets (t-test, $P=0.012$) are also significantly deviated to zero. But, neither $Z_{\Delta G_{open}}$ (t-test, $P=0.938$) nor Z_{tAI} values (t-test, $P=0.519$) in primate-specific targets have significant deviation to zero. This suggests selection of site accessibility and translation efficiency is more pronounced for miRNA targets with longer evolutionary history.

We then considered the functional importance of target gene. We used the protein complex size to denote gene's functional importance. We compared $Z_{\Delta G_{open}}$ and Z_{tAI} values of miRNA targets in genes with the highest 5% complex size to those with the lowest 5% complex size. The mean Z_{tAI} value of miRNA targets in genes with higher functionality is smaller than that of miRNA targets in genes with less functionality (Figure 7). We did not observe any significant difference between $Z_{\Delta G_{open}}$ values of miRNA targets in genes with different functional importance (Figure S5).

We finally considered the expression level and codon bias of target gene. We used *ENC* (Effective Number of Codons) to measure a gene's codon usage bias [46]. The higher a gene's codon usage bias, the lower the gene's *ENC*. We compared $Z_{\Delta G_{open}}$ and Z_{tAI} values of miRNA targets in genes the highest 5% expression level to those with the lowest 5% expression level. We also compared $Z_{\Delta G_{open}}$ and Z_{tAI} values of miRNA targets in genes with the highest 5% *ENC* to those with the lowest 5% *ENC*. But, we did not observe any significant difference in these comparisons (Figures S6, S7, S8 and S9).

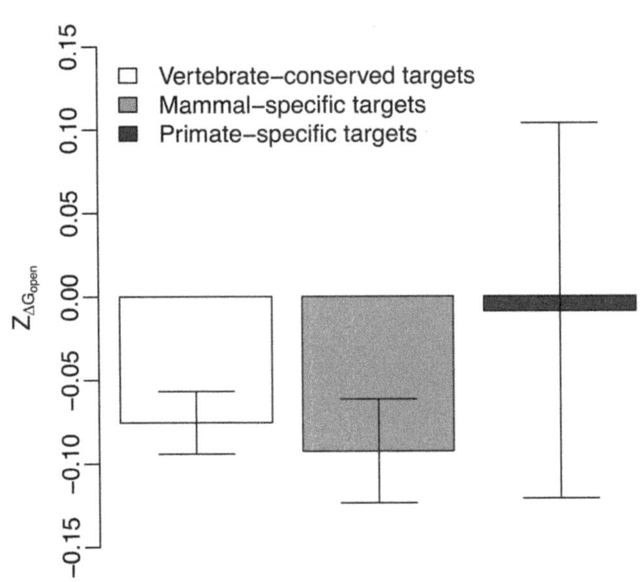

Figure 5. Comparison of the mean $Z_{\Delta G_{open}}$ between miRNA targets in genes at different conservation levels.

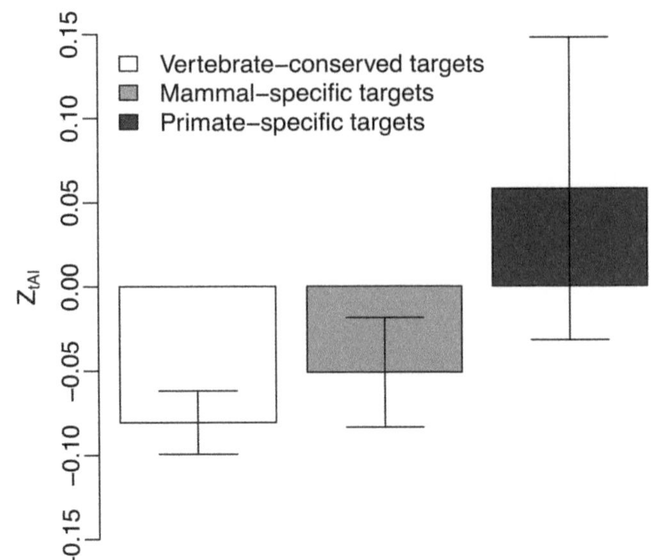

Figure 6. Comparison of the mean Z_{tAI} between miRNA targets in genes at different conservation levels.

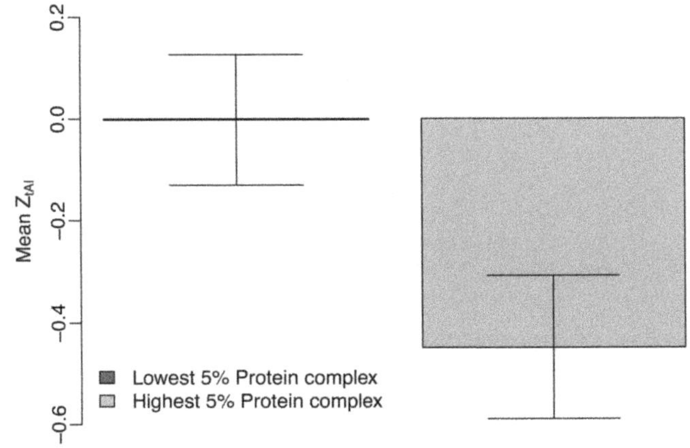

Figure 7. Comparison of the mean Z_{tAI} between miRNA targets in genes with the highest 5% and lowest 5% protein complex size.

Discussion

To investigate the biological basis of miRNA action when miRNA targets are located in protein coding region of mammalian transcripts, we have performed a genome scale analysis of site accessibility and translation efficiency near miRNA target region in the human genome. We have found both site accessibility and translation efficiency is selectively varied in the flank region of miRNA target sites (Figures 1 and 2). By focusing on *Z*-scores relative to the random expectation, our method has excluded some biases such as amino acid content or preferred codon usage that may cause the signal. Site accessibility in miRNA target region tends to be higher in real mRNA sequence than that expected from the permutated mRNA sequences (Figure 1). Our results are comparable to the observations made in some previous studies [44,45]. They had found site accessibility was the key factor in determining miRNA activity when miRNA target sites are located in 3′-UTR of human transcripts. Another two experimental studies had found active mRNA translation could impede miRNA association with target mRNAs in mammalian genomes [3,4]. When rare codons are introduced in the upstream region of miRNA target sites, miRNA activity could be restored [3]. In our results, we have found translation efficiency is decreased in the flank region of miRNA target sites (Figure 2). This confirms the importance of local translation efficiency in miRNA action for miRNA targets occurred in mammalian protein coding region. We have not observed any significant correlation between site accessibility and translation efficiency in the flank region of miRNA target sites (Pearson's product-moment correlation = 0.02, P = 0.27; Figure S10). Hence, site accessibility and local translation efficiency may be two independent factors in miRNA action when miRNA targets are located in mammalian protein coding regions.

In previous analysis, we had found site accessibility, but not local translation efficiency, was related to miRNA action when miRNA targets are located in protein coding region of plant genes [47,48]. This suggests site accessibility is a much general factor that affects miRNA action for different kinds of miRNA targets [44,45,47]. Comparing with site accessibility, local translation efficiency is only effective when miRNA targets are located in mammalian protein coding region [3,4,48]. Is it possible that the selection signal for reduced translation efficiency near miRNA target sites is an artifact caused by our method? To eliminate this possibility, for each miRNA target, we randomly picked a region of 21 nucleotides in length from the same mRNA sequence. We combined all these randomly selected gene regions as a dataset of randomized miRNA targets. We replicated our analysis of translation efficiency on all randomized miRNA targets. We did not observe any selection signal of reduced translation efficiency in all sliding windows (Figure S11). This implies that reduced translation efficiency near miRNA target region we have observed Figure 2 is a real signal. The reason why reduced translation efficiency is only important for miRNA targets in mammalian protein coding region is unknown. Some analysis had suggested most animal miRNAs regulated their target genes by translational repression, while plant miRNAs regulated their target genes mainly by RNA degradation [1,2]. This difference may explain the different importance of local translation efficiency in miRNA action between plants and mammals.

In our results, the region that tends to increase site accessibility is miRNA target region, which includes miRNA target sites, 17 flank upstream nucleotides and 10 downstream nucleotides (Figure 1). This region is the same as the one that we had observed in four plant genomes [29], and is comparable to those had been identified from experimental data [44,45]. We have observed that $Z_{\Delta G_{open}}$ value in most other windows is positive (Figure 1). This implies site accessibility in mRNA sequences is generally decreased. This is mainly caused by much tighter RNA structure in mRNA sequences (Figure S1). The latter result is comparable to the observations made by several previous studies that the overall mRNA stability is selectively increased in human [49,50].

We have observed that the region with decreased translation efficiency is located downstream of miRNA target sites (Figure 2). This region is different from the one suggested by Gu *et al.* [3]. They had employed model reporter constructs by introducing mutations into stop codons located upstream of miRNA target sites, which allowed translation to proceed through miRNA target sites [3]. They had found active translation could totally impede miRNA regulation. But, miRNA regulation can be restored when nine consecutive rare codons are introduced right upstream of miRNA target sites. In contrast, Lin *et al.* [4] had analyzed two naturally occurred miRNA target sites in viral protein coding sequences. They had suggested translation could modestly decrease miRNA regulation of those two naturally occurred miRNA targets. They had found the distribution of rare codons in the upstream region of those two miRNA targets is same as that of random sequences [4]. Our results are largely comparable to that observed in Lin *et al.* [4]. For naturally occurred miRNA targets in human protein coding sequences, we have observed reduced translation efficiency is selectively preferred near miRNA target region. But, translation efficiency in the region right upstream of miRNA target sites is not selectively varied. Instead, reduced translation efficiency has been observed in a nearby region located downstream of the miRNA target sites. A possible explanation is naturally occurred miRNA targets may use two separate regions to facilitate miRNA binding. Since the region right upstream of miRNA target sites has already been used to increase site accessibility, they may use the region downstream of the miRNA target region to slow down translational process near miRNA target sites. These two layers of selection may act together in the flank region of miRNA targets to ensure proper miRNA activity. It will be interesting to set up some experiments to validate this possibility.

Our results have suggested that the conservation level of target gene is a general factor that affects both site accessibility and translation efficiency in miRNA target region. We have found $Z_{\Delta G_{open}}$ and Z_{tAI} values are smaller in miRNA targets located in vertebrate-conserved genes and mammal-specific genes (Figures 5 and 6). But, miRNA targets in primate-specific genes have no obvious signal of selection for site accessibility and translation efficiency (Figures 5 and 6). These are consistent with the observations made by previous works [1,17,18,51]. They had suggested purifying selection was smaller in younger miRNA targets than that in conserved miRNA targets.

We have found $Z_{\Delta G_{open}}$ values of target regions in GC-rich genes are smaller than that in GC-poor genes (Figure 4). This finding can be explained by thermodynamic rules. Since GC pairs have three hydrogen bonds, GC-rich codons tend to fold in more stable RNA structures than AT-rich codons. Because loose RNA secondary structure is preferred in miRNA target region (Figure S1), GC-poor codons should be selectively preferred in miRNA target region to increase site accessibility, which is consistent with what we have observed in Figure 3. Assuming that selection targets the same higher site accessibility near all miRNA targets, we would expect that the increase of site accessibility is larger near miRNA targets located in GC-rich genes, simply because they start from a more-stable baseline. Whether selection actually targets the same higher site accessibility is not determined by our analysis. The actual larger decrease of $Z_{\Delta G_{open}}$ values for miRNA targets in GC-rich genes (Figure 4) implies this possibility.

In conclusion, we suggest site accessibility and translation efficiency have some effects on miRNA function when their targets are located in protein coding region of mammalian genes. Our results may help us better understand the process of miRNA action. It will also bring us important implications of protein coding sequence evolution in mammalian genome.

Materials and Methods

Data

Human PAR-CLIP data were downloaded from Supplementary Materials of Hafner *et al.* [14]. The PAR- CLIP data from Hafner *et al.* [14] contains genomic coordinates of DNA elements that were bound to human miRNAs in the experiment. Gene annotations of the human genome and mRNA sequences of all human transcripts were downloaded from Ensembl [52]. Overall, we extracted 4,344 miRNA targets that are located in human protein coding region.

To investigate the factors that may affect site accessibility and/or translation efficiency in the flank region of miRNA target sites, we considered the conservation level, the expression level and protein complex size of target gene. We downloaded a multiple alignment of 45 vertebrate genomes from UCSC genome server [53]. Using the 45-way alignment, we classified miRNA target genes into three groups (primate-specific genes, mammal-specific genes and vertebrate-conserved genes). We next downloaded expression data of human genes from Su *et al.* [54]. The expression level of each gene was quantified as the geometric mean of expression among different tissues. Finally, we downloaded protein complex data from the CORUM database [55]. For each gene, the complex size was measured as the number of proteins forming it and averaged across the number of complexes the gene participates.

Site accessibility

Site accessibility represents the difficulty in opening a segment of mRNA sequences around miRNA target sites for their binding with RNA-induced silencing complex (RISCs) [15]. We used ΔG_{open} to measure site accessibility for each miRNA target. ΔG_{open} is the difference between free energy of the secondary structure of miRNA target region and free energy of the secondary structure in which miRNA target sites are unpaired [45]. ΔG_{open} has been proved to be a good indicator of site accessibility [44,45]. We used a window of 48 nucleotides, including 21 nucleotides directly bound by miRNAs, 17 flank upstream and 10 flank downstream nucleotides, to represent a miRNA target region. Previous studies have observed a significant correlation between ΔG_{open} value of this region and miRNA activity [44,45]. We calculated ΔG_{open} for each miRNA target using *RNAddG4* program in *PITA* package [45] with default parameter settings. At the same time, the folding energy of local RNA secondary structure, ΔG_{local}, was also calculated for each miRNA target by RNAddG4 [45]. In *RNAddG4* [45], *RNAfold* [56] was used to calculate the free energy of RNA secondary structures. As suggested in previous studies [45,57–60], we used a segment of mRNA sequences, rather than the full-length mRNAs, as the input sequences in calculating free energy of local RNA secondary structure. For each miRNA target region, the input mRNA segment includes 48 nucleotides in miRNA target region and additional 140 flank upstream and downstream nucleotides. This is based on the fact that the probability of base pairing when nucleotides are separated by more than 140 nucleotides was low (data not shown) and it can substantially reduce computational complexity.

Translation efficiency

We used tRNA adaptation index (tAI) to quantify translation efficiency in the flank region of mRNA target sites. tAI is a measure of codon adaptation to the tRNA abundance in genome [61]. tAI has been widely used to estimate translation efficiency in human [31,61,62]. We calculatedtAI using codonR package [61], where tRNA copy numbers in the human genome were downloaded from the Genomic tRNA Database [63]. Since nine rare codons introduced in the upstream region of miRNA target sites are able to slow down local translational process and secure miRNA function [3], we calculated tAI in a window of nine consecutive codons in the flank region of miRNA target sites as a measure of local translation efficiency for each miRNA target.

mRNA randomization

We hypothesized that if site accessibility and/or local translation efficiency is important for miRNA action when miRNA target sites are located in human protein coding sequences, synonymous codons should be selected for site accessibility and/or local translation efficiency in human. To detect the selection signal on synonymous codons, we used a randomization method to compute the statistical deviation of site accessibility and/or translation efficiency between the real mRNA sequence and permuted sequences [29,47–49]. We randomly shuffled synonymous codons among sites for each mRNA transcript, while keeping the encoded protein sequences, gene's codon usage bias and gene's GC composition the same. We did not shuffle the codons that are directly targeted by miRNAs during mRNA randomization, since those nucleotides are crucial for proper miRNA function. We generated 1,000 such permuted mRNA sequences for all mRNA transcripts with miRNA target sites in their protein-coding region.

Z-score calculation

We used Z-score to determine the deviation of the real sequence from randomized sequences and estimate its statistical significance. For each miRNA target region, we calculated site accessibility in the real mRNA transcript $(\Delta G_{open})_N$ and each permutated mRNA sequence $(\Delta G_{open})_P$. Then, we calculated Z-score of site accessibility ($Z_{\Delta G_{open}}$) for each miRNA target region as formula 1.

$$Z_{\Delta G_{open}} = \frac{(\Delta G_{open})_N - \overline{(\Delta G_{open})_P}}{\sqrt{\sum_{i=1}^{n} \frac{((\Delta G_{open})_{P_i} - \overline{(\Delta G_{open})_P})^2}{n-1}}} \quad (1)$$

Here, $(\Delta G_{open})_N$ is site accessibility for the naturally occurring target region under consideration. $(\Delta G_{open})_{P_i}$ is site accessibility for the target region in i^{th} permuted sequence, and $\overline{(\Delta G_{open})_P}$ is the mean of $(\Delta G_{open})_{P_i}$ over all permuted sequences. The variable n represents the total number of permuted sequences, which is equal to 1,000 in our analysis.

Similarly, we calculated Z-score of local secondary structure ($Z_{\Delta G_{local}}$), Z-score of local GC content (Z_{GC}) and Z-score of local translation efficiency (Z_{tAI}) in the flank region of each miRNA target as formulas below.

$$Z_{\Delta G_{local}} = \frac{(\Delta G_{local})_N - \overline{(\Delta G_{local})_P}}{\sqrt{\sum_{i=1}^{n} \frac{((\Delta G_{local})_{P_i} - \overline{(\Delta G_{local})_P})^2}{n-1}}} \quad (2)$$

$$Z_{GC} = \frac{GC_N - \overline{GC_P}}{\sqrt{\sum_{i=1}^{n} \frac{(GC_{P_i} - \overline{GC_P})^2}{n-1}}} \quad (3)$$

$$Z_{tAI} = \frac{tAI_N - \overline{tAI_P}}{\sqrt{\sum_{i=1}^{n} \frac{(tAI_{P_i} - \overline{tAI_P})^2}{n-1}}} \quad (4)$$

The definitions for$(\Delta G_{local})_N$, GC_N, and tAI_N are analogous to$(\Delta G_{open})_N$, but refer to free energy of local RNA secondary structure, GC content and translation efficiency rather than to site accessibility. Similarly, the definitions for $(\Delta G_{local})_{P_i}$, GC_{P_i}, and tAI_{P_i} are analogous to $(\Delta G_{open})_{P_i}$, and the definitions for $\overline{(\Delta G_{local})_P}$, $\overline{GC_P}$, and $\overline{tAI_P}$ are analogous to $\overline{(\Delta G_{open})_P}$.

Supporting Information

Figure S1 The mean and standard error of $Z_{\Delta G_{local}}$ of each sliding window near miRNA target region in the human genome.

Figure S2 $Z_{\Delta G_{open}}$ in miRNA target region as a function of $Z_{\Delta G_{local}}$ in that region. Each point represents a miRNA target in human protein coding sequences.

Figure S3 $Z_{\Delta G_{local}}$ in miRNA target region as a function of Z_{GC} in that region. Each point represents a miRNA target in human protein coding sequences.

Figure S4 Comparison of the mean Z_{tAI} between miRNA targets in genes with the highest 5% and lowest 5% GC content.

Figure S5 Comparison of the mean $Z_{\Delta G_{open}}$ between miRNA targets in genes with the highest 5% and lowest 5% protein complex size.

Figure S6 Comparison of the mean $Z_{\Delta G_{open}}$ between miRNA targets in genes with the highest 5% and lowest 5% ENC.

Figure S7 Comparison of the mean Z_{tAI} between miRNA targets in genes with the highest 5% and lowest 5% ENC.

Figure S8 Comparison of the mean $Z_{\Delta G_{open}}$ between miRNA targets in genes with the highest 5% and lowest 5% expression level.

Figure S9 Comparison of the mean Z_{tAI} between miRNA targets in genes with the highest 5% and lowest 5% expression level.

Figure S10 $Z_{\Delta G_{open}}$ in miRNA target region as a function of Z_{tAI} in the window that is nine codons downstream of miRNA target sites. Each point represents a miRNA target in human protein coding sequences.

Figure S11 The mean and standard error of Z_{tAI} of each sliding window near randomized miRNA target region.

Author Contributions

Conceived and designed the experiments: WG XX TZ. Performed the experiments: WG XW CZ XX. Analyzed the data: WG XX TZ. Contributed reagents/materials/analysis tools: XW XX. Wrote the paper: WG XX TZ.

References

1. Bartel DP (2009) MicroRNAs: target recognition and regulatory functions. Cell 136: 215–233.
2. Brodersen P, Voinnet O (2009) Revisiting the principles of microRNA target recognition and mode of action. Nat Rev Mol Cell Biol 10: 141–148.
3. Gu S, Jin L, Zhang F, Sarnow P, Kay MA (2009) Biological basis for restriction of microRNA targets to the 3′ untranslated region in mammalian mRNAs. Nat Struct Mol Biol 16: 144–150.
4. Lin HR, Ganem D (2011) Viral microRNA target allows insight into the role of translation in governing microRNA target accessibility. Proc Natl Acad Sci U S A 108: 5148–5153.
5. Fang Z, Rajewsky N (2011) The impact of miRNA target sites in coding sequences and in 3′UTRs. PLoS One 6: e18067.
6. Forman JJ, Coller HA (2010) The code within the code: microRNAs target coding regions. Cell Cycle 9: 1533–1541.
7. Forman JJ, Legesse-Miller A, Coller HA (2008) A search for conserved sequences in coding regions reveals that the let-7 microRNA targets Dicer within its coding sequence. Proc Natl Acad Sci U S A 105: 14879–14884.
8. Reczko M, Maragkakis M, Alexiou P, Grosse I, Hatzigeorgiou AG (2012) Functional microRNA targets in protein coding sequences. Bioinformatics 28: 771–776.
9. Rigoutsos I (2009) New tricks for animal microRNAS: targeting of amino acid coding regions at conserved and nonconserved sites. Cancer Res 69: 3245–3248.
10. Schnall-Levin M, Rissland OS, Johnston WK, Perrimon N, Bartel DP, et al. (2011) Unusually effective microRNA targeting within repeat-rich coding regions of mammalian mRNAs. Genome Res 21: 1395–1403.
11. Schnall-Levin M, Zhao Y, Perrimon N, Berger B (2010) Conserved microRNA targeting in Drosophila is as widespread in coding regions as in 3′UTRs. Proc Natl Acad Sci U S A 107: 15751–15756.
12. Tay Y, Zhang J, Thomson AM, Lim B, Rigoutsos I (2008) MicroRNAs to Nanog, Oct4 and Sox2 coding regions modulate embryonic stem cell differentiation. Nature 455: 1124–1128.
13. Chi SW, Zang JB, Mele A, Darnell RB (2009) Argonaute HITS-CLIP decodes microRNA-mRNA interaction maps. Nature 460: 479–486.
14. Hafner M, Landthaler M, Burger L, Khorshid M, Hausser J, et al. (2010) Transcriptome-wide identification of RNA-binding protein and microRNA target sites by PAR-CLIP. Cell 141: 129–141.
15. Axtell MJ, Westholm JO, Lai EC (2011) Vive la difference: biogenesis and evolution of microRNAs in plants and animals. Genome Biol 12: 221.
16. Axtell MJ, Bowman JL (2008) Evolution of plant microRNAs and their targets. Trends Plant Sci 13: 343–349.
17. Fahlgren N, Jogdeo S, Kasschau KD, Sullivan CM, Chapman EJ, et al. (2010) MicroRNA gene evolution in Arabidopsis lyrata and Arabidopsis thaliana. Plant Cell 22: 1074–1089.
18. Cuperus JT, Fahlgren N, Carrington JC (2011) Evolution and functional diversification of MIRNA genes. Plant Cell 23: 431–442.
19. Saunders MA, Liang H, Li WH (2007) Human polymorphism at microRNAs and microRNA target sites. Proc Natl Acad Sci U S A 104: 3300–3305.
20. Itzkovitz S, Hodis E, Segal E (2010) Overlapping codes within protein-coding sequences. Genome Res 20: 1582–1589.
21. Vinogradov AE (2003) DNA helix: the importance of being GC-rich. Nucleic Acids Res 31: 1838–1844.

22. Warnecke T, Batada NN, Hurst LD (2008) The impact of the nucleosome code on protein-coding sequence evolution in yeast. PLoS Genet 4: e1000250.
23. Chamary JV, Hurst LD (2005) Biased codon usage near intron-exon junctions: selection on splicing enhancers, splice-site recognition or something else? Trends Genet 21: 256–259.
24. Chamary JV, Parmley JL, Hurst LD (2006) Hearing silence: non-neutral evolution at synonymous sites in mammals. Nat Rev Genet 7: 98–108.
25. Stoletzki N (2008) Conflicting selection pressures on synonymous codon use in yeast suggest selection on mRNA secondary structures. BMC Evol Biol 8: 224.
26. Parmley JL, Chamary JV, Hurst LD (2006) Evidence for purifying selection against synonymous mutations in mammalian exonic splicing enhancers. Mol Biol Evol 23: 301–309.
27. Warnecke T, Hurst LD (2007) Evidence for a trade-off between translational efficiency and splicing regulation in determining synonymous codon usage in Drosophila melanogaster. Mol Biol Evol 24: 2755–2762.
28. Wilke CO (2011) Transcriptional robustness complements nonsense-mediated decay in humans. PLoS Genet 7: e1002296.
29. Gu W, Zhou T, Wilke CO (2010) A universal trend of reduced mRNA stability near the translation-initiation site in prokaryotes and eukaryotes. PLoS Comput Biol 6: e1000664.
30. Kudla G, Murray AW, Tollervey D, Plotkin JB (2009) Coding-sequence determinants of gene expression in Escherichia coli. Science 324: 255–258.
31. Tuller T, Waldman YY, Kupiec M, Ruppin E (2010) Translation efficiency is determined by both codon bias and folding energy. Proc Natl Acad Sci U S A 107: 3645–3650.
32. Akashi H, Eyre-Walker A (1998) Translational selection and molecular evolution. Curr Opin Genet Dev 8: 688–693.
33. Duret L (2002) Evolution of synonymous codon usage in metazoans. Curr Opin Genet Dev 12: 640–649.
34. Ikemura T (1985) Codon usage and tRNA content in unicellular and multicellular organisms. Mol Biol Evol 2: 13–34.
35. Morton BR, Wright SI (2007) Selective constraints on codon usage of nuclear genes from Arabidopsis thaliana. Mol Biol Evol 24: 122–129.
36. Zhou T, Weems M, Wilke CO (2009) Translationally optimal codons associate with structurally sensitive sites in proteins. Mol Biol Evol 26: 1571–1580.
37. Komar AA, Lesnik T, Reiss C (1999) Synonymous codon substitutions affect ribosome traffic and protein folding during in vitro translation. FEBS Lett 462: 387–391.
38. Thanaraj TA, Argos P (1996) Ribosome-mediated translational pause and protein domain organization. Protein Sci 5: 1594–1612.
39. Hurst LD (2006) Preliminary assessment of the impact of microRNA-mediated regulation on coding sequence evolution in mammals. J Mol Evol 63: 174–182.
40. Hurst LD (2011) Molecular genetics: The sound of silence. Nature 471: 582–583.
41. Lin MF, Kheradpour P, Washietl S, Parker BJ, Pedersen JS, et al. (2011) Locating protein-coding sequences under selection for additional, overlapping functions in 29 mammalian genomes. Genome Res 21: 1916–1928.
42. Lindblad-Toh K, Garber M, Zuk O, Lin MF, Parker BJ, et al. (2011) A high-resolution map of human evolutionary constraint using 29 mammals. Nature 478: 476–482.
43. Brest P, Lapaquette P, Souidi M, Lebrigand K, Cesaro A, et al. (2011) A synonymous variant in IRGM alters a binding site for miR-196 and causes deregulation of IRGM-dependent xenophagy in Crohn's disease. Nat Genet 43: 242–245.
44. Hausser J, Landthaler M, Jaskiewicz L, Gaidatzis D, Zavolan M (2009) Relative contribution of sequence and structure features to the mRNA binding of Argonaute/EIF2C-miRNA complexes and the degradation of miRNA targets. Genome Res 19: 2009–2020.
45. Kertesz M, Iovino N, Unnerstall U, Gaul U, Segal E (2007) The role of site accessibility in microRNA target recognition. Nat Genet 39: 1278–1284.
46. Wright F (1990) The 'effective number of codons' used in a gene. Gene 87: 23–29.
47. Gu W, Wang X, Zhai C, Xie X, Zhou T (2012) Selection on Synonymous Sites for Increased Accessibility around miRNA Binding Sites in Plants. Molecular Biology and Evolution 29: 3037–3044.
48. Gu W, Zhai C, Wang X, Xie X, Parinandi G, et al. (2012) Translation Efficiency in Upstream Region of microRNA Targets in Arabidopsis thaliana. Evolutionary Bioinformatics 8: 565.
49. Chamary J, Hurst L (2005) Evidence for selection on synonymous mutations affecting stability of mRNA secondary structure in mammals. Genome Biology 6: R75.
50. Seffens W, Digby D (1999) mRNAs have greater negative folding free energies than shuffled or codon choice randomized sequences. Nucleic Acids Res 27: 1578–1584.
51. Ehrenreich IM, Purugganan M (2008) MicroRNAs in plants: Possible contributions to phenotypic diversity. Plant Signal Behav 3: 829–830.
52. Kersey PJ, Lawson D, Birney E, Derwent PS, Haimel M, et al. (2010) Ensembl Genomes: extending Ensembl across the taxonomic space. Nucleic Acids Res 38: D563–569.
53. Fujita PA, Rhead B, Zweig AS, Hinrichs AS, Karolchik D, et al. (2011) The UCSC Genome Browser database: update 2011. Nucleic Acids Research 39: D876–D882.
54. Su AI, Wiltshire T, Batalov S, Lapp H, Ching KA, et al. (2004) A gene atlas of the mouse and human protein-encoding transcriptomes. Proceedings of the National Academy of Sciences of the United States of America 101: 6062–6067.
55. Ruepp A, Waegele B, Lechner M, Brauner B, Dunger-Kaltenbach I, et al. (2010) CORUM: the comprehensive resource of mammalian protein complexes–2009. Nucleic Acids Res 38: D497–501.
56. Hofacker IL (2003) Vienna RNA secondary structure server. Nucleic Acids Res 31: 3429–3431.
57. Lu M, Zhang Q, Deng M, Miao J, Guo Y, et al. (2008) An Analysis of Human MicroRNA and Disease Associations. PLoS One 3: e3420.
58. Richter AS, Schleberger C, Backofen R, Steglich C (2010) Seed-based INTARNA prediction combined with GFP-reporter system identifies mRNA targets of the small RNA Yfr1. Bioinformatics 26: 1–5.
59. Ying X, Cao Y, Wu J, Liu Q, Cha L, et al. (2011) sTarPicker: a method for efficient prediction of bacterial sRNA targets based on a two-step model for hybridization. PLoS One 6: e22705.
60. Zhao Y, Samal E, Srivastava D (2005) Serum response factor regulates a muscle-specific microRNA that targets Hand2 during cardiogenesis. Nature 436: 214–220.
61. dos Reis M, Savva R, Wernisch L (2004) Solving the riddle of codon usage preferences: a test for translational selection. Nucleic Acids Res 32: 5036–5044.
62. Waldman YY, Tuller T, Shlomi T, Sharan R, Ruppin E (2010) Translation efficiency in humans: tissue specificity, global optimization and differences between developmental stages. Nucleic Acids Research 38: 2964–2974.
63. Chan PP, Lowe TM (2009) GtRNAdb: a database of transfer RNA genes detected in genomic sequence. Nucleic Acids Res 37: D93–97.

4

VirusFinder: Software for Efficient and Accurate Detection of Viruses and Their Integration Sites in Host Genomes through Next Generation Sequencing Data

Qingguo Wang[1], Peilin Jia[1,4], Zhongming Zhao[1,2,3,4]*

1 Department of Biomedical Informatics, Vanderbilt University School of Medicine, Nashville, Tennessee, United States of America, **2** Department of Psychiatry, Vanderbilt University School of Medicine, Nashville, Tennessee, United States of America, **3** Department of Cancer Biology, Vanderbilt University School of Medicine, Nashville, Tennessee, United States of America, **4** Center for Quantitative Sciences, Vanderbilt University School of Medicine, Nashville, Tennessee, United States of America

Abstract

Next generation sequencing (NGS) technologies allow us to explore virus interactions with host genomes that lead to carcinogenesis or other diseases; however, this effort is largely hindered by the dearth of efficient computational tools. Here, we present a new tool, VirusFinder, for the identification of viruses and their integration sites in host genomes using NGS data, including whole transcriptome sequencing (RNA-Seq), whole genome sequencing (WGS), and targeted sequencing data. VirusFinder's unique features include the characterization of insertion loci of virus of arbitrary type in the host genome and high accuracy and computational efficiency as a result of its well-designed pipeline. The source code as well as additional data of VirusFinder is publicly available at http://bioinfo.mc.vanderbilt.edu/VirusFinder/.

Editor: Dongxiao Zhu, Wayne State University, United States of America

Funding: This work was supported by National Institutes of Health grants (R01LM011177, P30CA68485, and P50CA095103). The funders had no role in study design, data collection and analysis, decision to publish, or preparation of the manuscript.

* E-mail: zhongming.zhao@vanderbilt.edu

Introduction

Viral infection, especially from tumorigenic viruses, is one of the leading causes of deaths worldwide. Some viruses, e.g. the hepatitis B virus (HBV), can fuse into a host genome ("integrated") to interrupt gene functions or induce chromosomal instability [1–4], while other viruses, e.g. the hepatitis C virus (HCV), rarely integrate into a host genome ("unintegrated"). Detecting the existence of viruses and, especially, their integration sites in host genomes is critical in understanding their molecular mechanisms in disease development.

With the rapid advances in next generation sequencing (NGS) technologies over the past several years and their increasingly widespread applications in clinical settings, recent large-scale investigations of virus-host interactions were carried out to shed light on virus-related cancers [2,3,5,6]. The strong demand for the NGS investigation of virus-host interactions is currently hindered by the lack of effective NGS tools for virus detection. Recent tools such as PathSeq [7], RINS [8], and READSCAN [9] apply computational subtraction to pathogen detection in NGS data. However, these tools do not identify virus integration events, which are important for studying tumorigenic viruses like HBV in carcinogenesis [2,3]. VirusSeq [10] uses whole transcriptome sequencing (RNA-Seq) data to detect virus integration sites in the human genome, yet is not practical for whole genome sequencing (WGS) data due to its high CPU requirements. Additionally, by including in its reference sequence 18 well-known viruses that have demonstrated potential to integrate into other genomes, VirusSeq is not able to identify integration sites of viruses other than the 18 predefined ones if without modification of its code. Another viral integration-detecting tool, ViralFusionSeq [11], can work with WGS data in addition to RNA-Seq, but is not applicable to samples infected with undiagnosed viruses; rather, it requires virus sequence in advance as a pre-defined input. All these tools were recently developed, and their computational capacities and accuracy remain largely uncertain.

To address the above limitations, here, we introduce VirusFinder for efficient and accurate identification of viruses and their integration sites (if present) from NGS data. Specifically, VirusFinder aims to: (i) detect the presence of viruses of arbitrary types in a host sample, whether they are integrated or unintegrated; (ii) detect virus insertion sites as long as the virus fuses into the host genome and its integration sites are captured by sequencing technologies; and (iii) work on WGS, RNA-Seq or targeted sequencing data. VirusFinder does not require virus sequences as a prerequisite input. Hence, it can not only work with NGS data with a specified virus type, like VirusSeq and ViralFusionSeq, but also be applied readily to samples infected with undiagnosed viruses, to which neither VirusSeq nor ViralFusionSeq is directly applicable.

Methods

Figure 1 illustrates the pipeline of VirusFinder, which overall follows a three-step procedure: (1) preprocessing, (2) virus detection, and (3) virus integration site detection. These steps are elaborated in detail in the text below.

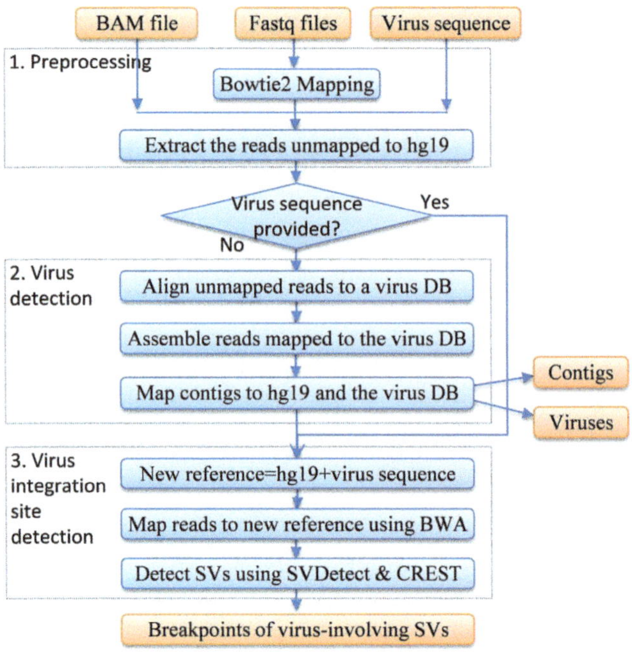

Figure 1. VirusFinder's pipeline to detect viruses and their integration sites in next generation sequencing (NGS) data. VirusFinder overall follows a three-step procedure: (1) preprocessing, (2) virus detection, and (3) virus integration site detection. The current version of VirusFinder (release 1.0) uses UCSC hg19 (http://hgdownload.cse.ucsc.edu/downloads.html#human) as reference human genome. DB: database. SVs: structural variants (especially inter-chromosomal translocations).

(1) Preprocessing

VirusFinder's input can either be raw sequencing reads (in Fastq format) or an alignment file (in BAM format). If user provides only raw sequencing reads, VirusFinder will first use the alignment tool Bowtie 2 [12] to map these reads to a human reference genome, which can either be NCBI build 37/36 (http://www.ncbi.nlm.nih.gov/projects/mapview/) or UCSC hg19/hg18 (http://hgdownload.cse.ucsc.edu/downloads.html#human). VirusFinder runs Bowtie 2 in its sensitive end-to-end mode, in which Bowtie 2 does not trim (or "soft clip") characters from short reads in order to achieve high alignment speed. With the alignment file generated by Bowtie 2 or provided by the user, VirusFinder then garners all reads unmapped to the human reference genome for downstream analysis.

Here in this step, user is allowed to provide the sequence of the virus being examined as an input parameter of VirusFinder. VirusFinder will skip step (2) of the pipeline if user provides the virus sequence.

(2) Virus detection

This step is used to detect the specific type(s) of virus(es) present in the sample. This step will be skipped if user supplies the virus sequence to VirusFinder. If the virus type is unknown, however, VirusFinder first aligns the unmapped reads collected in step (1) to a virus database (virus DB). The current version of VirusFinder (release 1.0) uses the same virus DB, virus.fa, as the one included with the RINS package (http://khavarilab.stanford.edu/resources.html) [8]. This virus DB contains viruses of all known classes (32,102 in total) [8]. User can replace virus.fa with an alternative virus DB, Genome Information Broker for Viruses (GIB-V) [13] (http://gib-v.genes.nig.ac.jp/), which collects 25,525 virus reference sequences, or a smaller set of viruses of user interest.

Next, VirusFinder *de novo* assembles the reads aligned to the virus DB into contigs and maps contigs to both the human genome and the virus DB. All contigs that are mapped to the human genome are discarded. The alignment scores of the nonhuman contigs, which align only to the virus DB, are then used to rank the viruses, to which they are mapped. The sequence of the top ranking virus is then applied to the next analysis step.

It may be worth mentioning that our virus detection method as described here used RINS [8] as a starting point. However, different from RINS that identifies viruses by recruiting all reads mapped to the virus DB, which can at the same time align to the human genome, VirusFinder utilizes only the reads mapped to the virus DB and unmapped to the human genome for virus detection. By using less reads than RINS and more importantly with a simplified pipeline, VirusFinder achieves significant speedup over RINS without sacrificing its accuracy. We have tested VirusFinder on more than 20 samples (including unpublished ones) infected with viruses of various types and VirusFinder detected correct virus types for all the test samples (see section below for results on publically available data).

(3) Virus integration site detection

VirusFinder combines the human reference genome with the virus sequence (designated as a separate pseudo-chromosome, chrVirus) identified in previous step (2) or provided by the user. It then uses the mapping tool BWA [14] to align the reads recruited in step (1) to this new reference. Another tool VirusSeq [10] also concatenates the human genome with virus sequences. But VirusSeq includes a fixed set of, i.e. 18, virus sequences in its reference genome and hence cannot be applied directly to detect virus insertion sites in samples infected with viruses other than the18 predefined ones. By concatenating the viruses detected in step (2) on the fly, VirusFinder is readily applicable to samples harboring viruses of arbitrary types (as long as they are represented in the virus DB).

From the resultant alignment file, VirusFinder calls inter-chromosomal structural variants (SVs) using CREST [15]. The breakpoints of the SVs that involve both the virus and human genome, if there are any, are then reported. CREST utilizes soft-clipped reads as breakpoint positions of SVs. On a WGS sample with a modest 30× coverage, CREST can take several days to complete. To speed up our pipeline, before executing CREST, we run a much faster SV-detecting tool SVDetect [16] on the alignment file to calculate potential regions harboring virus integration sites. Then we modified CREST to make it search primarily within the regions identified by SVDetect. By blending SVDetect with CREST, we are able to reduce the computational time of SV calling significantly from several days on a WGS sample to around an hour.

When the pipeline terminates, three files, 'virus.txt', 'contig.txt' and 'integration-sites.txt', are created in the working directory of VirusFinder. These files contain candidate viruses identified by VirusFinder, contigs mapped to these virus sequences, and detected virus insertion sites, respectively. For each virus insertion event, VirusFinder provides its breakpoints in both the virus sequence and the human genome. For detailed explanations of these files, please read our user manual in the Supplementary Material S1.

Table 1. Detection of viruses in seven NGS samples using VirusFinder and VirusSeq.

Sample	Sequencing technology	Virus	#Virus integration sites	VirusFinder	VirusSeq
HCC sample 198T	WGS	HBV	2	√	–
HCC sample 268T	WGS	HBV	3	√	–
HCC cell line HKCI-5a	RNA-Seq	HBV	3	√	×
HeLa cervical cancer cell line	RNA-Seq	HPV-18	1	√	×
MCC case 27	Targeted sequencing	MCV	1	√	√
MCC case 36	Targeted sequencing	MCV	2	√	√
An Indian patient with fever and acute encephalitis	Targeted sequencing	JEV	0	√	√

HCC: Hepatocellular carcinoma. MCC: Merkel cell carcinoma. WGS: whole genome sequencing. RNA-Seq: whole transciptome sequencing. √: detected. ×: failed. –: software did not end within allowable time.

(4) Software implementation

The entire pipeline of VirusFinder, from the initial preprocessing step to the final virus integration site detection, is fully automated. As far as we know, this is the first fully automatic pipeline combining virus detection (step 2) seamlessly with virus integration site identification (step 3) and, thus, the first NGS software enabling the automatic detection of virus integration sites in samples for which viruses may not necessarily be determined beforehand. The aforementioned tool VirusSeq provides both virus detection script and virus integration site identification script too. Unfortunately, in VirusSeq, they are separate programs and cannot work directly together. Another advantage of VirusFinder is that it is capable of analyzing large-scale NGS data efficiently by improving significantly its computational pipelines for viruses and their integration sites detection. VirusFinder further improves its analysis speed by blending a fast aligner Bowtie 2 in the time-consuming step (1) with a slower yet more sensitive aligner BWA on a smaller subset of reads in step (3).

VirusFinder was implemented in Perl programming language and has been tested on various Linux platforms. It depends on several third-party tools, including BLAST+ (or BLAST) [17], BLAT [18], SAMtools [19], and Trinity [20], in addition to the aforementioned Bowtie 2, BWA, SVDetect, and CREST. All these tools are publically available. Their download URLs and brief descriptions are provided in the Supplementary Material S1. Different from other tools, CREST requires the installation of a BLAT server. To ease the distribution of VirusFinder, we modified CREST into a standalone tool, which, together with several other software that CREST requires, is now included in the release package of VirusFinder. This removed completely the requirement to install a BLAT server on user's system.

Results

To evaluate VirusFinder, we compared VirusFinder with the aforementioned VirusSeq [10] and ViralFusionSeq [11], two publically available tools emerged recently for characterizing virus integration sites in host genomes.

Virus detection

We firstly tested the ability of VirusFinder to detect the presence of viruses in human samples. Table 1 lists the samples we collected for our benchmark experiment. These samples were infected with viruses of diverse types and six of them were confirmed to harbor

Table 2. Detection of virus integration sites in five NGS samples[a].

Sample	Integration sites	VirusFinder	VirusSeq	ViralFusionSeq
HCC sample 198T	chr5:1,269,387	chr5:1,269,387	–	chr5:1,269,387
	chr5:1,269,405	chr5:1,269,405		chr5:1,269,405
HCC sample 268T	chr5:1,292,391	chr5:1,292,391	–	chr5:1,292,391
	chr5:1,292,403	chr5:1,292,403		chr5:1,292,403
	chr19:30,298,787	chr19:30,298,787		chr19:30,298,787
HCC cell line HKCI-5[b]	N/A	chr7:98,532,319	chr7:98,532,182	~chr7: 98,532,184- 98,532,285
	N/A	chr16:30,407,194	chr16:30,408,118	~chr16:30,408,324
MCC case 27	chr9:121,417,276	chr9:121,417,092	chr9:121,417,087	×
MCC case 36	chr13:99,978,184	chr13:99,978,244	chr13:99,977,889	×
	chr13:97,820,362	chr13:97,820,192	chr13:97,820,189	

N/A: not available. –: software did not end within allowable time. ×: software failure. [a]Only the samples in Table 1 that harbor virus integration sites are included in this table (HeLa cervical cancer cell line was excluded from this table because a large chromosomal region in 8q24 instead of a precise virus insertion position was provided for this sample [7]). [b]It is the test data of ViralFusionSeq [11]. The virus integration sites of this sample were validated but are not publically available. ViralFusionSeq outputs human-virus fusion sequences instead of fusion breakpoints. Its predictions of virus integration sites for the first two samples, 198T and 268T, were taken from its published paper. When running ViralFusionSeq on the sample HCC cell line HKCI-5, we got the intermediate results that indicate a virus integration site around chr16:30408324. Its user manual provides another position, chr7: 98532184- 98532285, for this sample. Both loci were included here for the purpose of comparison.

Table 3. Comparison of computation time of three virus integration-detecting methods on whole genome sequencing (WGS) data[a].

Sample	Coverage	VirusFinder		ViralFusionSeq[b]		VirusSeq[c]	
		#CPUs	Time (days)	#CPUs	Time (days)	#CPUs	Time (days)
26T	65.5×	8	3.1	8	17.8	8	>7
71T	32.2×	8	1.9	8	11.5	8	>7
106T	44.8×	8	2.4	8	17.1	8	>12.5
180N	121.2×	8	7.3	8	>17.4	8	>12.5
186T	36.5×	8	2.0	8	13.0	8	>12.5
198T	34.4×	8	1.8	8	10.8	8	>12.5
200N	32.6×	8	1.9	8	11.5	8	>12.5
200T	31.7×	8	2.0	8	12.5	8	>12.5
268N	40.7×	8	2.7	8	14.5	8	>12.5
268T	34.1×	8	2.0	8	13.5	8	>9.9
Average			2.7		14.0		>11.1

[a]The computation time of the three methods on these samples were analyzed on Vanderbilt Advanced Computing Center for Research & Education (ACCRE, http://www.accre.vanderbilt.edu/), with the same configuration of CPUs in each node.
[b]ViralFusionSeq did not terminate successfully on sample 180N.
[c]We attempted to run VirusSeq three times on these WGS samples. The first trial failed because the size of its intermediate files exceeded our cluster quota. After getting more space, we reran VirusSeq. After non-stop running for a whole week, all our jobs were killed in server due to their exceeding allocated time – not realizing initially the long computation time of VirusSeq on WGS samples. In our latest trial of VirusSeq on February 13, 2013, we requested 35 GB memory, 8 CPUs, 30 days for each job and resubmitted our jobs to ACCRE. Seven jobs were scheduled to run first. After twelve and a half day, all these jobs were killed due to an unexpected internal network outage of ACCRE. Though we were not able to make VirusSeq terminate successfully on these WGS samples due to expensive computing, we may conclude from the data that VirusFinder runs much faster than VirusSeq.

virus integration sites. The first two samples in Table 1 were generated using WGS [2], the next two using RNA-Seq [7,11], and the remaining three targeted sequencing [21,22]. The first three samples are the complete test data set of ViralFusionSeq and the fourth one, HeLa cervical cancer cell line, was used to evaluate PathSeq [7].

As indicated in Table 1, VirusFinder identified the correct virus types in all the test samples. We also evaluated VirusFinder using additional 8 WGS samples (see the Computational efficiency section below) of different sequencing coverage, including 5 Hepatocellular carcinomas (HCCs) and 3 normal tissues. VirusFinder identified correct virus types for all these WGS samples and hence demonstrated its robustness in virus detection.

Table 1 shows that VirusSeq reported a false virus type for HCC cell line HKCI-5a. It also failed to detect the presence of HPV-18 virus in the HeLa cervical cancer cell line (VirusSeq failed probably because MosaikAligner, the mapping tool used in VirusSeq, extracted zero unaligned reads from the sequencing data of this cell line). ViralFusionSeq does not detect virus type and hence was not touched on here.

Virus integration site detection

We utilized the samples in Table 1 that harbor virus integration sites as test data to evaluate the virus insertion loci predicted by VirusFinder. Here, we excluded the HeLa cervical cancer cell line from this experiment, because a large chromosomal region in 8q24 instead of a precise virus insertion site was reported for this sample [7].

Table 2 presents the virus insertion loci detected by the three methods. The real virus insertion positions are in Column 2. From Table 2, we can see that on the two WGS samples, 198T and 268T, both VirusFinder and ViralFusionSeq pinpointed all the exact integration breakpoints reported in a recent study by Sung *et al.* [2]. On the two targeted sequencing samples, MCC cases 27 and 36, although the three integration positions predicted by VirusFinder are not as accurate as those on the WGS samples, they are comparable to or slightly more accurate than those detected by VirusSeq.

Computational efficiency

Finally, we compared the computation time of the 3 tools on 10 WGS samples infected with HBV virus (Table 3). All these samples are from the study by Sung *et al.* [2]. In their original study, Sung *et al.* validated 22 out of 176 WGS samples. Limited by computational resources, we picked 10 out of the 22 validated samples so that: (i) the hotspots of the integration breakpoints in both the human genome, e.g. genes *TERT*, *MLL4* and *CCNE1*, and the HBV virus genome, e.g. locus 1800, are covered; (ii) our test data includes not only tumor samples but also normal tissues – 200N, 268N, and 180N are three normals; (iii) different sequencing depths, which range from 31.7× to 121.2×, are represented; (iv) our test data includes the aforementioned two samples, 198T and 268T, so as to use them to compare VirusFinder with ViralFusionSeq (they were the test data of ViralFusionSeq).

The computation time of the three tools on these samples were analyzed on Vanderbilt Advanced Computing Center for Research & Education (ACCRE, http://www.accre.vanderbilt.edu/), with the same configuration of CPUs in each node. The memory we requested for each job of VirusFinder, ViralFusionSeq, and VirusSeq was 20 GB, 40 GB and 35 GB, respectively.

The average running time of these three tools is shown in Table 3. Based on 8 processors, VirusFinder and ViralFusionSeq took on average 2.7 and 14.0 days, respectively, on a WGS sample, while VirusSeq did not terminate within the allowable time (>11.1 days) on all the test samples. Among the 10 samples, 180N has the highest coverage (121.2×) and is the only normal sample validated to harbor a HBV fusion site [2]. Among the three methods, only VirusFinder successfully detected the HBV virus and its integration sites in 180N within the allocated time. This indicates that VirusFinder is capable of analyzing WGS samples with very high sequencing coverage. Both ViralFusionSeq and VirusSeq failed on this sample primarily due to their exceedingly high CPU requirements.

Discussion

With the increasing interest in applying NGS to investigate virus-host interactions in human cancer, new software tools emerged recently to detect viruses and their integration sites in the human genome. Unfortunately, these tools could hardly meet the challenges posed by the rapidly advancing NGS technologies of today, due to their limited capability and low computational efficiency. This is the main reason why we developed VirusFinder as introduced in this paper.

To the best of our knowledge, VirusFinder is the first fully automatic pipeline capable of characterizing integration loci of undiagnosed viruses of arbitrary types in NGS data. The results of our evaluation indicated that VirusFinder can accurately detect viruses and their integration sites in the human genome. The benchmark experiment on 10 WGS samples also demonstrated that VirusFinder is ideal for quick and accurate analysis of large-scale NGS data. These we believe will greatly benefit the studies that utilize NGS to investigate the etiologic association of viruses with disease, especially human cancer.

Acknowledgments

We are grateful to Dr. Eric J. Duncavage for sharing with us their data. We also thank Drs. Huy Vuong, Jingchun Sun, and the authors of PathSeq and RINS for their great assistance.

Author Contributions

Conceived and designed the experiments: QW ZZ. Performed the experiments: QW. Analyzed the data: QW PJ ZZ. Contributed reagents/materials/analysis tools: QW. Wrote the paper: QW PJ ZZ.

References

1. Lace MJ, Anson JR, Klussmann JP, Wang DH, Smith EM, et al. (2011) Human papillomavirus type 16 (HPV-16) genomes integrated in head and neck cancers and in HPV-16-immortalized human keratinocyte clones express chimeric virus-cell mRNAs similar to those found in cervical cancers. J Virol 85: 1645–1654.
2. Sung W-K, Zheng H, Li S, Chen R, Liu X, et al. (2012) Genome-wide survey of recurrent HBV integration in hepatocellular carcinoma. Nat Genet 44: 765–769.
3. Jiang Z, Jhunjhunwala S, Liu J, Haverty PM, Kennemer MI, et al. (2012) The effects of hepatitis B virus integration into the genomes of hepatocellular carcinoma patients. Genome Res 22: 593–601.
4. Lasithiotaki I, Antoniou K m., Derdas S p., Sarchianaki E, Symvoulakis E k., et al. (2013) The presence of Merkel cell polyomavirus is associated with deregulated expression of BRAF and Bcl-2 genes in non-small cell lung cancer. International Journal of Cancer. Doi: 10.1002/ijc.28062.
5. Ding D, Lou X, Hua D, Yu W, Li L, et al. (2012) Recurrent targeted genes of hepatitis B virus in the liver cancer genomes identified by a next-generation sequencing-based approach. PLoS Genet 8: e1003065.
6. Stransky N, Egloff AM, Tward AD, Kostic AD, Cibulskis K, et al. (2011) The mutational landscape of head and neck squamous cell carcinoma. Science 333: 1157–1160.
7. Kostic AD, Ojesina AI, Pedamallu CS, Jung J, Verhaak RGW, et al. (2011) PathSeq: software to identify or discover microbes by deep sequencing of human tissue. Nat Biotechnol 29: 393–396.
8. Bhaduri A, Qu K, Lee CS, Ungewickell A, Khavari PA (2012) Rapid identification of non-human sequences in high-throughput sequencing datasets. Bioinformatics 28: 1174–1175.
9. Naeem R, Rashid M, Pain A (2013) READSCAN: a fast and scalable pathogen discovery program with accurate genome relative abundance estimation. Bioinformatics 29: 391–392.
10. Yunxin Chen, Hui Yao, Thompson EJ, Tannir NM, Weinstein JN, et al. (2013) VirusSeq: software to identify viruses and their integration sites using next-generation sequencing of human cancer tissue. Bioinformatics 29: 266–267.
11. Li J-W, Wan R, Yu C-S, Co NN, Wong N, et al. (2013) ViralFusionSeq: accurately discover viral integration events and reconstruct fusion transcripts at single-base resolution. Bioinformatics 29: 649–651.
12. Langmead B, Salzberg SL (2012) Fast gapped-read alignment with Bowtie 2. Nat Methods 9: 357–359.
13. Hirahata M, Abe T, Tanaka N, Kuwana Y, Shigemoto Y, et al. (2007) Genome Information Broker for Viruses (GIB-V): database for comparative analysis of virus genomes. Nucleic Acids Res 35: D339–342.
14. Li H, Durbin R (2010) Fast and accurate long-read alignment with Burrows-Wheeler transform. Bioinformatics 26: 589–595.
15. Wang J, Mullighan CG, Easton J, Roberts S, Heatley SL, et al. (2011) CREST maps somatic structural variation in cancer genomes with base-pair resolution. Nat Meth 8: 652–654.
16. Zeitouni B, Boeva V, Janoueix-Lerosey I, Loeillet S, Legoix-né P, et al. (2010) SVDetect: a tool to identify genomic structural variations from paired-end and mate-pair sequencing data. Bioinformatics 26: 1895–1896.
17. Camacho C, Coulouris G, Avagyan V, Ma N, Papadopoulos J, et al. (2009) BLAST+: architecture and applications. BMC Bioinformatics 10: 421.
18. Kent WJ (2002) BLAT--the BLAST-like alignment tool. Genome Res 12: 656–664.
19. Li H, Handsaker B, Wysoker A, Fennell T, Ruan J, et al. (2009) The Sequence Alignment/Map format and SAMtools. Bioinformatics 25: 2078–2079.
20. Grabherr MG, Haas BJ, Yassour M, Levin JZ, Thompson DA, et al. (2011) Full-length transcriptome assembly from RNA-Seq data without a reference genome. Nat Biotechnol 29: 644–652.
21. Duncavage EJ, Magrini V, Becker N, Armstrong JR, Demeter RT, et al. (2011) Hybrid capture and next-generation sequencing identify viral integration sites from formalin-fixed, paraffin-embedded tissue. J Mol Diagn 13: 325–333.
22. Patowary A, Chauhan RK, Singh M, Kv S, Periwal V, et al. (2012) De novo identification of viral pathogens from cell culture hologenomes. BMC Res Notes 5: 11.

Exome Sequencing of 47 Chinese Families with Cone-Rod Dystrophy: Mutations in 25 Known Causative Genes

Li Huang[1]◔, Qingyan Zhang[2]◔, Shiqiang Li[1]◔, Liping Guan[2], Xueshan Xiao[1], Jianguo Zhang[2], Xiaoyun Jia[1], Wenmin Sun[1], Zhihong Zhu[2], Yang Gao[1], Ye Yin[2], Panfeng Wang[1], Xiangming Guo[1], Jun Wang[2]*, Qingjiong Zhang[1]*

1 State Key Laboratory of Ophthalmology, Zhongshan Ophthalmic Center, Sun Yat-sen University, Guangzhou, Guangdong, China, **2** BGI-Shenzhen, Shenzhen, Guangdong, China

Abstract

Objective: The goal of this study was to identify mutations in 25 known causative genes in 47 unrelated Chinese families with cone-rod dystrophy (CORD).

Methods: Forty-seven probands from unrelated families with CORD were recruited. Genomic DNA prepared from leukocytes was analyzed by whole exome sequencing. Variants in the 25 genes were selected and then validated by Sanger sequencing.

Results: Fourteen potential pathogenic mutations, including nine novel and five known, were identified in 10 of the 47 families (21.28%). Homozygous, compound heterozygous, and hemizygous mutations were detected in three, four, or three families, respectively. The 14 mutations in the 10 families were distributed among *CNGB3* (three families), *PDE6C* (two families), *ABCA4* (one family), *RPGRIP1* (one family), *RPGR* (two families), and *CACNA1F* (one family).

Conclusions: This study provides a brief view on mutation spectrum of the 25 genes in a Chinese cohort with CORD. Identification of novel mutations enriched our understanding of variations in these genes and their associated phenotypes. To our knowledge, this is the first systemic exome-sequencing analysis of all of the 25 CORD-associated genes.

Editor: Andreas R. Janecke, Innsbruck Medical University, Austria

Funding: This study was supported by the National Natural Science Foundation of China (81170881, U1201221, http://www.nsfc.gov.cn/Portal0/default152.htm), Guangdong Translational Medicine Public Platform, the "985 project" of Sun Yat-sen University, and the Fundamental Research Funds of the State Key Laboratory of Ophthalmology. The funders had no role in study design, data collection and analysis, decision to publish, or preparation of the manuscript.

Competing Interests: The authors have declared that no competing interests exist.

* E-mail: zhangqji@mail.sysu.edu.cn (QJZ); wangj@genomics.cn (JW)

◔ These authors contributed equally to this work.

Introduction

Cone-rod dystrophy (CORD) refers to a series of hereditary retinal disorders with a predominantly cone involvement [1]. Rod impairment may occur at the same time as the cone impairment or appear later. Patients with CORD usually have reduced visual acuity, photophobia, and color vision defects.

CORD may be transmitted as an autosomal dominant (adCORD), autosomal recessive (arCORD), or X-linked trait (xlCORD). To date, mutations in at least 25 genes have been reported to be associated with different forms of CORD, including the following: aryl hydrocarbon receptor interacting protein-like 1 (*AIPL1*) [1]; the cone-rod homeobox containing gene (*CRX*) [2]; guanylate cyclase activator 1A (*GUCA1A*) [3]; guanylate cyclase 2D (*GUCY2D*) [4]; PITPNM family member 3 (*PITPNM3*) [5]; prominin 1 (*PROM1*) [6]; peripherin 2 (*PRPH2*) [7]; regulating synaptic membrane exocytosis 1 (*RIMS1*) [8]; sema domain, immunoglobulin domain (Ig), transmembrane domain (TM) and short cytoplasmic domain, (semaphorin) 4A (*SEMA4A*) [9]; unc-119 homolog (*UNC119*) [10]; ATP-binding cassette, sub-family A (ABC1) member 4 (*ABCA4*) [11]; ADAM metallopeptidase domain 9 (*ADAM9*) [12]; chromosome 8 open-reading frame 37 (*C8ORF37*) [13]; calcium channel voltage-dependent alpha 2/delta subunit 4 (*CACNA2D4*) [14]; cadherin-related family member 1 (*CDHR1*) [15]; ceramide kinase-like (*CERKL*) [16]; cyclic nucleotide gated channel beta 3 (*CNGB3*) [17]; cyclin M4 (*CNNM4*) [18]; potassium channel subfamily V member 2 (*KCNV2*) [19]; phosphodiesterase 6C, cGMP-specific, cone, alpha prime (*PDE6C*) [20]; retina and anterior neural fold homeobox 2 (*RAX2*) [21]; retinol dehydrogenase 5 (*RDH5*) [22]; retinitis pigmentosa GTPase regulator interacting protein 1 (*RPGRIP1*) [23]; calcium channel voltage-dependent L type alpha 1F subunit (*CACNA1F*) [24]; and retinitis pigmentosa GTPase regulator (*RPGR*) [25] (RetNet: https://sph.uth.edu/Retnet/). Of the 25 genes, mutations in the first 10 genes are responsible for adCORD, the next 13 for arCORD, and the last two for xlCORD. The associated genomic information of the 25 genes is listed in Table S1.

In our previous study on CORD, mutations were only detected in 7 of 130 (5.38%) Chinese families with CORD by using cycle sequencing of all coding exons of five genes (*CRX*, *GUCY2D*,

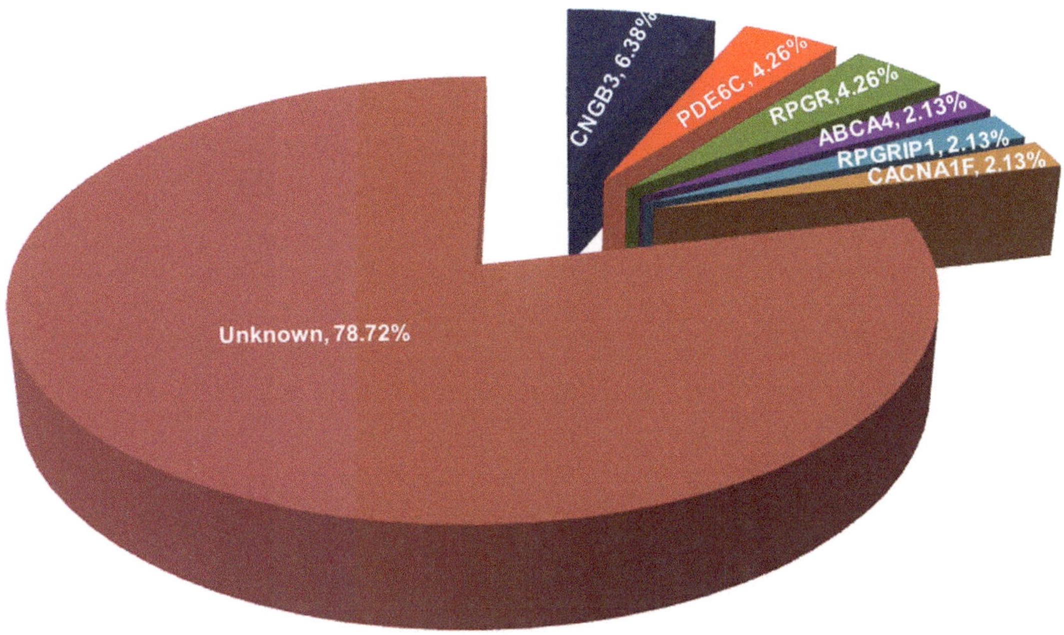

Figure 1. Prevalence of mutations in the investigated genes in our cohort of 47 CORD patients.

GUCA1A, PRPH2, and *KCNV2*) as well as of all exons harboring reported mutations in other 17 CORD-associated genes [26,27,28,29]. Of the seven families, all mutations were identified in genes responsible for adCORD but none in genes for arCORD and xlCORD. The genetic cause for most families (the remaining 123 of 130 (94.62%) Chinese families) was still unknown. In order to identify the additional cause of most CORD and to disclose further the mutation spectrum and frequency of the 25 genes, whole exome sequencing was used to screen for mutations in 47 unrelated Chinese families with CORD.

Materials and Methods

Patients

Forty seven probands from unrelated families with CORD were recruited from the Eye Hospital of Zhongshan Ophthalmic Center, Sun Yat-sen University. Patients with identified mutation who were included in our previous study were excluded from this one. Written informed consents were obtained from the participants or their guardians before the study, which was conforms to the tenets of the Declaration of Helsinki and follows the Guidance of Sample Collection of Human Genetic Diseases (863-plan) by the Ministry of Public Health of China. This study was approved by the Institute Review Board of the Zhongshan Ophthalmic Center. Genomic DNA was prepared from the blood leukocytes as previous described [30].

Exome Sequencing

Exome sequencing was completed by using a commercial service from BGI Shenzhen (http://www.genomics.cn/index.php). The exome sequencing, genotype calling, and SNP calling were in the same way as the methods reported before [31]. In brief, exome capture was carried out by using a NimbleGen SeqCap EZ Exome (44 M) array. Exon-enriched DNA fragments were sequenced by the Illumina Genome Analyzer II. The average sequencing depth was set to 60-fold. SOAP aligner was used to set the sequencing reads to UCSC hg19 [32,33]. The likelihood of possible genotypes in the target regions was calculated using SOAPsnp [34]. Variants in all the 25 genes detected by exome sequencing were selected for validation. Exome sequencing dataset of the patients with identified mutations in this study have been deposited to NIH (http://www.ncbi.nlm.nih.gov/biosample: accession number SAMN01997562 to SAMN01997571).

Sanger Sequencing

Sanger sequencing was used to validate variants in the 25 genes that resulted from exome sequencing, including heterozygous variants in the adCORD genes, homozygous or compound heterozygous variants in the arCORD genes, or hemizygous variants in the xlCORD genes. Primers (Table S2) used to amplify the regions containing the variants were designed by primer design tool Primer3 (http://frodo.wi.mit.edu/primer3/) [35]. A touch-down polymerase chain reaction (PCR) was used to amplify the fragments with variants, as previously reported [36], and the amplicons were analyzed with an ABI BigDye Terminator cycle sequencing kit v3.1 (Applied Biosystems, Foster City, CA) on an ABI3100 Genetic Analyzer (Applied Biosystems). Sequencing results from patients and controls were compared using the SeqManII program of the Lasergene package (DNAStar Inc, Madison, WI). Detected variants were further sequenced in the available family members. Novel variants were further evaluated in 192 control individuals. The description of mutations was in accordance with the nomenclature for the description of sequence variants [37](HGVS: http://www.hgvs.org/mutnomen/). The conservation of a variation was evaluated by Phastcons_score (http://varianttools.sourceforge.net/Annotation/PhastCons) [38], the effect of a missense variation was analyzed by using SIFT [39] (http://sift.jcvi.org/) and Polyphen-2 [40] (http://genetics.bwh.harvard.edu/pph2/) online tools, and the effect of splicing site changes was predicted by Berkeley Drosophila Genome Project (BDGP) [41] (http://www.fruitfly.org/).

Considering that CORD-causing mutations are rare and the presence of the normal carriers of arCORD gene mutations, we assumed that the affected individuals were likely homozygous or

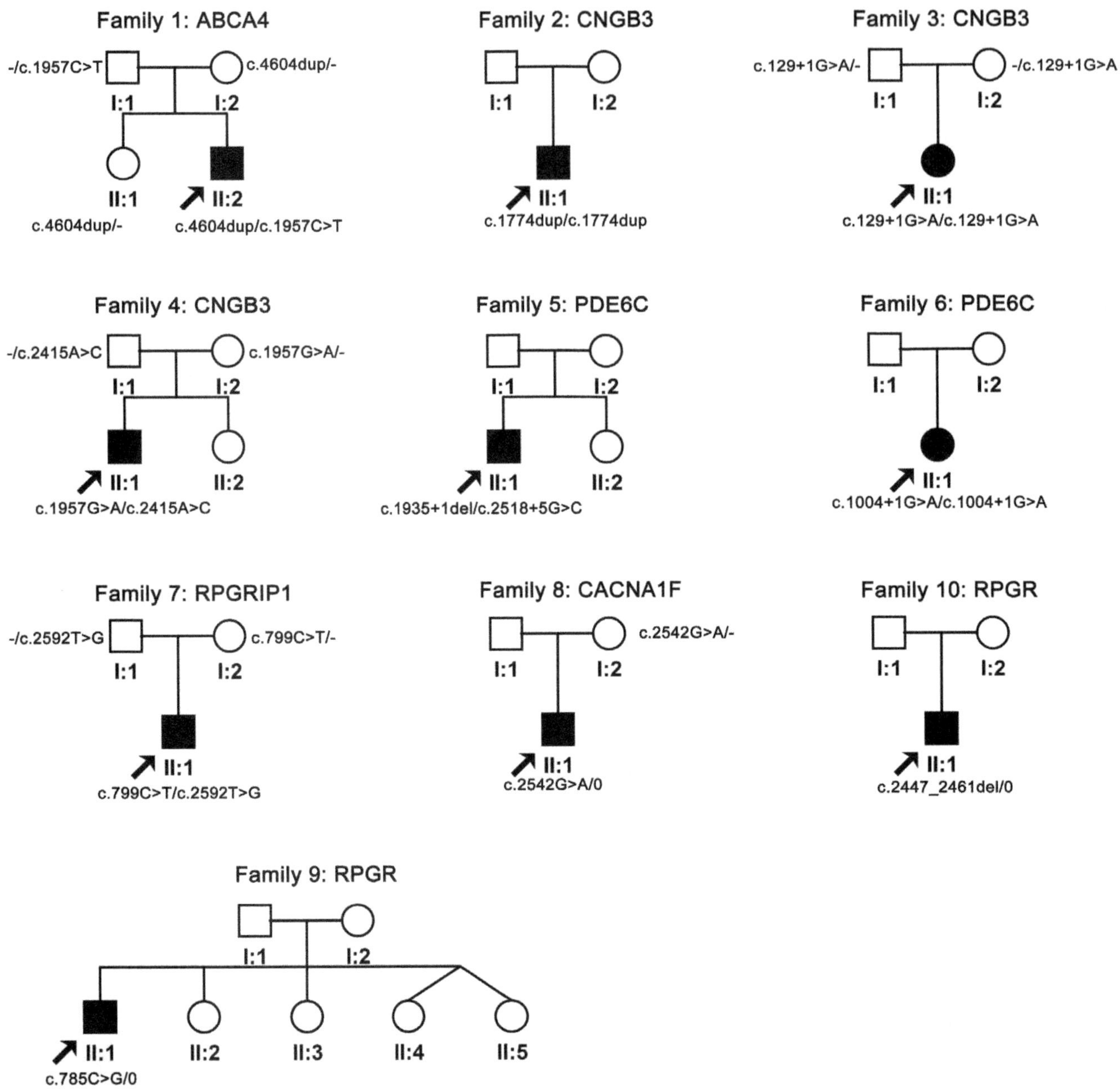

Figure 2. Pedigrees of the 10 families with mutations. The family numbers and their corresponding mutations were shown just above the pedigrees.

compound heterozygous, so variants absent in the dbSNP134, 1000Genome or with allelic frequencies ≤0.006 were considered to be potentially pathogenic (frequency of heterozygote carriers calculated based on a disease incidence of 1:40,000, under the hypothesis that a unique arCORD gene would explain the remaining 40% of cases [42]).

Results

Whole exome sequencing identified 14 potential pathogenic mutations in 10 of the 47 (21.28%) families with CORD (Table 1), including seven homozygous or compound heterozygous mutations in four (*ABCA4, CNGB3, PDE6C,* and *RPGRIP1*) of the 13 genes associated with arCORD, and three hemizygous mutations in the two genes (*RPGR* and *CACNA1F*) associated with xlCORD. Of the 14 mutations, nine were novel. The 14 mutations in the 10 families involved six of the 25 CORD-associated genes, including *CNGB3* (three families), *PDE4C* (two families), *RPGR* (two families), *ABCA4* (one family), *RPGRIP1* (one family), and *CACNA1F* (one family) (Figure 1), respectively. Sanger sequencing confirmed the 14 mutations in the 10 families (Figure S1). Segregation analysis was available for five of the 10 families in where the mutations co-segregated with the disease in the family (Figure 2). No potential pathogenic mutation was identified in the other 19 genes of the 47 families.

Clinical data of the 10 probands with potential pathogenic mutations are listed in Table 2. All probands with identified mutations had an early onset severe form of retinal dystrophy with predominantly cone involvement. Fundus changes were mainly in

Table 1. Potential pathogenic mutations detected in 10 of the 47 families.

Family ID	Gene	Variations		Status	Bioinformation analysis				Allele frequency in		Reference
		DNA	Protein		SIFT	Polyphen-2	Splice	Phastcons _score	patients	controls	
Family 1	ABCA4	c.4604dup	p.T1537Nfs*18	hetero	–	–	–	0.997	1/94	0/384	novel
Family 1	ABCA4	c.1957C>T	p.R653C	hetero	D	PD	–	1.000	1/94	NA	[38]
Family 2	CNGB3	c.1774dup	p.E592Gfs*44	homo	–	–	–	1.000	2/94	0/384	novel
Family 3	CNGB3	c.129+1G>A	–	homo	–	–	DSA	1.000	2/94	0/384	novel
Family 4	CNGB3	c.2415A>C	p.E805D	hetero	D	PD	–	1.000	1/94	NA	rs186448979#
Family 4	CNGB3	c.1957G>A	p.A653T	hetero	tolerated	benign	–	0.000	1/94	0/384	novel
Family 5	PDE6C	c.1935+1del	–	hetero	–	–	DSA	1.000	1/94	0/384	novel
Family 5	PDE6C	c.2518+5G>C	NA	hetero	–	–	DSA	0.112	1/94	0/384	novel
Family 6	PDE6C	c.1004+1G>A	–	homo	–	–	DSA	1.000	2/94	0/384	novel
Family 7	RPGRIP1	c.2592T>G	p.Y864*	hetero	–	–	–	0.994	1/94	0/384	novel
Family 7	RPGRIP1	c.799C>T	p.R267*	hetero	–	–	–	1.000	1/94	NA	[36]
Family 8	CACNA1F	c.2542G>A	p.G848S	hemi	tolerated	benign	–	1.000	1/94	0/384	novel
Family 9	RPGR	c.785C>G	p.A262G	hemi	tolerated	benign	–	0.002	1/94	NA	[39]
Family10	RPGR	c.2447_2461del	p.G816_E820del	hemi	–	–	–	NA	1/94	NA	[40]

Note: D = damaging; PD = probably damaging; DSA = donor site abolished.
#The variation was found in 1000 Genomes database with the Global minor allele frequency (MAF) of G = 0.001/3 so that the pathogeneity of the variants in this family need to be clarified further.

the macular regions, showing mild pigmentary changes and a loss of foveal reflex, as well as attenuated retinal arteries in rare cases.

Discussion

Based on an initial screening of exome sequencing and the subsequent confirmation of Sanger sequencing, potential pathogenic mutations were identified in 10 of the 47 (21.28%) families with CORD, involving 14 mutations in six of the 25 CORD-associated genes. For the 47 families in this study, the contributions of causative mutations in individual genes are as follows: *CNGB3* (6.38%), *PDE6C* (4.26%), *RPGR* (4.26%), *ABCA4* (2.13%), *RPGRIP1* (2.13%), and *CACNA1F* (2.13%).

The frequency of mutations detected in this study is significantly higher than in our previous study, in which mutations were only detected in seven of 130 (5.38%) Chinese families with CORD through the process of sequencing the coding exons of five genes and all exons with reported mutations in 17 other genes [26,27,28,29]. In addition, in this study, mutations in seven of the 10 families were found in arCORD-associated genes but no mutations were detected in the adCORD-associated genes; in contrast, mutations in all seven families from our previous study were found in adCORD-associated genes. This difference may

Table 2. Clinical data of the probands with CORD and identified potential pathogenic mutations.

Family ID	Gene	Nucleotide changes	Sex	Age (year) at		First	Best visual acuity		Fundus changes		ERG responses from	
				exam	onset	symptom	right	left	right	left	rod	cone
Family 1	ABCA4	c.[1957C>T];[4604dup]	M	7.0	6.0	PV, PP	0.20	0.15	MA, TPOD, ARA	MA, TPOD, ARA	Normal	Reduced
Family 2	CNGB3	c.[1774dup];[1774dup]	M	6.5	EC	PV,NYS	0.20	0.20	MA, TPOD	MA, TPOD	Mildly reduced	Extinguished
Family 3	CNGB3	c.[129+1G>A];[129+1G>A]	F	5.0	0.3	PV,NYS, PP	0.20	0.20	MA, TPOD, ARA	MA, TPOD, ARA	Normal	Extinguished
Family 4	CNGB3	c.[1957G>A];[2415A>C]	F	4.5	0.5	PP,NYS	0.10	0.20	ARA	ARA	Mildly reduced	Severely reduced
Family 5	PDE6C	c.[1935+1del];[2518+5G>C]	M	7.0	EC	NYS	0.05	0.05	High myopic	High myopic	Normal	Severely reduced
Family 6	PDE6C											

c.[1004+1G>A];[1004+1G>A]F2.0ECPV,PP,NYSPOPONANAMildly reducedExtinguishedFamily 7RPGRIP1c.[799C>T];[2592T>G]M3.6FMBPV, PP, NYSNANAARA,CRDARA,CRDExtinguishedExtinguishedFamily 8CACNA1Fc.[2542G>A]; [0]M2.3NAPP, NYSNANAMAMASeverely reduced*ExtinguishedFamily 9RPGRc.[785C>G]; [0]M28.0ECPV, PP0.100.10MA, TPOD, ARAMA, TPOD, ARAModerately reducedExtinguishedFamily 10RPGRc.[2447_2461del]; [0]M9.0ECPV, PP0.200.20MAMANormalExtinguished

Note: F = female; M = male; EC = early childhood; FMB = first few months after birth; PV = poor vision; PP = photophobia; NYS = nystagmus; PO = pursuing object; NA = not available; MA = macular atrophy; TPOD = temporal pallor of optic disc; ARA = attenuated retinal arteries; CRD = carpet-like retinal degeneration.
*This patient did not have the "electronegative" ERG in the standard combined response.

partly have resulted from 1) several genes associated with adCORD having been systemically analyzed before and all patients with mutations in these genes being excluded from this study; and 2) selective analysis of exons with reported mutations perhaps failing to identify both mutations in arCORD associated genes if one mutation in the exons is not analyzed. The real proportion of mutations in the 25 genes should be higher than 21.28% in Chinese patients with CORD if mutations identified in previous study are taken into account [26,27,28,29]. Further analysis of the results from exome sequencing for those patients without identified mutations in the 25 genes may provide useful clues in the identification of new genes responsible for CORD.

Previously studies on an individual gene or a set of genes have revealed a different frequency of mutation in CORD-associated genes in different populations. Sanger sequencing of 10 adCORD-associated genes identified mutations in 25 of 52 (48%) German families with adCORD, and mutations were found in *GUCY2D* (24%), *PRPH2* (12%), *GUCA1A* (8%), *CRX* (4%), and *PROM1* (2%) [43]. In other studies for individual genes, mutations in *CRX* are responsible for a 4.76% proportion of CORD [26], *GUCY2D* for 11.11% of CORD in a Japanese population [44] and 9.09% in a Spanish one [45], *GUCA1A* for 16.67% of CORD in a German one [46], *PRPH2* for 11% of CORD in another German population [43], *AIPL1* for 1.82% of CORD in an American sample [1], *PROM1 for* 0.93% of CORD in a Dutch population [16], *SEMA4A* for 8.00% of CORD in a Pakistani group [9], and *UNC119 for* 5.00% of CORD in another American population [10].

For genes known to be associated with arCORD, their mutations might be responsible for nearly 40% of arCORD [42], whereas mutations in ABCA4 are most frequent in European and American populations, ranging from 16.13% to 65.00% [47,48,49,50]. However, ABCA4 mutations are very rare in Chinese families with CORD. The mutation frequency of CNGB3 [49], *PDE6C* [20,49], and *PRGRIP1* [23] in other populations are relatively rare, as seen in Chinese sample. For other CORD-associated genes without identified mutations in this study, mutations in other populations are also rare, as with *RAX2* in 0.62% of CORD [21], *CERKL* in 1.85% of Canadian CORD [16], or *ADAM9* [12], *RDH5* [22], *CDHR1* [15], and *C8ORF37* [13] in a few reports based on an analysis of a limited number of families.

Supporting Information

Figure S1 Sequence chromatography. Forteen sequence changes detected in the probands with CORD are shown (left column) compared with corresponding normal sequences (right column). Some known mutations were not verified in the normal controls, so the normal sequences are absent.

Table S1 Genomic information of the 25 genes referred in this study. This table listed the accession numbers of the genomic DNA, mRNA, and protein for each of the 25 genes. The information is based on human genome reference GRCh37.p10.

Table S2 Sequences of primers used in this study. This table listed 52 primers used to amplify the genomic fragments with variants detected by exome sequencing.

Acknowledgments

The authors are grateful to the patients for their participation.

Author Contributions

Conceived and designed the experiments: QJZ. Performed the experiments: LH QYZ SL LG XX JZ XJ WS ZZ YG YY PW XG JW. Analyzed the data: LH QYZ LG JZ WS YY JW QJZ. Contributed reagents/materials/analysis tools: LH SL XX JZ XJ WS YG PW XG JW QJZ. Wrote the paper: LH QJZ.

References

1. Sohocki MM, Perrault I, Leroy BP, Payne AM, Dharmaraj S, et al. (2000) Prevalence of AIPL1 mutations in inherited retinal degenerative disease. Mol Genet Metab 70: 142–150.
2. Freund CL, Gregory-Evans CY, Furukawa T, Papaioannou M, Looser J, et al. (1997) Cone-rod dystrophy due to mutations in a novel photoreceptor-specific homeobox gene (CRX) essential for maintenance of the photoreceptor. Cell 91: 543–553.
3. Payne AM, Downes SM, Bessant DA, Taylor R, Holder GE, et al. (1998) A mutation in guanylate cyclase activator 1A (GUCA1A) in an autosomal dominant cone dystrophy pedigree mapping to a new locus on chromosome 6p21.1. Hum Mol Genet 7: 273–277.
4. Kelsell RE, Gregory-Evans K, Payne AM, Perrault I, Kaplan J, et al. (1998) Mutations in the retinal guanylate cyclase (RETGC-1) gene in dominant cone-rod dystrophy. Hum Mol Genet 7: 1179–1184.
5. Kohn L, Kadzhaev K, Burstedt MS, Haraldsson S, Hallberg B, et al. (2007) Mutation in the PYK2-binding domain of PITPNM3 causes autosomal dominant cone dystrophy (CORD5) in two Swedish families. Eur J Hum Genet 15: 664–671.
6. Yang Z, Chen Y, Lillo C, Chien J, Yu Z, et al. (2008) Mutant prominin 1 found in patients with macular degeneration disrupts photoreceptor disk morphogenesis in mice. J Clin Invest 118: 2908–2916.
7. Fishman GA, Stone EM, Alexander KR, Gilbert LD, Derlacki DJ, et al. (1997) Serine-27-phenylalanine mutation within the peripherin/RDS gene in a family with cone dystrophy. Ophthalmology 104: 299–306.
8. Kelsell RE, Gregory-Evans K, Gregory-Evans CY, Holder GE, Jay MR, et al. (1998) Localization of a gene (CORD7) for a dominant cone-rod dystrophy to chromosome 6q. Am J Hum Genet 63: 274–279.
9. Abid A, Ismail M, Mehdi SQ, Khaliq S (2006) Identification of novel mutations in the SEMA4A gene associated with retinal degenerative diseases. J Med Genet 43: 378–381.
10. Kobayashi A, Higashide T, Hamasaki D, Kubota S, Sakuma H, et al. (2000) HRG4 (UNC119) mutation found in cone-rod dystrophy causes retinal degeneration in a transgenic model. Invest Ophthalmol Vis Sci 41: 3268–3277.
11. Cremers FP, van de Pol DJ, van Driel M, den Hollander AI, van Haren FJ, et al. (1998) Autosomal recessive retinitis pigmentosa and cone-rod dystrophy caused by splice site mutations in the Stargardt's disease gene ABCR. Hum Mol Genet 7: 355–362.
12. Parry DA, Toomes C, Bida L, Danciger M, Towns KV, et al. (2009) Loss of the metalloprotease ADAM9 leads to cone-rod dystrophy in humans and retinal degeneration in mice. Am J Hum Genet 84: 683–691.
13. Estrada-Cuzcano A, Neveling K, Kohl S, Banin E, Rotenstreich Y, et al. (2012) Mutations in C8orf37, encoding a ciliary protein, are associated with autosomal-recessive retinal dystrophies with early macular involvement. Am J Hum Genet 90: 102–109.
14. Wycisk KA, Budde B, Feil S, Skosyrski S, Buzzi F, et al. (2006) Structural and functional abnormalities of retinal ribbon synapses due to Cacna2d4 mutation. Invest Ophthalmol Vis Sci 47: 3523–3530.
15. Ostergaard E, Batbayli M, Duno M, Vilhelmsen K, Rosenberg T (2010) Mutations in PCDH21 cause autosomal recessive cone-rod dystrophy. J Med Genet 47: 665–669.
16. Littink KW, Koenekoop RK, van den Born LI, Collin RW, Moruz L, et al. (2010) Homozygosity mapping in patients with cone-rod dystrophy: novel mutations and clinical characterizations. Invest Ophthalmol Vis Sci 51: 5943–5951.
17. Michaelides M, Aligianis IA, Ainsworth JR, Good P, Mollon JD, et al. (2004) Progressive cone dystrophy associated with mutation in CNGB3. Invest Ophthalmol Vis Sci 45: 1975–1982.
18. Polok B, Escher P, Ambresin A, Chouery E, Bolay S, et al. (2009) Mutations in CNNM4 cause recessive cone-rod dystrophy with amelogenesis imperfecta. Am J Hum Genet 84: 259–265.
19. Wu H, Cowing JA, Michaelides M, Wilkie SE, Jeffery G, et al. (2006) Mutations in the gene KCNV2 encoding a voltage-gated potassium channel subunit cause "cone dystrophy with supernormal rod electroretinogram" in humans. Am J Hum Genet 79: 574–579.

20. Thiadens AA, den Hollander AI, Roosing S, Nabuurs SB, Zekveld-Vroon RC, et al. (2009) Homozygosity mapping reveals PDE6C mutations in patients with early-onset cone photoreceptor disorders. Am J Hum Genet 85: 240–247.
21. Wang QL, Chen S, Esumi N, Swain PK, Haines HS, et al. (2004) QRX, a novel homeobox gene, modulates photoreceptor gene expression. Hum Mol Genet 13: 1025–1040.
22. Wada Y, Abe T, Sato H, Tamai M (2001) A novel Gly35Ser mutation in the RDH5 gene in a Japanese family with fundus albipunctatus associated with cone dystrophy. Arch Ophthalmol 119: 1059–1063.
23. Hameed A, Abid A, Aziz A, Ismail M, Mehdi SQ, et al. (2003) Evidence of RPGRIP1 gene mutations associated with recessive cone-rod dystrophy. J Med Genet 40: 616–619.
24. Jalkanen R, Mantyjarvi M, Tobias R, Isosomppi J, Sankila EM, et al. (2006) X linked cone-rod dystrophy, CORDX3, is caused by a mutation in the CACNA1F gene. J Med Genet 43: 699–704.
25. Demirci FY, Rigatti BW, Wen G, Radak AL, Mah TS, et al. (2002) X-linked cone-rod dystrophy (locus COD1): identification of mutations in RPGR exon ORF15. Am J Hum Genet 70: 1049–1053.
26. Huang L, Xiao X, Li S, Jia X, Wang P, et al. (2012) CRX variants in cone-rod dystrophy and mutation overview. Biochem Biophys Res Commun 426: 498–503.
27. Huang L, Li S, Xiao X, Jia X, Sun W, et al. (2013) Novel GUCA1A mutation identified in a Chinese family with cone-rod dystrophy. Neurosci Lett 541: 179–183.
28. Xiao X, Guo X, Jia X, Li S, Wang P, et al. (2011) A recurrent mutation in GUCY2D associated with autosomal dominant cone dystrophy in a Chinese family. Mol Vis 17: 3271–3278.
29. Huang L, Li S, Xiao X, Jia X, Wang P, et al. (2013) Screening for variants in 20 genes in 130 unrelated patients with cone-rod dystrophy. Mol Med Rep 7: 1779–1785.
30. Wang Q, Wang P, Li S, Xiao X, Jia X, et al. (2010) Mitochondrial DNA haplogroup distribution in Chaoshanese with and without myopia. Mol Vis 16: 303–309.
31. Li Y, Vinckenbosch N, Tian G, Huerta-Sanchez E, Jiang T, et al. (2010) Resequencing of 200 human exomes identifies an excess of low-frequency non-synonymous coding variants. Nat Genet 42: 969–972.
32. Li R, Li Y, Kristiansen K, Wang J (2008) SOAP: short oligonucleotide alignment program. Bioinformatics 24: 713–714.
33. Li R, Yu C, Li Y, Lam TW, Yiu SM, et al. (2009) SOAP2: an improved ultrafast tool for short read alignment. Bioinformatics 25: 1966–1967.
34. Li R, Li Y, Fang X, Yang H, Wang J, et al. (2009) SNP detection for massively parallel whole-genome resequencing. Genome Res 19: 1124–1132.
35. Rozen S, Skaletsky H (2000) Primer3 on the WWW for general users and for biologist programmers. Methods Mol Biol 132: 365–386.
36. Li L, Xiao X, Li S, Jia X, Wang P, et al. (2011) Detection of variants in 15 genes in 87 unrelated Chinese patients with Leber congenital amaurosis. PLoS One 6: e19458.
37. den Dunnen JT, Antonarakis SE (2000) Mutation nomenclature extensions and suggestions to describe complex mutations: a discussion. Hum Mutat 15: 7–12.
38. Siepel A, Bejerano G, Pedersen JS, Hinrichs AS, Hou M, et al. (2005) Evolutionarily conserved elements in vertebrate, insect, worm, and yeast genomes. Genome Res 15: 1034–1050.
39. Kumar P, Henikoff S, Ng PC (2009) Predicting the effects of coding non-synonymous variants on protein function using the SIFT algorithm. Nat Protoc 4: 1073–1081.
40. Adzhubei IA, Schmidt S, Peshkin L, Ramensky VE, Gerasimova A, et al. (2010) A method and server for predicting damaging missense mutations. Nat Methods 7: 248–249.
41. Reese MG, Eeckman FH, Kulp D, Haussler D (1997) Improved splice site detection in Genie. J Comput Biol 4: 311–323.
42. den Hollander AI, Black A, Bennett J, Cremers FP (2010) Lighting a candle in the dark: advances in genetics and gene therapy of recessive retinal dystrophies. J Clin Invest 120: 3042–3053.
43. Kohl S, Kitiratschky V, Papke M, Schaich S, Sauer A, et al. (2012) Genes and mutations in autosomal dominant cone and cone-rod dystrophy. Adv Exp Med Biol 723: 337–343.
44. Ito S, Nakamura M, Nuno Y, Ohnishi Y, Nishida T, et al. (2004) Novel complex GUCY2D mutation in Japanese family with cone-rod dystrophy. Invest Ophthalmol Vis Sci 45: 1480–1485.
45. Garcia-Hoyos M, Auz-Alexandre CL, Almoguera B, Cantalapiedra D, Riveiro-Alvarez R, et al. (2011) Mutation analysis at codon 838 of the Guanylate Cyclase 2D gene in Spanish families with autosomal dominant cone, cone-rod, and macular dystrophies. Mol Vis 17: 1103–1109.
46. Kitiratschky VB, Behnen P, Kellner U, Heckenlively JR, Zrenner E, et al. (2009) Mutations in the GUCA1A gene involved in hereditary cone dystrophies impair calcium-mediated regulation of guanylate cyclase. Hum Mutat 30: E782–796.
47. Ducroq D, Rozet JM, Gerber S, Perrault I, Barbet D, et al. (2002) The ABCA4 gene in autosomal recessive cone-rod dystrophies. Am J Hum Genet 71: 1480–1482.
48. Fishman GA, Stone EM, Eliason DA, Taylor CM, Lindeman M, et al. (2003) ABCA4 gene sequence variations in patients with autosomal recessive cone-rod dystrophy. Arch Ophthalmol 121: 851–855.
49. Thiadens AA, Phan TM, Zekveld-Vroon RC, Leroy BP, van den Born LI, et al. (2012) Clinical course, genetic etiology, and visual outcome in cone and cone-rod dystrophy. Ophthalmology 119: 819–826.
50. Maugeri A, Klevering BJ, Rohrschneider K, Blankenagel A, Brunner HG, et al. (2000) Mutations in the ABCA4 (ABCR) gene are the major cause of autosomal recessive cone-rod dystrophy. Am J Hum Genet 67: 960–966.

6

Comparison of Sequencing Platforms for Single Nucleotide Variant Calls in a Human Sample

Aakrosh Ratan[1]◔, Webb Miller[1], Joseph Guillory[2], Jeremy Stinson[2], Somasekar Seshagiri[2], Stephan C. Schuster[1*◔]

1 Center for Comparative Genomics and Bioinformatics, Pennsylvania State University, University Park, Pennsylvania, United States of America, **2** Department of Molecular Biology, Genentech Inc., South San Francisco, California, United States of America

Abstract

Next-generation sequencings platforms coupled with advanced bioinformatic tools enable re-sequencing of the human genome at high-speed and large cost savings. We compare sequencing platforms from Roche/454(GS FLX), Illumina/HiSeq (HiSeq 2000), and Life Technologies/SOLiD (SOLiD 3 ECC) for their ability to identify single nucleotide substitutions in whole genome sequences from the same human sample. We report on significant GC-related bias observed in the data sequenced on Illumina and SOLiD platforms. The differences in the variant calls were investigated with regards to coverage, and sequencing error. Some of the variants called by only one or two of the platforms were experimentally tested using mass spectrometry; a method that is independent of DNA sequencing. We establish several causes why variants remained unreported, specific to each platform. We report the indel called using the three sequencing technologies and from the obtained results we conclude that sequencing human genomes with more than a single platform and multiple libraries is beneficial when high level of accuracy is required.

Editor: Zhaoxia Yu, University of California Irvine, United States of America

Funding: This study is based on sequencing data generated for the Southern African genomes Project [5]. The 454 data was generated at Penn State University and supported by Roche/454 via the donation of sequencing reagents. We like to thank Timothy Harkins and Kevin McCarren for making SOLiD data for the KB1 genome available to this group of investigators. The sequence data for Illumina GA IIx and HiSeq 2000 was generated at Penn State University and supported through internal funds. The funders had no role in study design, data collection (other than what has been stated above) and analysis, decision to publish, or preparation of the manuscript. This project is funded, in part, under a grant by the Pennsylvania Department of Health using Tobacco CURE Funds to AR. The Pennsylvania Department of Health specifically disclaims responsibility for any analyses, interpretations or conclusions.

Competing Interests: Some of the authors are employed by a commercial company, Genentech Inc.

* E-mail: scs@bx.psu.edu

◔ These authors contributed equally to this work.

Introduction

The Human Genome Project [1,2] published the first working draft of the human reference sequence in 2000. That sequence was generated in its entirety by capillary sequencing; all subsequent genomes of human individuals [3–9], except one [4], have relied on next-generation sequencing (NGS) platforms. Starting in 2005, 454/Roche [10], and subsequently Illumina [11] and SOLiD/ABI [12] entered the market with technology that ultimately aims to re-sequence human genomes for less than $1000, which would transform the field of personalized medicine. While several new entrants may have the future potential to change the sequencing landscape yet again, the current sequencing market is dominated by these three matured platforms. With the introduction of these technologies, reports of biases in all platforms [13], as well as efforts to monitor [14], remove [15], or compensate for them arose. Initially, the three sequencing approaches were evaluated in targeted regions (up to 4 Mbp) [16] where they were compared with Sanger-generated sequences for validation. A recent study compared the accuracy of the SNP calls and the quality of the short-reads from the three platforms in an E. coli sample [17], while another study compared single nucleotide variants from Illumina data with the data from Complete Genomics, another entrant in the sequencing landscape [18].

Here we compare three platforms namely 454/Roche GS FLX, Illumina HiSeq 2000 and ABI SOLiD 3 ECC in their ability to identify the single-nucleotide substitutions in the same human individual. In contrast to previous studies that generated a saturating level of redundant coverage to eliminate low coverage as a factor in SNP calling, we sequenced the individual's DNA to read-depths that allows for variant detection in each corresponding dataset with sufficient confidence. This allows us to assess the performance and biases of each sequencing platform, as it would affect the completeness of the variant detection in a genome-sequencing project. In this study we show that both Illumina HiSeq 2000 and SOLiD 3 ECC sequencing exhibit variation in coverage with GC content, and report on the probable reasons why certain SNPs are missed by each of the technologies. We also discuss a method to calculate the uniquely mappable regions of a reference genome, which can then be used to filter SNPs and improve the quality of the variant calls, thereby avoiding the generation of false positive variants. This approach is especially useful for 454 GS FLX sequences analyzed using the software Newbler [3], which does not utilize the concept of mapping quality [19].

Results

Generation and Alignment of Reads

We sequenced the genomic DNA from a human DNA sample called KB1 [5]; to 10.04 fold coverage using 454 GS FLX sequencer, 58.89 fold coverage using Illumina HiSeq 2000 sequencer and 78.63 fold coverage using SOLiD 3 ECC technology (Table 1). All three platforms are expected to exhibit different error characteristics and therefore should complement one another to yield the most accurate set of human single nucleotide variants. Since the dominant type of error for SOLiD and Illumina reads is substitutions, it is possible to compare the sequences using similar alignment criteria and software. Using the software BWA [20] (version 0.5.9rc), we aligned the SOLiD and Illumina reads to the human reference GRCh37, henceforth referred to as hg19. In contrast to randomly dispersed Illumina and SOLiD sequencing templates, the array-based pyrosequencing technology used by 454 generates sequences with homopolymer errors (indels in runs of the same nucleotide). The assembly/mapping software Newbler aligns these data in a format called flow-space in an attempt to correct these systematic errors associated with pyrosequencing. We used Newbler version 2.3 to align the single-end fragment reads to hg19 with the default parameters. Only the two platforms 454 and Illumina exclude the low-quality reads using filters prior to reporting the final set of reads to the user; a fact that is reflected in a higher alignment rate for these two technologies. In contrast, the SOLiD system uses the alignment to a reference as a mean to determine reads of sufficient quality (Table 1).

Coverage Distribution and Variation

The depth-of-coverage distributions of the reference genome from data from the three sequencing platforms are shown in Figure 1. The coverage distributions are bimodal (Figure 1), with the two modes that at first appeared to reflect the coverage on the autosomes and the sex chromosomes. However, we found that removing the sex chromosomes from the analysis did not eliminate the bimodal behavior. Since, earlier studies reported a decrease in coverage in A/T rich regions with Illumina sequencing [21], we investigated the correlation between the GC content and coverage for the three platforms, for potentially influencing bimodal behavior. Figure 2 shows a significant variation in coverage with GC content; coverage by Illumina and SOLiD sequences is notably lower in G/C rich regions. This variation with GC content, along with the expected haploid coverage on sex chromosomes, explains the observed bimodal behavior of these distributions. Despite sharing the emulsion PCR approach in the sequencing template preparation with the SOLiD 3 system, only the Roche/454 FLX sequencing chemistry seemed to be immune to this bias, as is demonstrated with only a minor correlation between GC content and coverage for 454 reads (Figure 2). The behavior exhibited by Illumina HiSeq sequences is in stark contrast to the behavior of GA II reads, which exhibit a lower coverage in A/T rich regions.

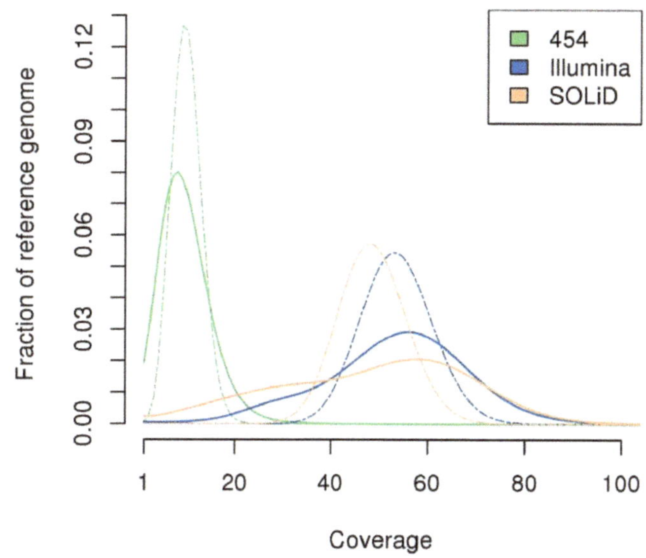

Figure 1. Depth of coverage distribution for the three platforms. The y-axis indicates the fraction of the bases in the reference sequence that has a particular coverage. This does not include secondary alignments and potential PCR duplicates. The dashed lighter curves depict the coverage distribution as calculated using a Poisson model for each sequencing technology.

Detection of SNPs and Indels

A large portion of the genomic regions requires local realignments due to the presence of indels in the sequenced genome when compared to the reference genome. Indels can lead to alignment artifacts where a lot of bases around the indel do not agree with the reference and can masquerade as SNPs. We used the realignment tool in GATK [22] version 1.2–29 to realign the sequences from the Illumina and SOLiD dataset; followed by use of SAMtools version 0.1.16 to call variants. As described in the

Table 1. Sequencing and alignment statistics. Coverage is calculated with and without the putative PCR duplicates.

	454	Illumina	SOLiD
Number of reads generated	83,331,227	1,867,073,052	6,905,193,148
Number of bases generated	29,246,232,549	188,349,876,745	397,681,271,500
Read lengths	350 avg. single-end	101 paired-end	50 paired-end, 75 single-end
Number of reads aligned	82,310,265 (98.77%)	1,751,042,389 (93.79%)	4,429,505,837 (64.15%)
Number of bases aligned	28,732,501,185 (98.24%)	168,495,777,999 (89.46%)	224,998,686,646 (56.58%)
Coverage	10.04/9.78 X	58.89/55.06 X	78.63/53.20 X
Duplicate reads	2,211,903	115,528,614	1,216,108,795
Reference bases covered	2,781,827,482	2,858,458,440	2,850,277,778

The number of aligned reads includes the duplicate reads.

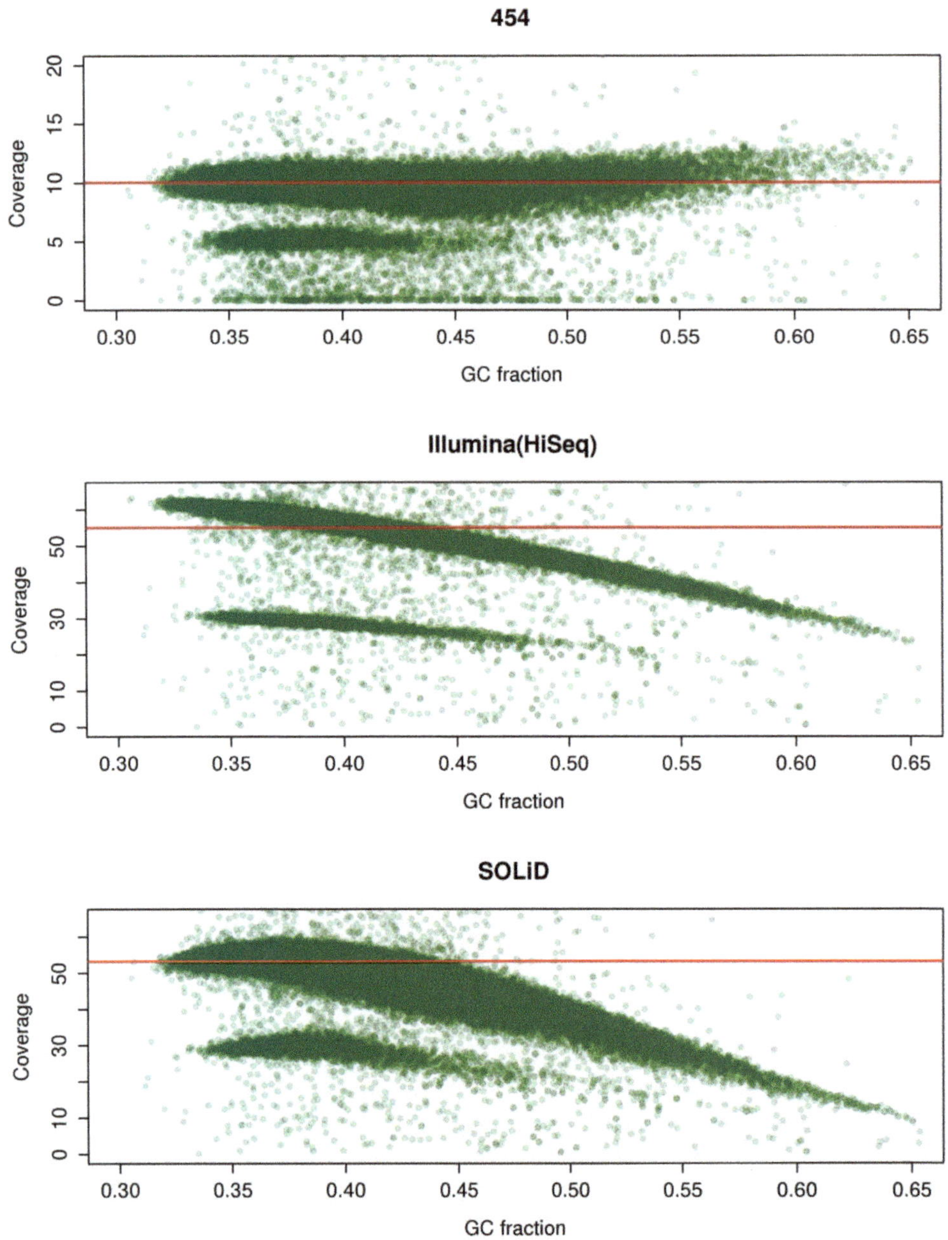

Figure 2. Variation of coverage with GC content in the three sequencing technologies. The red line shows the mean coverage across the whole genome. Each point on the plot reflects the mean coverage and fraction of GC content in 50 kbp non-overlapping window. The y-axis shows the coverage whereas the x-axis shows the fraction of C, G nucleotides in the window. This does not include secondary alignments and potential PCR duplicates.

previous section, Newbler handles the 454 data in flow-space and we used Newbler version 2.3 to call variants in it (See Methods). While the three sequencing approaches resulted in a similar number of single substitution variants, namely 4,331,131 variants for the 454 data, 4,691,363 variants using the Illumina data, and 4,145,208 variants using the SOLiD sequences, the combination resulted in a total of 5,252,985 potential variant locations. However, only a common set of 3,401,954 variant locations was shared between all three technologies, whereas only one or two of the platforms supported the remaining 1,851,031 locations (Figure 3a). As for indels, we found 614,794 indels using the 454 data, 554,138 small indels using the Illumina data and 303,937 potential indels using the SOLiD data.

Reasons for Failure of Detection of Substitutions

The 1,851,031 discrepant locations between 454 FLX, Illumina HiSeq and SOLiD 3 sequences allow for the study the false positive and the false negative rates for the obtained variant calls. Considering the SNP calls from each platform as independent evidence, we use the locations where two platforms agree to study the false negatives for the third one. This allows us to quantify and understand the reasons why this SNP was not called in the dataset sequenced using the third platform. Similarly, at the locations where only one platform calls a variant, we identify and study the false positives for that platform, barring exceptions that are explained later in the text.

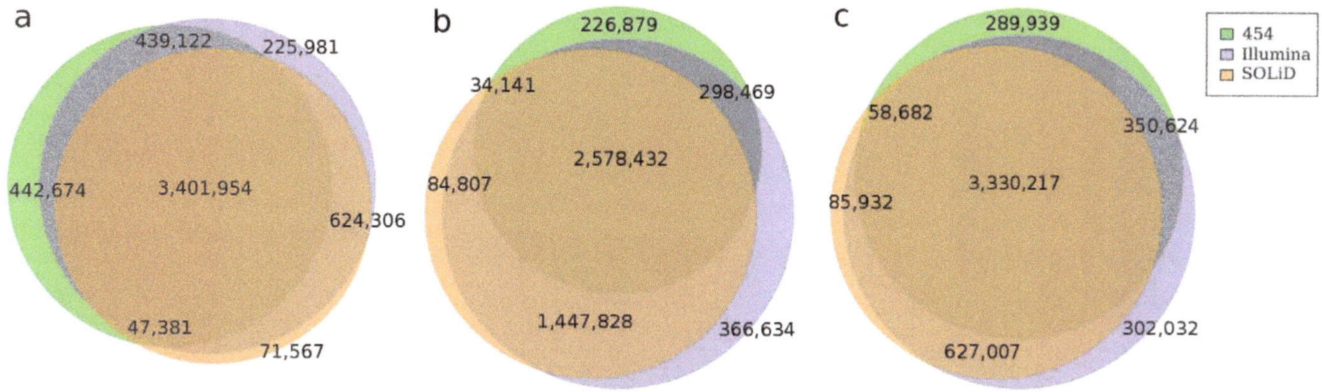

Figure 3. Venn diagram showing the overlap in the SNP calls made using data from the three sequencing technologies. We display the sizes of each of the seven categories of overlaps among the variant calls in the three technologies. (a) depicts the overlaps when all substitution calls are used, (b) depicts the overlaps when all calls from Illumina and SOLiD are used but only the high-confidence subset of the 454 dataset is used, and (c) depicts the overlaps when only the variants in the uniquely alignable regions of the reference sequence are used.

The reasons why a SNP is not detected by one sequencing technology, whereas it is reported by another, can be broadly divided into three categories:

- **Issues related to coverage**: These can be further subdivided into complete lack of coverage, low coverage (which is not enough to call a SNP based on predefined criteria), and higher-than-expected coverage (based on a model used to separate SNPs from structural variants and assembly errors) at the candidate location.
- **Issues with the alternate allele**: Most software tools (including SAMtools and Newbler) require observing the alternate allele at least twice or more, before they consider the location as a potential variant. These can be further subdivided into instances where the alternate allele is not seen at all and others, when the alternate allele is not seen a sufficient number of times.
- **Issues with the variant calling**: These refer to the situations where the alternate allele is seen a requisite number of times, but the SNP is not called due to other reasons. These reasons may include proximity to many other SNPs, proximity to a high quality indel, existence in a non-uniquely alignable region, and a huge deviation from the expected diploid behavior of the sample for the data aligned using BWA. For the reads aligned using Newbler, the reasons include the location being in a non-uniquely alignable region and other alignment errors that arise due to the unique error-profile of the 454 reads.

We investigated the alignments at the 439,122 locations that were called as putative variants by using 454 and Illumina sequences, but not using SOLiD sequences (Figure 4a i). We assigned each location to a particular category based on the reason why it was not called a SNP. We found that the variant allele was observed in the SOLiD reads in 64% of these cases, but the SNP was filtered away for various reasons. 27% of the locations were filtered away due to a low SNP quality (defined as the Phred-scaled likelihood that the called genotype is identical to the reference), 18% of them were filtered away due to a low RMS (root mean square) mapping quality (reflecting the limitation of shorter reads) and another 19% were filtered away as the variant allele was not seen enough number of times. Coverage related issues (no coverage, too little coverage or more than expected coverage) were responsible for another 19% of the locations. The alternate allele was not seen at all, despite adequate coverage at the site, for the remaining 17% locations.

For the 71,567 locations that were called using the SOLiD sequences (but not by others), we looked at the alignments for both the 454 dataset and the Illumina datasets. At about 15% of these locations (Figure 4a ii), the alternate allele was seen just once in the 454 dataset and at about another 16% of them, the coverage of 454 reads was not enough to call a SNP. For another 21% of the locations the SNP was not called by Newbler, even though the allele was seen multiple times in the pairwise alignments between the reference and the 454 reads, with most of them being associated with homopolymer errors. On the other hand at 25% of these locations the SNP was seen in the Illumina dataset (Figure 4a iii), but it was filtered away due to a lower SNP quality (15%), or because lower mapping quality (9%). Another 14% of these locations did not have sufficient coverage with Illumina reads to be considered in SNP calling. Considering the locations where both 454 and Illumina had little, no, or higher than expected coverage, and where the alternate allele was seen at least once in either 454 or Illumina dataset as true SNPs, we expect 14,707 of the 71,567 locations to be false-positives for the SOLiD calls.

When we looked at the 47,381 locations that were called a SNP using 454 and SOLiD reads, we found that primary reason (at 60% of the locations) these were not called a SNP with Illumina reads had to do with the coverage (Figure 4b i). 57% of the locations were in regions where the coverage was more than expected (signaling a putative structural variant), whereas there was little of no coverage for the remaining 3%. We used a Poisson distribution with the same mean value to calculate the coverage threshold to filter variants, but this data suggests that a gamma distribution with more weight on more tails is probably a better model for Illumina data. The second largest contributor was low SNP quality (22% of the locations), which is the result of an observed deviation from the expectation that both allele should be seen approximately the same number of times on a heterozygous location.

We found 225,981 locations that were called as putative variants using Illumina reads only. Looking at the alignments for the SOLiD reads at those locations (Figure 4b ii), we found that for 22% of them we saw the alternate allele a sufficient number of times, but it was filtered away either due to low RMS mapping quality or a low SNP quality. Another 16% of the locations were

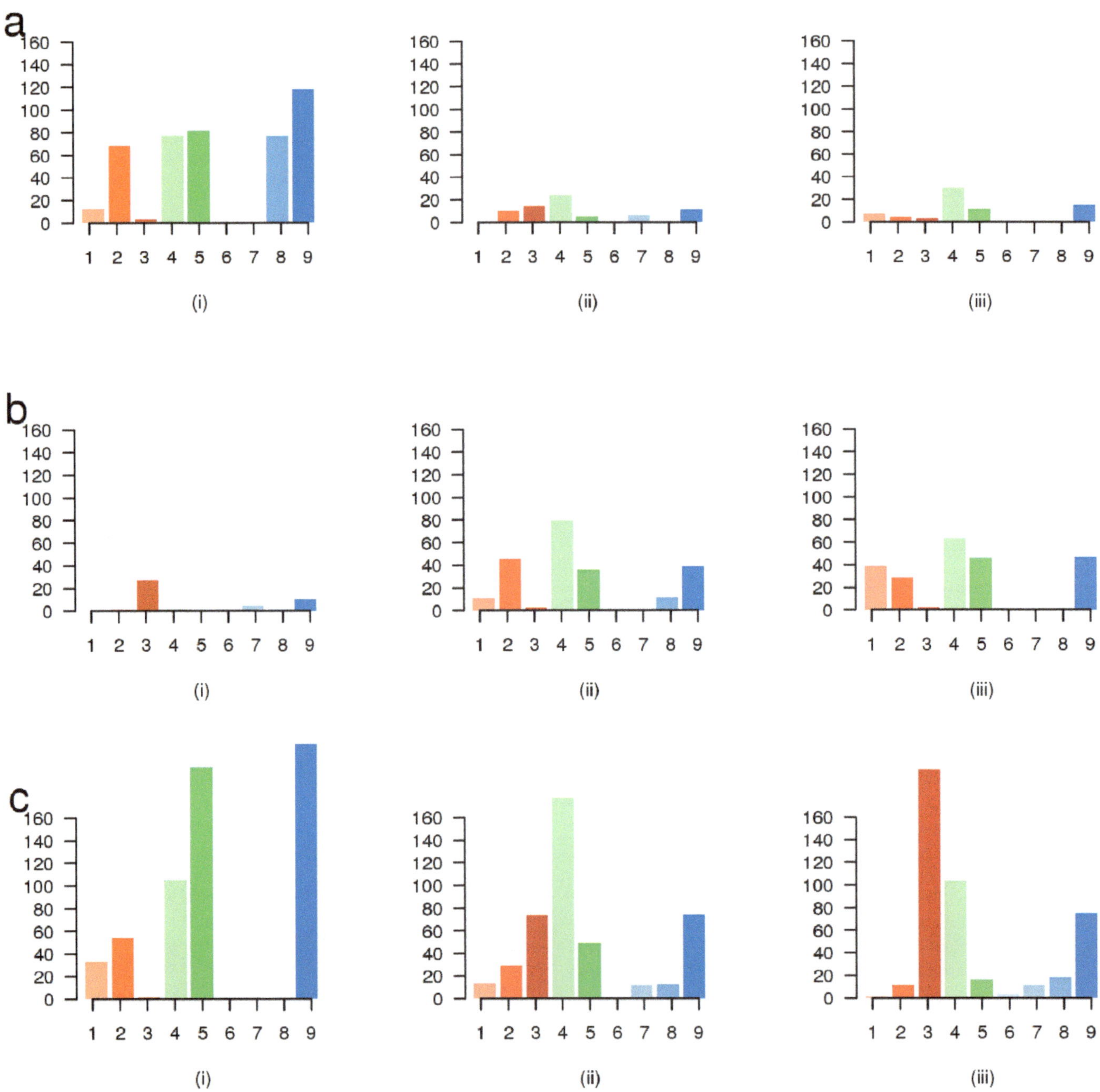

Figure 4. Discrepant SNP calls from each platform. The categories on the x-axis are (1) no coverage at location (2) not enough coverage at location (3) more than expected coverage (4) alternate allele not seen (5) alternate allele seen just once (6) too many SNPs around location (7) close to a high-quality indel (8) low RMS mapping quality (9) low SNP quality. The y-axis depicts the number of locations (frequency) in each category. a) Comparison of SOLiD generated sequences with other sequences based on SNP calls and alignments. (i) SNPs called using 454 and Illumina sequences but not called using SOLiD reads. (ii) SNPs called only by SOLiD sequences. We investigate why they were not called using Illumina alignments. (iii) SNPs called only by SOLiD sequences. We investigate why they were not called using 454 alignments. b) Comparison of Illumina generated sequences with other sequences based on SNP calls and alignments. (i) SNPs called using 454 and SOLiD reads but not called using Illumina reads. (ii) SNPs called only by Illumina sequences. We investigate why they were not called using SOLiD alignments. (iii) SNPs called only by SOLiD sequences. We investigate why they were not called using 454 alignments. c) Comparison of 454 generated sequences with other sequences based on SNP calls and alignments. (i) SNPs called using SOLiD and Illumina reads but not called using 454 reads. (ii) SNPs called only by 454 sequences. We investigate why they were not called using SOLiD alignments. (iii) SNPs called only by 454 sequences. We investigate why they were not called using Illumina alignments.

filtered away as the variant was not seen sufficient number of times. The alternate allele was not seen at all for 35% of the locations and the coverage was deemed insufficient for the remaining 27% of the locations. This was very different from 454 (Figure 4b iii), where the alternate allele was not seen at all for about 28% of the locations, only seen once for about 20% of the locations, and for another 21% the allele was seen multiple times but was not called as a variant by Newbler. Using the same criterion as for SOLiD data, we expect 31,696 of the calls to be false-positives for the Illumina data.

Among the 624,306 locations that were called a SNP using both Illumina and SOLiD reads, but not using 454 reads, 225,443 (36%) had the alternate allele multiple times in the pairwise alignments of the reads with the reference (Figure 4c i). However, they were not flagged as variants in the later stages of processing, with most of them being associated with homopolymer errors. There were another 205,060 SNPs (33%) where we saw the alternate allele once, not twice or more as required by Newbler to call it a SNP. As for the 442,674 locations that were only called as variants using 454 sequences, SOLiD (Figure 4c ii) and Illumina (Figure 4c iii) did not call 19% of them due to low RMS mapping quality or low SNP quality. 45% of the variant locations were not considered in Illumina reads as the coverage exceeded the expectations, and another 23% were not called as the variant was not seen enough times. In contrast, only 17% of the locations had more than expected coverage in SOLiD reads, but 40% of the variants were not called as the variant allele was not seen enough times. We expect 75,695 calls to be false-positives for sequences generated using 454 instruments for this sample using the same criterion as that used for Illumina and SOLiD. However a large proportion of these putative false-positive calls were in locations where both Illumina and SOLiD saw the alternate allele multiple times, but the SNP was filtered away as the depth-coverage at that location was higher than expected. This could be an artifact of our SNP-calling pipeline, which used Poisson distributions to set the coverage thresholds for the SNP calls, or a consequence of structural variants present in the target genome. This also means that false-positives for 454 sequences are indicative of certain biases in Illumina and SOLiD reads along with errors in the 454 reads.

454 Data and Newbler-specific Variant Calling Variation

Newbler labels a subset of its variant calls as "high-confidence". These calls require more evidence than that required for the calls used in the analysis summarized above. We therefore decided to use the high-confidence subset of the 454 FLX calls to investigate whether this increases the concordance rate of 454 data with the other two technologies. Newbler called 3,137,921 substitutions when the high-confidence subset was used, resulting in a reduction from 442,674 454-specific calls (with default parameters) to now only 226,879. The more stringent setting, however, significantly increased the number of SNPs that were called by Illumina and SOLiD, but not by 454, to 1,447,828. It also decreased the number of SNPs called by all three technologies to 2,578,432 (Fig. 3b). Therefore, even though using the high-confidence subset of SNP calls seems to reduce the false-positive rate of calls (as evident by a decrease in the number of calls made only by 454 reads), it significantly increases the false-negative rate for variant calls.

BWA and many other aligners assign a mapping quality to each read, which is a scaled error probability that the alignment for the read is wrong. A read with a mapping quality of zero does not contribute to SNP calls. Newbler on the other hand aligns the reads to the reference and if it finds more than one alignment that passes all its alignment thresholds, then it tags the read as a "repeat", and that read is not used in SNP calling. Both methods attempt to reduce the false-positive rate of SNP calls, while attempting to use as much of the data as possible.

We normally post-process the SNP calls to restrict the calls to the uniquely alignable regions of the reference genome, for the reads of a certain length and error profile. This method was used in Schuster et al. 2010 [5] to call the variants and explicitly discussed in Koehler et al. 2010 [23], which termed the uniquely alignable region of the reference as the "uniqueome". We calculated the uniquely mappable region for 454 reads of average length 350 bp, 76 bp paired-end Illumina reads and 75 bp single-end/50 bp paired-end reads for SOLiD reads, and then filtered to throw away the SNP calls that did not lie in these regions. The results are shown in Figure 3c, where we see a substantial decrease in the calls made only by 454 sequencing, a marginal decrease in the number of calls made only by SOLiD sequencing, and an increase in Illumina only calls.

Our study highlights some of the issues encountered when these technologies are used for whole-genome sequencing of samples. We realize that the results of our analyses reflect a cumulative effect of library preparation, base callers, sequencing bias, alignment tools and variant callers. All platforms have their own biases, and use of more than one technology can be used to overcome the limitations that are posed by a single sequencing technology.

Comparison of Indels

Insertions and deletions are harder to call and analyse, compared to single nucleotide substitutions. Calculation of indel boundaries requires computationally expensive local realignments, and even then the exact start and stop of an indel can vary between algorithms and samples, making any comparison based on coordinates difficult. We had fragment reads from 454, and paired-end reads from Illumina and SOLiD (with different insert length distributions); each technology is best suited for different lengths and ranges of indels making any meaningful comparison difficult. Prior report on indels in whole-genome sequences have varied greatly; from 135,262 in the Han Chinese genome, to 823,396 indels in the Venter genome, showing the limitations faced in calling these variants.

Newbler called 614,794 indels in KB1 with the 454 data, 225,980 of which were homozygous and 282,909 of them were tagged as high confidence. Intersection of indel intervals in dbSNP 132 with these indel calls results in 263,262 (42.82%) overlaps and another 67,895 (11.04%) indels are found with +/−10 bp of the indels in dbSNP. We called 554,138 indels using the Illumina data, and 303,937 indels with the SOLiD data. 440,514 (79.49%) of the indels called using the Illumina data and 255,232 (83.97%) of the indels called using the SOLiD data were also found in dbSNP132. These numbers increased to 456,328 (82.35%) and 259,333 (85.32%) when an overlap +/−10 bp around indels was allowed.

In terms of overlap between the technologies, out of the possible 842,281 intervals, 177,772 of the indels were called using the sequences from all the technologies. 223,029 of them were supported by 2 of the 3 technologies, whereas the remaining had support from just one of them.

Experimental Validation of SNPs

We randomly selected 300 SNPs from each of the six sets in Figure 3a, i.e., where at least one of the three platforms (454 GS FLX, Illumina HiSeq 2000, SOLiD 3) disagreed on the computed genotype. For this purpose a sequence interval of 50 bp flanking a potential variant on each side was used for primer design using the MassARRAY® Assay Design Software (Sequenom, Inc.). These SNPs were assayed using mass spectrometry based genotyping technology and the results are shown in Figure 5. The validation rate was 50% for variants supported only by 454, 69% for variants only supported by Illumina, and 87% for the ones supported only by SOLiD. The validation rate for variants supported by more than one platform was higher when compared to the rate from the individual platforms, e.g. the validation rate for variants supported by 454 and Illumina was 78% whereas the validation rate for variants supported only by 454 was 50% and the validation rate for variants supported only by Illumina was 69%. The validation rate for SNPs called using 454 and SOLiD was 80% and it was 89% for variants called by Illumina and SOLiD. The longer reads generated by 454 resulted in some SNPs with 50 bp flanks (as required by the Sequenom assay) in repeat regions that have a higher dimer/hairpin potential when compared to the flanks for SNPs in the other two platforms. As a result, we saw significantly more primer design and assay failures for the 454 derived SNPs when compared to SNPs in the two short read technologies. Furthermore, the number of assays and primers that failed was also notably higher for locations that had been called by only one of the sequencing technologies.

Discussion

With the ultimate goal to sequence and analyze a human genome as fast and cost-effective as possible, it is highly desirable to simplify sampling and library preparation steps, but also to carry out DNA-sequencing on a single technology platform. In fact, with the availability of a high quality reference genome, re-sequencing with "low-cost per base" technologies has been at the center of recent efforts of most academic and commercial sequencing endeavors. However, with hundreds of human genomes currently being deciphered, the question of completeness relative to today's technological possibilities has become secondary to other pursuits. Faced with increasing evidence that human genomics does not provide allelic association to critical human phenotypes at predicted rates, the question is being raised whether missed genetic variants might be causal due to unmapped regions of the genome. Further, understanding interactions of so far unrecognized genetic variations with small molecule drugs is of increasing interest to pharmaceutical and diagnostic industry. Our analysis on the false negative rate of SNP identification suggests that a significant number of variants are not reported using a single sequencing platform, limiting insights which could enable a more complete understanding of the human genome. On the other hand, false positives from a single sequencing technology lead to a continuous degradation of the quality of SNP databases via the inclusion of non-existing genetic variants. Using a non-sequencing approach we report that the validation of SNP calls is significantly improved for variants that are reported using more than a single sequencing platform.

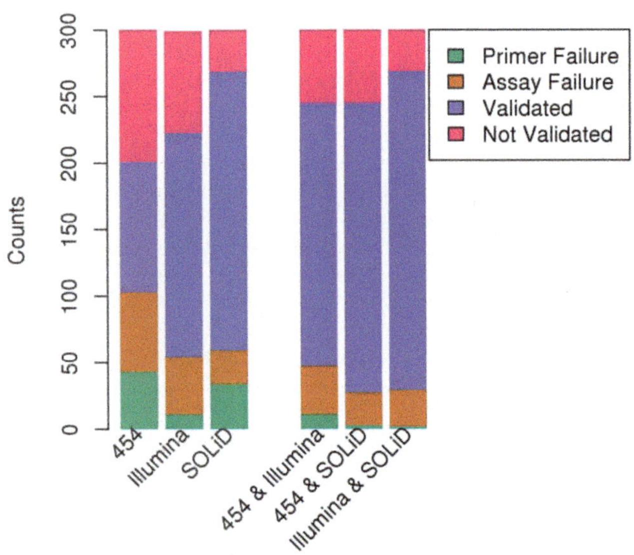

Figure 5. SNP Validation using Mass spectroscopy. Validation of 300 putative SNP locations from each of the six sets of SNP calls in Figure 3a, where not all three technologies agree on the computed genotype. The categories on x-axis are "454" (SNPs called by 454 only), "Illumina" (SNPs called by Illumina only), "SOLiD" (SNPs called by SOLiD only), "454 & Illumina" (SNPs called by 454 and Illumina), "454 & SOLiD" (SNPs called by 454 and SOLiD), "Illumina & SOLiD" (SNPs called by Illumina and SOLiD). The color categories include "Primer Failure" (Primer extension failure), "Assay Failure" (Assay Failure), "Validated" and "Not Validated".

We realize that every lab and sequencing platform has a "best practice" protocol from sample-prep to the bioinformatics pipeline, and that all the steps heavily influence any conclusion that one attempts to derive in such an experiment. In this study we show evidence that each of the three applied sequencing technologies contributes, at the respective coverage of each dataset, an additional unique 71,000 to 443,000 variants (1.4–8.9%) of the total of 5 million found in the human individual KB1. Remarkably, at least 1.4% of these technology-dependent variants would have gone unnoticed, even if the genome were sequenced on two of the three platforms. Furthermore as evident using Sequenom mass spectrometry, the validation rates for variants using two platforms are improved by more than 10% over the validation rates for the individual platforms. It therefore seems highly beneficial to sequence at least reference genomes from each geographic region by the multi-platform approach presented here. A larger number of multi-platform human reference genomes would not only minimize systematic technological biases, but also reduce ethnical biases from today's human genome reference sequence.

In this comparative study we looked at nine factors that affect the ability of a sequencing platform to accurately call variants in the human genome (see Figure 4 a–c). That include coverage related issues, allele related issues where the alternate allele was not seen multiple times and SNP related issues where the SNPs were filtered away in an attempt to reduce the false-positives. Among all the factors impacting the probability of a variant to be called, one of the most important one is the unbiased distribution of reads across a genome. As shown in Figure 2, the Roche/454 data is much more uniformly aligned, independent from the GC content of the genome than the short read platforms, resulting in comparable number of variant calls despite a much lower overall coverage. This trend however is counteracted by the 454 Roche specific error model that is prone to inaccurately assess the length of a given homopolymer, leading to potential false-positive SNPs calls, as shown by calls made only by 454 sequencing (Figure 3a). In addition to the platform specific errors, systematic errors are introduced through the varying algorithms and parameters used in the process of variant calling. This becomes apparent when the Roche/454 dataset is reanalyzed to include only the "high-confidence" variants, which reduces the total number of detected Newbler SNPs by 21%. The Roche/454 specific SNPs are reduced this way by more than 51% (226,879 calls, down from 442,674) (Figure 3b), however at the cost of increasing potential false negative prediction, as is apparent in a largely increased number of calls that are seen only by the Illumina and SOLiD platform (1,447,828 calls, up from 624,306). In order to avoid the conflict of Newbler's two modes of that either over-predicts false-negative or under-predicts true positive variants, we calculated the uniquely mappable region of the human genome for each dataset specific on the platform's read length (Roche/454 length 350 bp,

Illumina 76/101 bp paired-end reads and SOLiD 75 bp single-end, 50 bp paired-end reads). This approach is highly successful in combining the sensitivity of the variant detection for the Roche/454 data, while significantly reducing the rate of false negative prediction, as shown in Figure 3c. In our approach, the total number of variants agreed upon by all three platforms remains by ~3.3 million, while the number of Roche/454 specific variants is lowered to 302,032.

Interestingly, the lower GC bias as seen in 454 reads is not observed for the Life Technologies/SOLiD platform, despite the fact it also uses emulsion-PCR for template amplification. The performance of the SOLiD system was greatly aided in this comparison by its much higher coverage when compared to the 454 or Illumina data set.

An ideal setup for a high quality human or other mammalian genome project therefore would include all three sequencing platforms and thus allow for the generation of the most complete SNP sets and genome sequences. While we find that the three technology platforms complement one another, efforts are underway to improve currently used chemistries to combine advantages from multiple platforms into a single one.

Methods

Sequencing

454 Sequencing on the GS FLX. The protocols and details of library construction and sequencing on four Roche/454 GS FLX instruments using Titanium chemistry, for a total of 72 runs have been previously described in Schuster et al, 2010.

Illumina Sequencing on the HiSeq 2000. The KB1 library was prepared from genomic DNA using the Multiplexing Sample Preparation Oligonucleotide Kit from Illumina according to manufacturers instructions. The library was subjected to 5 cycles of PCR enrichment prior to cluster generation and then sequenced on the HiSeq 2000 according to manufacturers instructions.

SOLiD sequencing using SOLiD 3 ECC platform. SOLiD sequencing of KB1 was performed using recent developments in ligation-based sequencing. Specifically, improvements in paired-end sequencing have allowed us to sequence longer read lengths in both the forward (75 bp) and reverse direction (30 bp). The longer paired end reads were combined with a mate pair (2×50, 1.5 kb insert size) sequencing approach to achieve ~60× paired coverage. In addition, an increase in accuracy was obtained by using Exact Call Chemistry (ECC) for mate pair sequencing, as well as the 75 base pair forward tag.

ECC is based on standard "error correcting codes" commonly used in modern communication and data storage systems. This approach works by transforming data and augmenting it with redundancy to make it more resistant to measurement error. SOLiD and its ligation-based sequencing approach has the unique ability to use ECC by applying an additional, second sequencing probe set and repeating a sequencing primer for redundancy (Figure 5). This reduces measurement error and improves accuracy with minimal impact on sequencing time. Furthermore, ECC allows one to decode SOLiD sequencing data directly to base space without using a reference by using Bayesian inference. All of the analysis on the data was performed in base space and more details about the involved chemistry can be found at http://www3.appliedbiosystems.com/cms/groups/global_marketing_group/documents/generaldocuments/cms_091372.pdf.

Alignment of Reads

454. We used Newbler version 2.3 to align the 454 reads to the human reference genome (hg19) using the default parameters.

Illumina. We aligned the Illumina reads to the human reference genome (hg19) using BWA version 0.5.9rc, allowing up to 2 differences for reads of length 36 bp, up to 4 differences for reads of length 76 bp and up to 5 differences for reads of length 101 bp. The base quality deteriorates towards the 3′ end of the read, so we ran BWA with an option '–q 20′ to allow trimming of the read down to 35 bp.

SOLiD. We aligned the SOLiD reads to the human reference genome (hg19) using BWA version 0.5.9rc, allowing up to 3 differences in reads of length 50 bp and up to 4 differences in reads of length 76 bp. All the other parameters and arguments used were the same as that used for alignment of the Illumina reads.

Coverage

Coverage from the three technologies is defined as:

C = (number of distinct bases aligned)/(number of non-N bases in hg19 = 2,861,343,702).

Removal of Duplicate Reads

We used the MarkDuplicates tool offered as part of the Picard command-line suite (http://picard.sourceforge.net) to mark the duplicate reads in the Illumina and SOLiD datasets. Newbler takes care of the duplicate reads internally by requiring that variants be confirmed from non-duplicate reads. However, we did calculate the number of duplicate reads in the 454 dataset (Table 1) to compare them to Illumina and SOLiD, by calculating the number of reads that had the same alignment coordinates on the reference.

Variant Calling

In previous studies [16], Poisson distribution has been assumed as the model of depth coverage and we used it to set the coverage thresholds to filter SNPs.

454. We used Newbler version 2.3 to align and call SNPs from the 454 reads. The SNPs (available in the file 454AllDiffs.txt produced by Newbler) were analysed and were kept if the coverage at the SNP location was between 2 and 30. We deemed the SNPs as homozygous if more than 80% of the reads supported the alternate allele. We filtered to only keep the homozygous SNPs from the sex chromosomes and the mitochondria.

Illumina. We used SAMtools version 0.1.16 to call the variants in the Illumina reads. We required a minimum coverage of 4, a maximum coverage of 60 and a minimum quality of 20 for the SNPs and indels that were found to be on the autosomes. We reduced the maximum coverage requirement to 45 for the sex chromosomes and increased it to 10,000 for the mitochondrial DNA. Only homozygous SNP and indels calls were kept from the sex chromosomes and mtDNA.

SOLiD. We used SAMtools version 0.1.16 to call the variants in the Illumina reads. We required a minimum coverage of 4, a maximum coverage of 100 and minimum quality of 20 for the SOLiD reads that mapped to the autosomes. The maximum coverage limit was reduced to 60 for the sex chromosomes, and increased to 10,000 for the mitochondrial DNA and only homozygous SNP and indel calls were kept from these chromosomes.

Variation with GC Content

We calculated the mean coverage and the GC content in non-overlapping windows of 50 kbps and that information was used to plot Figure 2.

Selection of SNPs for Validation

We collated the SNPs belonging to each of the six categories in Figure 5, where the three technologies did not agree on the call. We removed the SNPs where the 50 bp flanks for the SNPs aligned to more than one location on hg19 with greater than 95% identity. We randomly selected 300 SNPs from each of the categories for validation.

Single Copy Intervals in hg19

We used a "self-masking" process to identify the regions in the reference genome where reads should align uniquely. The process breaks the reference genome into smaller fragments of length equal to the length of the reads, and adjacent fragments overlap each other by half the read length. These fragments are aligned to each reference chromosome, with alignment parameters selected to allow differences with a certain mutation rate. Any reference position appearing in no more than two alignments is considered uniquely mappable, since we only expect it to align to the two fragments that include it. We used LASTZ [24] and utilized the dynamic masking option to mask the reference genome. Initially the reference sequence is marked entirely as non-masked, and counters associated with each base are set to zero. Scoring parameters are set to reflect those that will be used later to map reads. As alignments are found, the corresponding counter is incremented for every base in the alignment. Whenever a base's counter exceeds 2, that base is soft-masked and removed from the alignment seeding tables. This latter action prevents highly repetitive regions from overwhelming the computational process.

Sequenom Genotyping

Whole Genome Amplification (WGA). We amplified genomic DNA samples using the REPLI-g Whole Genome Amplification Midi Kit (Qiagen, Valencia CA). Amplified DNA was cleaned up as per manufacturer's recommendation.

Sequenom validation. The predicted SNPs were validated on the Sequenom Mass Spectrometry platform using the iPLEX Gold Chemistry (Sequenom, San Diego CA). Primers for validating the SNPs were made using iPLEX Assay Design 3.0 software (Sequenom, San Diego, CA). The software creates a pair of primers that will allow amplification of an approximately 100 base pair PCR product that encompasses the region where the SNP of interest occurs. It also creates an extension primer that is complementary to the region of interest and extends up to the penultimate base right before the SNP position that is being interrogated. The PCR and extension primers used in this study are listed in Table S1. The primers used in the study were synthesized at Integrated DNA technologies in Iowa. PCR reactions with the appropriate template DNA and 100 nM of each PCR primer was set up and brought to 94°C and held for 2 minutes at 94°C followed by 35 cycles of 94°C for 15 sec, 56°C for 30 sec and 70°C for 1 min. A final extension setup at 70°C for 8 min concluded the PCR. Excess primers and unincorporated deoxynucleotides were removed from PCR reactions by adding 0.3 units of shrimp alkaline phosphatase (Sequenom, San Diego, CA), 0.3 to each reaction and incubating at 37°C for 40 min. The shrimp alkaline phosphatase was heat inactivated for 5 min at 85°C. The PCR reactions products were then subject to single base pair extension using the extension primer Thermosequenase and iPLEX nucleotides (Sequenom, San Diego, CA). Extension primers were used at a final concentration of 0.2 μM in a 10 μl reaction. Extension reactions were initially incubated at 94°C for 30 sec followed by 40 cycles of 94°C for 5 sec, 5 cycles of 54°C for 5 sec and 80°C for 5 sec. After this, primer extension reaction products were desalted with SpectroCLEAN resin (Sequenom, San Diego, CA). Ten nanoliters of the extension reaction was dispensed on a 384-format SpectroCHIP (Sequenom) prespotted with 3-hydroxypicolinic acid using a MassARRAY nanodispenser (Sequenom, San Diego, CA). Data was collected using a matrix-assisted laser desorption ionization/time of flight mass spectrometer (Sequenom, San Diego, CA). Primer extension data was analysed using MassARRAY Typer 3.4 software (Sequenom, San Diego, CA).

Supporting Information

Table S1 The putative allele, the flanking sequence in the genome, PCR and extension primers used in this study are listed in this spreadsheet. The sheet "Only 454" lists the details for locations that were only called using 454 generated sequences, "Only Illumina" refers to the details for locations that were only called by sequences generated using Illumina HiSeq 2000 sequencing, and "Only SOLiD" refers to the details for the locations that were only called by sequences generated using SOLiD 3 ECC technology. "454 and Illumina", "454 and SOLiD", and "Illumina and SOLiD" refer to details for locations that were called by only two of the three sequencing platforms that were used in this study.

Acknowledgments

This study is based on sequencing data generated for the Southern African genomes Project [5]. We would like to thank Timothy Harkins and Kevin McCarren for making SOLiD data for the KB1 genome available to this group of investigators.

Author Contributions

Conceived and designed the experiments: AR WM SS SCS. Performed the experiments: JG JS. Analyzed the data: AR WM SS SCS. Contributed reagents/materials/analysis tools: AR SCS. Wrote the paper: AR SS SCS.

References

1. Venter JC, Adams MD, Myers EW, Li PW, Mural RJ, et al. (2001) The sequence of the human genome. Science (New York, NY) 291: 1304–1351. doi:10.1126/science.1058040.
2. Lander ES, Linton LM, Birren B, Nusbaum C, Zody MC, et al. (2001) Initial sequencing and analysis of the human genome. Nature 409: 860–921. doi:10.1038/35057062.
3. Wheeler D a, Srinivasan M, Egholm M, Shen Y, Chen L, et al. (2008) The complete genome of an individual by massively parallel DNA sequencing. Nature 452: 872–876. doi:10.1038/nature06884.
4. Levy S, Sutton G, Ng PC, Feuk L, Halpern AL, et al. (2007) The diploid genome sequence of an individual human. PLoS biology 5: e254. doi:10.1371/journal.pbio.0050254.
5. Schuster SC, Miller W, Ratan A, Tomsho LP, Giardine B, et al. (2010) Complete Khoisan and Bantu genomes from southern Africa. Nature 463: 943–947. doi:10.1038/nature08795.
6. Tong P, Prendergast JGD, Lohan AJ, Farrington SM, Cronin S, et al. (2010) Sequencing and analysis of an Irish human genome. Genome biology 11: R91. doi:10.1186/gb-2010-11-9-r91.
7. Kim J-I, Ju YS, Park H, Kim S, Lee S, et al. (2009) A highly annotated whole-genome sequence of a Korean individual. Nature 460: 1011–1015. doi:10.1038/nature08211.
8. Ahn S-M, Kim T-H, Lee S, Kim D, Ghang H, et al. (2009) The first Korean genome sequence and analysis: full genome sequencing for a socio-ethnic group. Genome research 19: 1622–1629. doi:10.1101/gr.092197.109.

9. Wang J, Wang W, Li R, Li Y, Tian G, et al. (2008) The diploid genome sequence of an Asian individual. Nature 456: 60–65. doi:10.1038/nature07484.
10. Margulies M, Egholm M, Altman WE, Attiya S, Bader JS, et al. (2005) Genome sequencing in microfabricated high-density picolitre reactors. Nature 437: 376–380. doi:10.1038/nature03959.
11. Bentley DR, Balasubramanian S, Swerdlow HP, Smith GP, Milton J, et al. (2008) Accurate whole human genome sequencing using reversible terminator chemistry. Nature 456: 53–59. doi:10.1038/nature07517.
12. Valouev A, Ichikawa J, Tonthat T, Stuart J, Ranade S, et al. (2008) A high-resolution, nucleosome position map of C. elegans reveals a lack of universal sequence-dictated positioning. Genome Research: 1051–1063. doi:10.1101/gr.076463.108.3.
13. Dohm JC, Lottaz C, Borodina T, Himmelbauer H (2008) Substantial biases in ultra-short read data sets from high-throughput DNA sequencing. Nucleic acids research 36: e105. doi:10.1093/nar/gkn425.
14. Lassmann T, Hayashizaki Y, Daub CO (2010) SAMStat: monitoring biases in next generation sequencing data. Bioinformatics 27: 130–131. doi:10.1093/bioinformatics/btq614.
15. Schwartz S, Oren R, Ast G (2011) Detection and Removal of Biases in the Analysis of Next-Generation Sequencing Reads. PLoS ONE 6: e16685. doi:10.1371/journal.pone.0016685.
16. Shen Y, Sarin S, Liu Y, Hobert O, Pe'er I (2008) Comparing platforms for C. elegans mutant identification using high-throughput whole-genome sequencing. PloS one 3: e4012. doi:10.1371/journal.pone.0004012.
17. Suzuki S, Ono N, Furusawa C, Ying B-W, Yomo T (2011) Comparison of Sequence Reads Obtained from Three Next-Generation Sequencing Platforms. PLoS ONE 6: e19534. doi:10.1371/journal.pone.0019534.
18. Lam HYK, Clark MJ, Chen R, Chen R, Natsoulis G, et al. (2012) Performance comparison of whole-genome sequencing platforms. Nature biotechnology 30: 78–82. doi:10.1038/nbt.2065.
19. Li H, Ruan J, Durbin R (2008) Mapping short DNA sequencing reads and calling variants using mapping quality scores. Genome research 18: 1851–1858. doi:10.1101/gr.078212.108.
20. Li H, Durbin R (2009) Fast and accurate short read alignment with Burrows-Wheeler transform. Bioinformatics (Oxford, England) 25: 1754–1760. doi:10.1093/bioinformatics/btp324.
21. Harismendy O, Ng PC, Strausberg RL, Wang X, Stockwell TB, et al. (2009) Evaluation of next generation sequencing platforms for population targeted sequencing studies. Genome biology 10: R32. doi:10.1186/gb-2009-10-3-r32.
22. McKenna A, Hanna M, Banks E, Sivachenko A, Cibulskis K, et al. (2010) The Genome Analysis Toolkit: a MapReduce framework for analyzing next-generation DNA sequencing data. Genome research 20: 1297–1303. doi:10.1101/gr.107524.110.
23. Koehler R, Issac H, Cloonan N, Grimmond SM (2010) The Uniqueome: A mappability resource for short-tag sequencing. Bioinformatics (Oxford, England): 1–2. doi:10.1093/bioinformatics/btq640.
24. Harris RS (2007) Improved pairwise alignment of genomic DNA. The Pennsylvania State University,University Park, Pennsylvania, USA.

7

Genotype-Phenotype Correlation of 2q37 Deletions Including *NPPC* Gene Associated with Skeletal Malformations

Elisa Tassano[1*§], Jens Buttgereit[2,3§], Michael Bader[3§], Margherita Lerone[4§], Maria Teresa Divizia[4§], Renata Bocciardi[5§], Flavia Napoli[7§], Giovanna Pala[7§], Frédérique Sloan-Béna[6§], Stefania Gimelli[6§], Giorgio Gimelli[1§]

1 Laboratorio di Citogenetica, Istituto G. Gaslini, Genova, Italy, **2** Experimental and Clinical Research Center (ECRC), a joint institution of the Max Delbrück Center for Molecular Medicine (MDC) and the Charité Medical Faculty, Berlin, Germany, **3** Max Delbrück Center for Molecular Medicine (MDC), Campus Berlin-Buch, Berlin, Germany, **4** Laboratorio di Genetica Molecolare, Istituto G. Gaslini, Genova, Italy, **5** Dipartimento di Neuroscienze, Riabilitazione, Oftalmologia, Genetica e Scienze Materno-Infantili, (DiNOGMI), and Laboratorio di Genetica Molecolare, Istituto G. Gaslini, Genova, Italy, **6** Service of Genetic Medicine, University Hospitals of Geneva, Geneva, Switzerland, **7** Dipartimento di Pediatria, Istituto G. Gaslini, Università di Genova, Genova, Italy

Abstract

Coordinated bone growth is controlled by numerous mechanisms which are only partially understood because of the involvement of many hormones and local regulators. The C-type Natriuretic Peptide (CNP), encoded by *NPPC* gene located on chromosome 2q37.1, is a molecule that regulates endochondral ossification of the cartilaginous growth plate and influences longitudinal bone growth. Two independent studies have described three patients with a Marfan-like phenotype presenting a *de novo* balanced translocation involving the same chromosomal region 2q37.1 and overexpression of NPPC. We report on two partially overlapping interstitial 2q37 deletions identified by array CGH. The two patients showed opposite phenotypes characterized by short stature and skeletal overgrowth, respectively. The patient with short stature presented a 2q37 deletion causing the loss of one copy of the *NPPC* gene and the truncation of the *DIS3L2* gene with normal CNP plasma concentration. The deletion identified in the patient with a Marfan-like phenotype interrupted the *DIS3L2* gene without involving the *NPPC* gene. In addition, a strongly elevated CNP plasma concentration was found in this patient. A possible role of *NPPC* as causative of the two opposite phenotypes is discussed in this study.

Editor: Bin He, Baylor College of Medicine, United States of America

Funding: This work was supported by "Cinque per mille dell'IRPEF- Finanziamento della ricerca sanitaria" and "Finanziamento Ricerca Corrente, Ministero Salute (contributo per la ricerca intramurale)". The funders had no role in study design, data collection and analysis, decision to publish, or preparation of the manuscript.

Competing Interests: The authors have declared that no competing interests exist.

* E-mail: eli.tassano@gmail.com

§ These authors contributed equally to this work.

Introduction

During embryonic development and in postnatal life, longitudinal bone growth occurs at the cartilaginous growth plates located at the ends of the vertebrae and long bones. This process involves systemic hormones and local regulators that are finely tuned to allow harmonious and coordinated bone growth. Among these, CNP (C-type natriuretic peptide) plays an essential role in regulating bone growth by influencing cartilage homeostasis, proliferation and differentiation of osteoblasts and osteoclasts, and endochondral bone growth [1–4]. It is encoded by the *NPPC* gene located on chromosome 2q37.1 and exerts its effect through the interaction with the natriuretic peptide receptor B (NPR2 or NPRB) [5]. Knock-out mice for the CNP coding gene showed marked dwarfism due to impaired endochondral ossification [6–8].Targeted expression of CNP in chondrocytes rescued the dwarfism of CNP deficient mice, indicating that CNP acts as a local regulator of longitudinal bone growth [7]. Up to now no mutation or loss of the *NPPC* gene associated with dwarfism has been reported in humans. Homozygous mutations of the *NPR2* gene have been identified in patients affected by acromesomelic dysplasia or Maroteaux type (MIM 602875), while heterozygous mutations of the same receptor gene have been associated with short stature [9–10]. Conversely, overexpression of the *NPPC* gene determines skeletal overgrowth in transgenic mice [3] and in humans [11–13]. A recent genome-wide association study suggested that variation in the CNP signaling pathway, involving the *NPPC* and *NPR3* genes, plays an important role in determining human body height [14].

We report on two different patients with *de novo* 2q37 deletions. Patient 1 is affected by a pathological short stature and presents a 2q37.1 deletion of 1.9 Mb, while Patient 2 has a 2q37.1q37.3 deletion of 4.5 Mb associated with a Marfan-like phenotype. We discuss the possible role of *NPPC* in determining the two opposite phenotypes.

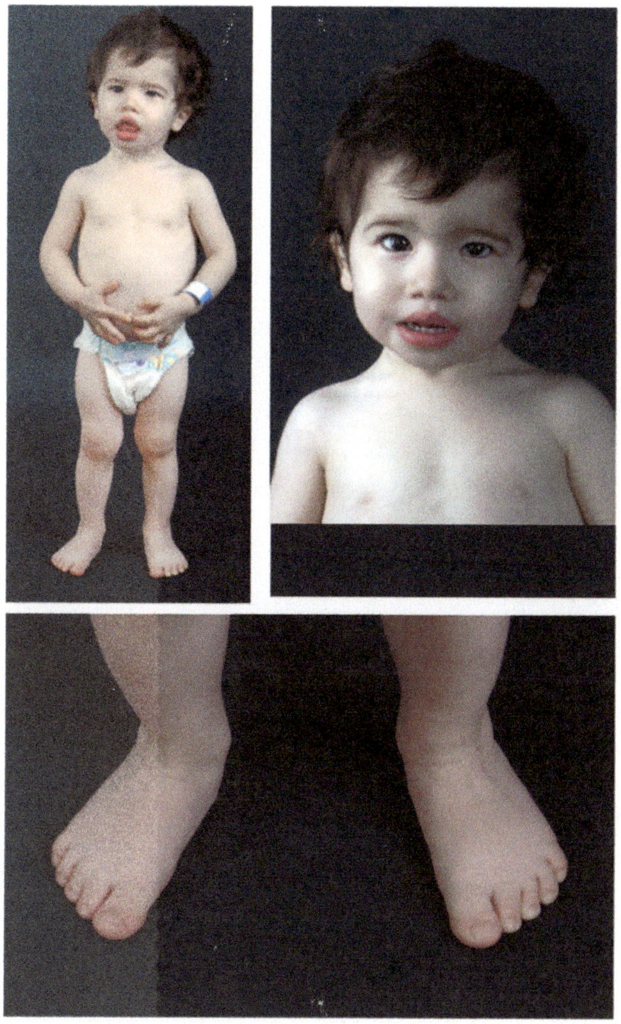

Figure 1. Clinical signs of the patient 1. The patient 1 at 3 years of age: short stature, prominent forehead, strabismus, epicanthus, low-set-ears, depressed nasal bridge, short phyltrum, V-shaped mouth. Mild brachydactyly, cutaneous syndactyly of 2nd and 3rd toes.

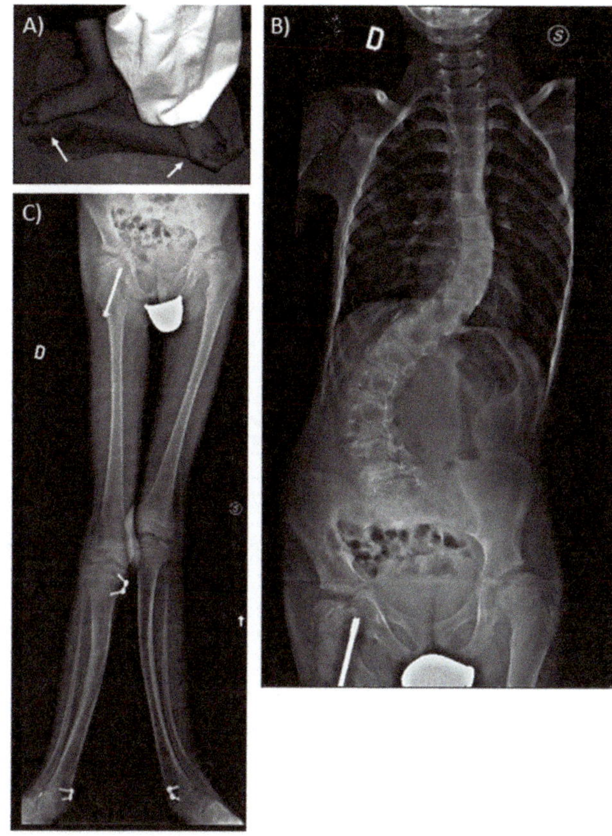

Figure 2. Clinical signs of the patient 2. A) Details of feet, showing very long hallux bilaterally B) X-rays, showing severe scoliosis. C) Long and curved lower limbs.

Results

Clinical Report

Patient 1. The patient, a 7 year-old-girl, is the second child of healthy non-consanguineous parents; parents' height was normal (father's height was 180,8 cm and mother's height 157,4 cm). In the family, a maternal first cousin had severe mental delay with abnormal behaviour and a paternal granduncle showed bilateral upper limb malformation. The patient has a healthy 9-year-old sister. Ultrasound examination revealed intrauterine growth retardation and polyhydramnios. The patient was born at 37 weeks of gestation by spontaneous delivery.

Birth weight was 2275 g (<3th centile), length 42 cm (<3th centile) and the occipitofrontal circumference (OFC) 34.5 cm (75th centile); Apgar score was 9 and 10 at 1 and 5 minutes, respectively. At the age of 20 months, growth parameters were weight 7620 g (<3th centile), height 73 cm (<3th centile), OFC 46.5 cm (50th–75th centile). Clinical examination showed facial dysmorphisms including pyriform skull with prominent forehead, upward slanting eyebrows, strabismus, epicanthus, low-set ears, depressed nasal bridge, short philtrum, V-shaped mouth and short upper lip. Feet showed mild brachydactyly, cutaneous syndactyly of 2nd and 3rd toes bilaterally, axial and limb girdle hypotonia, mild lower limb dysmetria (left>right), joint laxity. Growth has always been quite poor. At 6 years 5 months height was 103.4 cm (3th centile). Psychomotor and language development were delayed. Ophthalmological examination showed refractive errors and, at the age of two months, peripheral hearing loss was confirmed by auditory brainstem response (ABR test). Chest X-ray revealed 13 rib pairs and brain MRI showed a corpus callosum dysgenesis. Echocardiogram showed mild mitral valve insufficiency. Arginine stimulation test showed a normal GH response, thus ruling out GH deficiency. Photographs of the patient is shown in Figure 1.

Patient 2. The patient is a 12-year-old boy born to non-consanguineous parents. No family history of congenital anomalies or mental retardation was reported. He has a healthy 4 year-old sister. The patient was born at full term after an uneventful pregnancy by spontaneous vaginal delivery. Exposure to drugs or infection during the pregnancy was denied. At birth, weight was 3.300 kg (25th–50th centile), length 54 cm (90–95th centile) and occipitofrontal circumference (OFC) 34 cm (10th–25th centile). Psychomotor development was normal. At the age of 12 years, weight was 30 kg (5th centile), height 165 cm (>97th centile) and OFC 53.7 cm (>50th centile). Marfanoid habitus, arachnodactyly of hands and feet with very long hallux bilaterally associated and mild syndactyly of 3rd, 4th and 5th fingers, and severe scoliosis were observed (Fig. 2). Echocardiography showed patent ductus arteriosus.

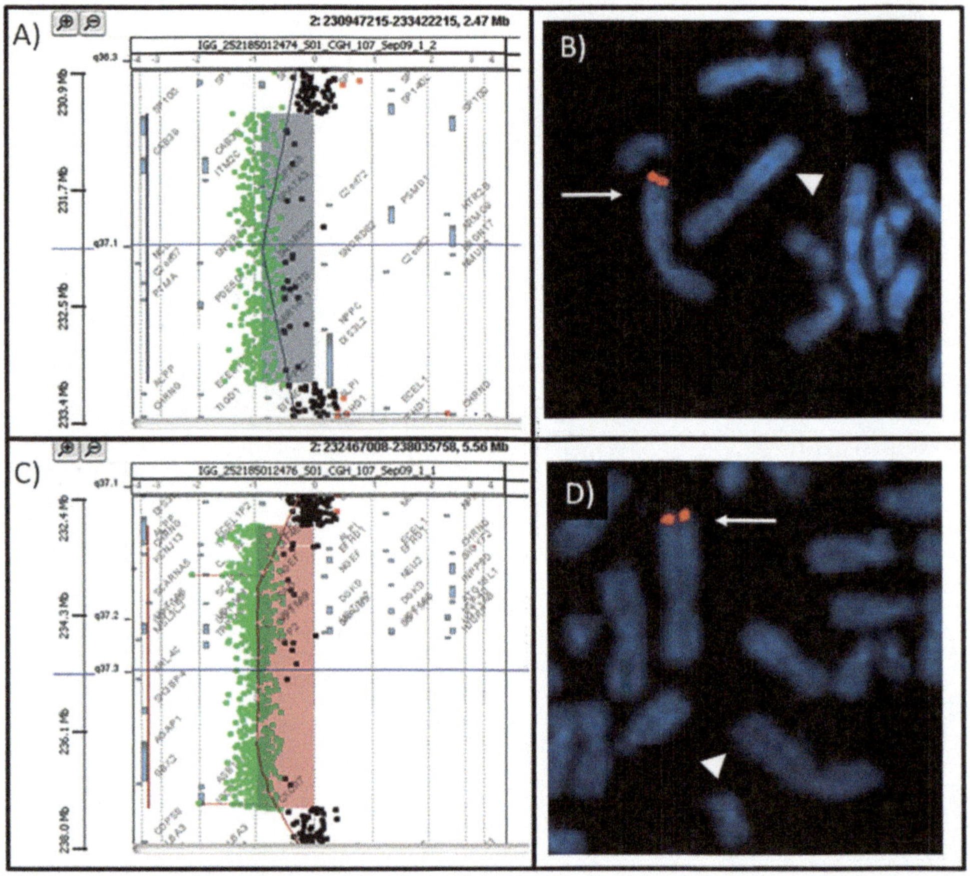

Figure 3. Array-CGH and FISH results. A) Result of array-CGH analysis of chromosome 2 with Agilent Human Genome CGH microarray Kit G3 400 in patient 1. The 1.914 Mb interstitial q37.1 deleted region of chromosome 2 extends between oligomers A_18_P13670199 (231,264,956 bp) and A_16_P00618306 (233,178,325 bp) flanked by oligomers A_16_P00615757 (231,257,468 bp) and A_16_P00618312 (233,181,399 bp) (UCSC Genome Browser, http://genome.ucsc.edu/; February 2009). B) FISH with BAC clones RP11-395A23 (AC010149) (chr2:231,304,236–231,476,367). The arrowhead indicates the deleted chromosome 2. D) Result of array-CGH analysis of chromosome 2 with Agilent Human Genome CGH microarray Kit G3 400 in patient 2. The 4.515 Mb interstitial deletion at bands q37.1q37.3 of chromosome 2 was comprised between oligomers A_16_P16076619 (232,963,736 bp) and A_16_P36124457(237,479,062 bp) flanked by oligomers A_16_P16076610 (232,954,321 bp) and A_16_P36124475 (237,483,914 bp). C) FISH with RP11-485M18 (AC079400)(chr2:236,766,818-236,919,215). The arrowhead shows the deleted chromosome 2.

Cytogenetic and Array-CGH Analysis

Cytogenetic analysis, performed on GTG-banded metaphases from cultured lymphocytes of both patients and their parents showed normal karyotypes. Considering the phenotypic abnormalities of the patients, array CGH analysis (Agilent G3 400 K) was performed.

In patient 1 array CGH analysis revealed a *de novo* interstitial deletion of ~1.9 Mb at 2q37.1 band (chr2:231,264,596–233,178,325)×1 (Fig. 3, A). The presence of the deletion was confirmed by FISH with BAC clone RP11-395A23 (AC010149) (Fig. 3, B).

At least 20 annotated genes were included within the deleted region. Among those genes, the *NPPC* gene had been previously associated with overgrowth and skeletal anomalies [11–13]. The deletion also leads to the truncation of *DIS3L2* locus in the intronic region comprised between exons 14 and 15 (Fig. 4).

In patient 2, a *de novo* interstitial 2q37.1q37.3 deletion of ~4.5 Mb (chr2:232,963,736 -237,479,062)×1 was identified by array CGH (Fig. 3, C). The deletion was confirmed by FISH with BAC clone RP11-485M18 (AC079400)(Fig 3, D). The deleted region contained 45 annotated genes and its proximal (centromeric) breakpoint interrupted the *DIS3L2* gene in intron 6 (Fig 4).

The two rearrangements shared an overlapping region of 214 kb containing 8 exons of *DIS3L2* locus (from exon 7 to exon 14) and a gene coding for hsa-miR-562 (Fig 4). The microRNA sequence was within intron 9 of *DIS3L2* and it was deleted in both patients. No additional imbalance was detected in patients 1 and 2.

Gene expression analysis

CNP concentration was measured in plasma samples from the two patients, their parents and six female and six male volunteers. For each family member, two double determinations were performed and, for controls, only one double determination. CNP-22 plasma concentration was significantly elevated in patient 2 compared to controls as showed in Fig. 5, A. In this patient, the increased plasmatic levels of CNP were correlated with overexpression of the corresponding gene at mRNA level, as assessed by RT-qPCR (Fig 5, B). Conversely, patient 1 showed normal CNP-22 concentration and normal expression of *NPPC* mRNA, despite complete deletion of one copy of the gene (Fig. 6). The presence of mutations in the promoter region and in the coding sequence of *NPPC* was excluded for both patients (data not shown).

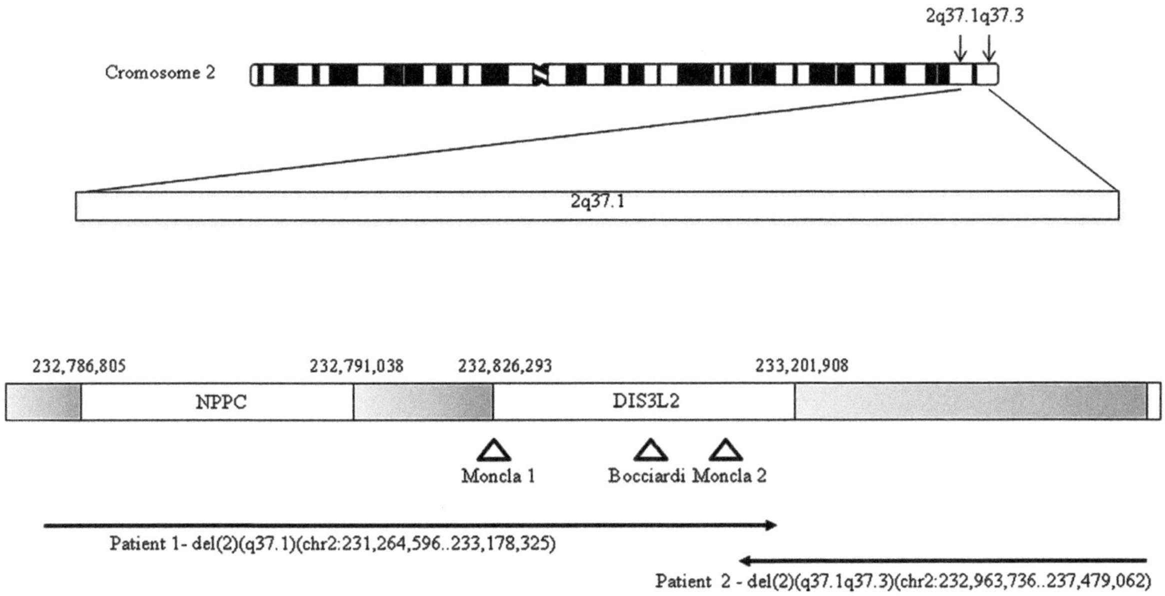

Figure 4. Schematic representation of the genomic region spanning the described deletions according to the UCSC Genome Browser Feb 2009 (GRCh37/hg19). Several annotated genes and ESTs are present, the two deletions are indicated by black arrows (patient 1 and patient 2), whereas triangles indicate the position of the translocation breakpoints previously published [11–12].

Discussion

The control mechanisms that allow for coordinated bone growth throughout the body are only partially understood, due to the involvement of many systemic hormones and local regulators. The *NPPC* gene codes for CNP, a molecule that regulates endochondral ossification of the cartilaginous growth plate and, hence, influences longitudinal bone growth. The CNP/NPR2 system has only recently proved to be an important regulator of human growth. Overexpression of *NPPC* has been shown to cause skeletal overgrowth. On the contrary, disruption of *Nppc* or *Npr-2* genes proved to lead to dwarfism and severe skeletal malformations in mice [7–8].

Moreover, two independent studies have described three patients characterized by postnatal skeletal overgrowth, Marfanoid habitus, arachnodactyly with abnormally long halluces, and marked overgrowth of the long bones, carrying *de novo* balanced translocations involving the same chromosomal region in 2q37 and three different chromosome partners [11–12]. The three breakpoints mapped around the 5′ end of the *DIS3L2* gene (Fig. 4), however this locus was not considered as a candidate.

Recently, germline homozygous mutations in *DIS3L2* have been associated with Perlman syndrome, a rare overgrowth and cancer susceptibility disorder [15–16]. However, considering its wide expression, the *DIS3L2* gene does not represent a convincing candidate for a bone affecting phenotype.

The peculiar skeletal phenotype of the translocated patients has been found to an up-regulation of the *NPPC* gene both at mRNA and at circulating peptide level [11–12]. As the gene was not disrupted by the translocations nor mutated in the proximal promoter region or in the coding sequence, it was assumed that rearrangements caused a separation of the *NPPC* gene unit from a negative, cis-acting, long-distance regulatory sequence.

Interestingly, a genome-wide association study on northwestern Europeans found that genetic variants close to the *NPPC* gene were significantly linked with body height [14].

Here we report on two patients with a deletion in 2q37 locus identified by array CGH.

Patient 1 showed short stature, cranial asymmetry, moderate psychomotor delay, and a *de novo* 2q37.1 interstitial deletion encompassing many genes including *NPPC* and the first 14 exons of *DIS3L2*. *NPPC* expression and CNP circulating peptide levels were normal.

Mice with targeted disruption of CNP (*Nppc–/–*) displayed striking narrowing of the growth plate of vertebrae and long bones compared to wild type mice and showed severe dwarfism as a result of impaired endochondral ossification [7]. Similarly, the "long bone abnormality" (*lbab/lbab*) mice, mutants due to a spontaneous autosomal recessive hypomorphic mutation in the *Nppc* gene, developed severe dwarfism characterized by short tails and extremities, while heterozygous mice did not differ from wild-type mice. These studies confirmed that haploinsufficiency of the *NPPC* gene does not exist in mice [17]. On the contrary, the effects of haploinsufficiency of this gene are evident in humans which could be due to differences between species or some other unknown mechanism(s). CNP is especially crucial for skeletal growth spurt that occurs in early life, is expressed in the growth plate cartilage, and works as an autocrine/paracrine regulator [17]. In the study by Moncla et al. [12], the *NPPC* gene was not expressed in normal lymphoblasts under normal conditions while it is expressed 6-fold in chondrocytes of his patient P1. We think that "normal" expression of *NPPC* in our patient 1was due to a mild overexpression of the unique gene copy and it could interpreted as an attempt to functional recovery.

The second patient of our study is a 12-year-old boy with marfanoid habitus associated with arachnodactyly of hands and feet, mild syndactyly, and severe scoliosis. He is affected by an interstitial 2q37.1q37.3 deletion of 4.5 Mb leading to overexpression of the NPPC gene. The deletion disrupts the DIS3L2 gene at the level of exon 6. The phenotype is very similar to that of the other three patients previously described [11–12]. Interestingly, in two cases (patient 2, reported by Bocciardi [11] and patient 2 reported by Moncla [12]), the breakpoints were located into the

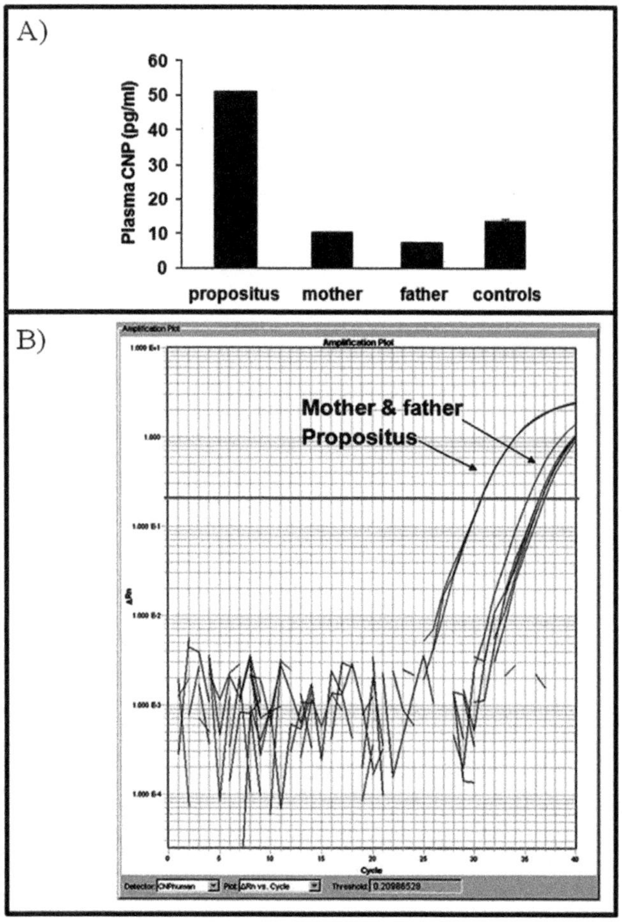

Figure 5. Expression and biochemical results. A) CNP plasma concentration in patient 2, his parents and three controls. The concentration of the peptide was increased more than 4-fold in the propositus compared to parents and unrelated controls (n = 3). B) Analysis of NPPC expression by real-time PCR analysis in lymphoblasts from patient and his parents. The expression of NPPC in the propositus was remarkably higher compared to those of his father and mother.

DIS3L2 gene, while in patient 1 described by Moncla et al. [12] the translocation breakpoint was about 18 Kb centromeric to DIS3L2.

The overlapping region of 214 kb shared by the two deletions contained an additional gene coding for a microRNA named hsa-miR-562 which is expressed only in kidney and colon and regulates EYA1, a critical gene for renal development. The haploinsufficiency of miR-562 contributes to the etiology of Wilms tumor by promoting deregulation of EYA1 [18].Considering the putative role of miR-562 in sporadic Wilms tumor and its specific expression pattern, we couldn't speculate a possible role of this microRNA in determining the phenotypes observed in our patients.

About the other genes present in the two deletions, we could retain that the ones included in patient 1 deletion may have caused the additional dysmorphisms other than the bone phenotype, while the 45 genes in patient 2 deletion appear to have had no significant effects on the phenotype.

To our knowledge, this is the first report of *NPPC* heterozygous deletion associated with short stature. The description of a second case (patient 2) with an opposite phenotype confirms a correlation between the loss of long range control elements which regulate *NPPC* functional activity and skeletal overgrowth. Moreover, these results highlight the possibility of identifying novel therapeutic strategies for the treatment of dwarfism or skeletal overgrowth.

Materials and Methods

Ethics Statement

The current study was performed using peripheral blood of the members of the two families treated at the Istituto Giannina Gaslini, Genova, Italy. The parents of the patients gave written informed consent allowing molecular and genetic studies. We didn't request approval by Review Board of our institution, because our study request only classical and molecular cytogenetic analyses. For cytogenetics analyses are sufficient only written informed consent of the parents (DM 21 dicembre 2007). The informed consents of the parents were previously authorized by the Review Board of our institution. We didn't conduct research outside our country of residence. We didn't approach the local authorities before beginning work on this study. The full name of the ethics committee of our institution is Comitato di Etica per la Ricerca Scientifica Biomedica, per la Buona Pratica Clinica e per la Sperimentazione dei Farmaci. The parents of the two patients allowed us to publish the descriptive details of their children's malformations.

The parents of the individuals in this manuscript have given written informed consent (as outlined in PLOS consent form) to publish these case details.

Clinical material

The current study was performed using peripheral blood of the members of the two families treated at the Istituto Giannina Gaslini, Genova, Italy.

Cytogenetic, array-CGH and fluorescence *in situ* hybridization analyses. Cytogenetic analysis was performed on GTG-banded metaphases at a resolution of 400–550 bands. Chromosome preparations were made from cultured lymphocytes of patients and their parents.

Fluorescence in situ hybridization (FISH) analysis performed using conventional protocols. BAC clones were selected from the human library RPCI-11 according to the UCSC Genome Browser (http://genome.ucsc.edu/; GRCh37/hg19, February 2009). BAC probes were directly labeled with Spectrum-Orange by nick translation (Vysis/Abbott, Downers Grove, IL).

Array-CGH analyses were performed using the Agilent Human Genome CGH Microarray Kit G3 400 (Agilent Technologies, Palo Alto, USA) with 5.3 Kb overall median probe spacing. Labeling and hybridization were performed following the protocols provided by the manufacturer. Data analysis was performed using the Agilent Genomic Workbench Lite Edition Software 6.5.0.18(2) with the following settings for aberration calling: ADM-2 algorithm (threshold 6) with a moving average of 500 KB and visual inspection of the log2 ratios. DNA sequence information was according to the UCSC Genome Browser (http://genome.ucsc.edu/; GRCh37/hg19, February 2009).

Expression analysis

For CNP radio-immunoassays, blood samples from the patient, his parents and healthy volunteers were withdrawn in pre-chilled tubes containing aprotinin. After centrifugation at 1.200 rpm for 10 min at 4°C, plasma was collected and stored at −80°C until measurement. Plasma samples were purified on Sep-Pak C18 cartridges (Waters) and CNP content was determined using a RIA kit (Phoenix Pharmaceuticals, Belmont, CA; RKU-012-03) according to the manufacturer's protocol. For analysis of CNP

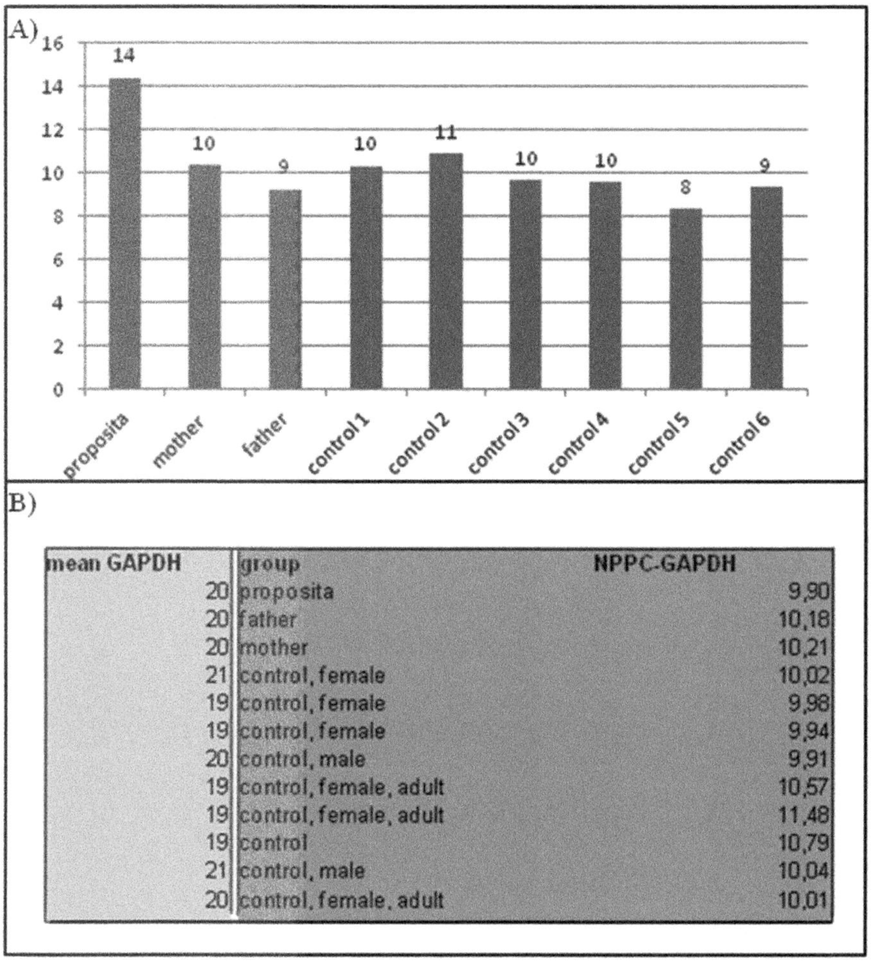

mean GAPDH	group	NPPC-GAPDH
20	proposita	9,90
20	father	10,18
20	mother	10,21
21	control, female	10,02
19	control, female	9,98
19	control, female	9,94
20	control, male	9,91
19	control, female, adult	10,57
19	control, female, adult	11,48
19	control	10,79
21	control, male	10,04
20	control, female, adult	10,01

Figure 6. Expression and biochemical results. A) CNP plasma concentration in patient 1, his parents and six controls. The concentration of the peptide was normal compared to parents and unrelated controls (n = 6). B) Analysis of *NPPC* expression by real-time PCR analysis in lymphoblasts from patient, his parents, and ten controls. The expression of NPPC in the proposita was normal compared to those of his parents and unrelated controls (n = 10).

and GAPDH mRNA expression in lymphoblast cells, total RNA was isolated using the RNeasy Mini Kit (Qiagen) according to the manufacturer's protocol. Following DNAse treatment, first strand cDNA was synthesized using the Advantage-RT-for-PCR kit (Clontech). Expression of CNP and GAPDH mRNAs was analyzed by real-time PCR (TaqMan® gene expression master mix; Applied Biosystems) using ABI 7900 HT device.

The following primers were used:

1. NPPC forward: 5′-AGCGTGGGCTCGCCTT-3′
2. NPPC reverse: 5′-CTTGTTGGCTCCTTTGTATTTGC-3′
3. NPPC probe: 5′-6-FAM-TGCAAGAGCACCCCAACGCG-TAMRA- 3′
4. GAPDH forward: 5′-CCACTCCTCCACCTTTGAC -3′
5. GAPDH reverse: 5′-ACCCTGTTGCTGTAGCCA -3′
6. GAPDH probe: 5′-6-FAM-TTGCCCTCAACGAC-CACTTTGTC-TAMRA- 3′

Reactions included 40 cycles at 96°C for 30 sec and 60°C for 1 min. Sample volume was 20 μl.

Acknowledgments

We thank the patient's parents for their gracious participation and support. We are grateful to Marco Bertorello and Corrado Torello for their technical assistance.

Author Contributions

Conceived and designed the experiments: ET SG GG. Performed the experiments: ET JB MB FSB SG GG. Analyzed the data: ET JB MB ML MTD RB FSB SG GG. Contributed reagents/materials/analysis tools: ET JB MB FSB SG GG. Wrote the paper: ET SG GG. Clinical studies: FN GP.

References

1. Pejchalova K, Krejci P, Wilcox WR (2007) C-natriuretic peptide: an important regulator of cartilage. Mol Genet Metab 92:210–215.
2. Yasoda A, Ogawa Y, Suda M, Tamura N, Mori K, et al. (1998) Natriuretic peptide regulation of endochondral ossification. Evidence for possible roles of the

C-type natriuretic peptide/guanylyl cyclase-B pathway. J Biol Chem 273:11695–11700.
3. Yasoda A, Komatsu Y, Chusho H, Miyazawa T, Ozasa A, et al. (2004) Overexpression of CNP in chondrocytes rescues achondroplasia through a MAPK-dependent pathway. Nat Med 10:80–86.
4. Silberbach M, Roberts CT Jr (2001) Natriuretic peptide signalling: molecular and cellular pathways to growth regulation. Cell Signal 13:221–231.
5. Potter LR, Abbey-Hosch S, Dickey DM (2006) Natriuretic peptides, their receptors, and cyclic guanosine monophosphate-dependent signaling functions. Endocr Rev F27:47–72.
6. Matsukawa N, Grzesik WJ, Takahashi N, Pandey KN, Pang S, et al. (1999) The natriuretic peptide clearance receptor locally modulates the physiological effects of the natriuretic peptide system. Proc Natl Acad Sci USA 96:7403–7408.
7. Chusho H, Tamura N, Ogawa Y, Yasoda A, Suda M, et al. (2001) Dwarfism and early death in mice lacking C-type natriuretic peptide. Proc Natl Acad Sci USA 98:4016–4021.
8. Tsuji T, Kunieda T (2005) A loss-of-function mutation in natriuretic peptide receptor 2 (Npr2) gene is responsible for disproportionate dwarfism in cn/cn mouse. J Biol Chem 280:14288–14292.
9. Bartels CF, Bukulmez H, Padayatti P, Rhee DK, van Ravenswaaij-Arts C, et al. (2004) Mutations in the transmembrane natriuretic peptide receptor NPR-B impair skeletal growth and cause acromesomelic dysplasia, type Maroteaux. Am J Hum Genet 75:27–34.
10. Olney RC, Bükülmez H, Bartels CF, Prickett TC, Espiner EA, et al. (2006) Heterozygous mutations in natriuretic peptide receptor-B (NPR2) are associated with short stature. J Clin Endocrinol Metab 91:1229–1232.
11. Bocciardi R, Giorda R, Buttgereit J, Gimelli S, Divizia MT, et al. (2007) Overexpression of the C-type natriuretic peptide (CNP) is associated with overgrowth and bone anomalies in an individual with balanced t(2;7) translocation. Hum Mutat 28:724–731.
12. Moncla A, Missirian C, Cacciagli P, Balzamo E, Legeai-Mallet L, et al. (2007) A cluster of translocation breakpoints in 2q37 is associated with overexpression of NPPC in patients with a similar overgrowth phenotype. Hum Mutat 28:1183–1188.
13. Bocciardi R, Ravazzolo R (2009) C-type natriuretic peptide and overgrowth. Endocr Dev 14:61–66.
14. Estrada K, Krawczak M, Schreiber S, van Duijn K, Stolk L, et al. (2009) A genome-wide association study of northwestern Europeans involves the C-type natriuretic peptide signaling pathway in the etiology of human height variation. Hum Mol Genet 18:3516–3524.
15. Astuti D, Morris MR, Cooper WN, Staals RH, Wake NC, et al. (2012) Germline mutations in DIS3L2 cause the Perlman syndrome of overgrowth and Wilms tumor susceptibility. Nat Genet 44:277–284.
16. Higashimoto K, Maeda T, Okada J, Ohtsuka Y, Sasaki K, et al. (2013) Homozygous deletion of DIS3L2 exon 9 due to non-allelic homologous recombination between LINE-1s in a Japanese patient with Perlman syndrome. Eur J Hum Genet. Mar 13. (doi: 10.1038/ejhg.2013.45).
17. Kondo E, Yasoda A, Tsuji T, Fujii T, Miura M, et al. (2012) Skeletal Analysis of the Long Bone Abnormality (lbab/lbab) Mouse, A Novel Chondrodysplastic C-Type Natriuretic Peptide Mutant. Calcif Tissue Int 90:307–318.
18. Drake KM, Ruteshouser EC, Natrajan R, Harbor P, Wegert J, et al. (2009) Loss of heterozygosity at 2q37 in sporadic Wilms' tumor: putative role for miR-562. Clin Cancer Res. 15:5985–5992.

8

A Comparison of 100 Human Genes Using an *Alu* Element-Based Instability Model

George W. Cook[1], Miriam K. Konkel[1], Jerilyn A. Walker[1], Matthew G. Bourgeois[1], Mitchell L. Fullerton[2], John T. Fussell[3], Heath D. Herbold[4], Mark A. Batzer[1]*

1 Department of Biological Sciences, Louisiana State University, Baton Rouge, Louisiana, United States of America, **2** Department of Bioengineering, Clemson University, Clemson, South Carolina, United States of America, **3** Electrochemical Materials, Louisiana Business and Technology Center, Baton Rouge, Louisiana, United States of America, **4** Albemarle Corporation, Pasadena, Texas, United States of America

Abstract

The human retrotransposon with the highest copy number is the *Alu* element. The human genome contains over one million *Alu* elements that collectively account for over ten percent of our DNA. Full-length *Alu* elements are randomly distributed throughout the genome in both forward and reverse orientations. However, full-length widely spaced *Alu* pairs having two *Alus* in the same (direct) orientation are statistically more prevalent than *Alu* pairs having two *Alus* in the opposite (inverted) orientation. The cause of this phenomenon is unknown. It has been hypothesized that this imbalance is the consequence of anomalous inverted *Alu* pair interactions. One proposed mechanism suggests that inverted *Alu* pairs can ectopically interact, exposing both ends of each *Alu* element making up the pair to a potential double-strand break, or "hit". This hypothesized "two-hit" (two double-strand breaks) potential per *Alu* element was used to develop a model for comparing the relative instabilities of human genes. The model incorporates both 1) the two-hit double-strand break potential of *Alu* elements and 2) the probability of exon-damaging deletions extending from these double-strand breaks. This model was used to compare the relative instabilities of 50 deletion-prone cancer genes and 50 randomly selected genes from the human genome. The output of the *Alu* element-based genomic instability model developed here is shown to coincide with the observed instability of deletion-prone cancer genes. The 50 cancer genes are collectively estimated to be 58% more unstable than the randomly chosen genes using this model. Seven of the deletion-prone cancer genes, *ATM, BRCA1, FANCA, FANCD2, MSH2, NCOR1 and PBRM1*, were among the most unstable 10% of the 100 genes analyzed. This algorithm may lay the foundation for comparing genetic risks posed by structural variations that are unique to specific individuals, families and people groups.

Editor: Gil Ast, Tel Aviv University, Israel

Funding: This research was supported by the National Institutes of Health (http://www.nih.gov/)RO1 GM59290 (MAB), and the Louisiana Board of Regents Governor's Biotechnology Initiative GBI (http://web.laregents.org/programs/special-programs/) (2002-005) (MAB). The funders had no role in study design, data collection and analysis, decision to publish, or preparation of the manuscript.

Competing Interests: Also note that three of the authors of this paper, George W. Cook, John T. Fussell, and Heath D. Herbold, are employees of commercial entities. George and Heath are employees of Albemarle Corporation and John is an employee of Electrochemical Materials, LLC. Each of these three author's participation in the preparation of this manuscript was conducted as independent researchers and their contributions in its preparation were made wholly outside of their respective employment. Furthermore, Albemarle Corporation and Electrochemical Materials, LLC are in no way attached to any consultancy, patents, products in development or marketed products etc. related to this manuscript.

* E-mail: mbatzer@lsu.edu

Introduction

The draft human genome is interspersed with approximately 45% of mobile element related repetitive sequence [1]. Advanced sequence analyses indicate that the repeat related portion of the genome may be as high as 69% [2]. Retrotransposons, which reproduce through a copy and paste mechanism, have generated the majority of this repetition. The human retrotransposon with the highest copy number is the *Alu* element. *Alu* elements have populated the human genome with over one million copies and account for over 10 percent of all human DNA [3].

Both by insertion and by recombination, *Alu* elements spawn genetic disease [4–7]. Over 100 studies link Alu elements to deletion-related diseases (Table S1). It has been suggested that the most damaging impact of mobile elements may not be their insertion into genes, but their potential interactions with each other. Such interactions could result in deletions, duplications, inversions and a host of more complex genomic structural changes [8–11]. *Alus* have also been associated with copy number variation breakpoints [12,13]. The incidence of *Alu-Alu* interactions is further supported by studies highlighting *Alu-Alu* gene conversion events [14,15]. The homogenization of neighboring *Alu* sequences in ostensibly healthy subjects is consistent with the theory that *Alu-Alu* interactions routinely occur in healthy cells [16,17].

Recombinant inverted *Alu* pairs have been shown to be unstable in genetically engineered yeast experiments when separated by up to 100 base pair (bp) and are potential sources of chromosome instability when separated by up to 350,000 bp in humans [18–20]. Furthermore, fusions of inverted *Alu* pairs previously separated by 1–5 kb have been recently identified at the break-points of high copy number loci in cancer cells [13].

Previously we reported that full-length inverted *Alu* pairs (represented by the letter, I) were statistically underrepresented in the human genome when compared to full-length direct oriented *Alu* pairs (represented by the letter, D). The term, *Alu* pair exclusions (APEs), was used to describe this human I:D *Alu* pair imbalance [18]. In this study we find that the lower number of inverted *Alu* pairs (when compared to direct *Alu* pairs) applies to all combinations of human *Alu* sizes. Additionally, we characterize this human *Alu* pair I:D imbalance and construct a model for estimating relative human genome instability based upon the premise that the human *Alu* pair I:D imbalance is generated as a consequence of inverted *Alu* pair instability.

This newly developed *Alu* induced instability model was used to compare the relative instabilities of 50 human cancer genes with 50 randomly selected genes from the human genome to experimentally validate the model. The cancer genes considered in this study were selected for their potential susceptibility to deletions based upon previous studies [21–23]. This selection criterion was adopted in order to maximize the model's opportunity to distinguish between these two groups of genes. Taken together, the model estimates that the deletion-prone cancer genes are 58% more unstable than the randomly chosen genes.

Results

Each human gene resides within a unique landscape of *Alu* elements. The structures of these landscapes vary in attributes that include *Alu* density, clustering and orientation. Adding further to *Alu* landscape complexity is the number of exons and their spacings. Within these backdrops inverted *Alu* pairs are statistically less numerous than direct oriented *Alu* pairs. It has been hypothesized that this imbalance is primarily the consequence of deletions generated by interactions between inverted *Alu* pairs [18].

This hypothesis was tested by construction of an algorithm designed to estimate the risk that a gene's *Alu* landscape could potentially impose upon its coding sequence. The coding sequence risk was estimated by multiplying two independent probabilities. The first probability, the *Alu*-induced deletion risk, is the probability of the occurrence of an *Alu*-induced deletion. This deletion probability was estimated by characterization of the human genome-wide inverted *Alu* pair to direct *Alu* pair imbalance. The imbalance is described by the ratio of inverted to direct *Alu* pairs, I:D. In this study, the statistically significant departure of the I:D ratio below unity ($p<0.05$) is assumed to be a consequence of deletions that remove inverted *Alu* pairs from the genome. The predicted likelihood of a deletion arising from the instability of a given *Alu* pair is derived from the genome-wide I:D imbalance pattern. This likelihood is estimated as a function of three parameters which are discussed later in this section. The second probability, the *Alu*-induced deletion size risk, is the risk that once a deletion is formed, it will be of sufficient size to extend into the coding region of the gene being evaluated. Deletion size risk is estimated using an algorithm constructed from recent studies describing the human indel size frequency distribution. Each of these two probabilities is discussed in greater detail later in this section.

This *Alu* element-based instability model was used to compare the relative stabilities of 50 human cancer genes with 50 randomly selected genes from the human genome. The cancer genes considered in this study were selected for their potential susceptibility to deletions [21–23]. This methodology was utilized to increase the likelihood that the model would be able to discriminate between these two groups of genes.

Two-hit potential of *Alu* elements

The instability model assumes that each end of an *Alu* element is vulnerable to a double-strand break, DSB. These DSB sites are identified from the proposed DNA conformations associated with two mechanisms that have been suggested to explain human inverted *Alu* pair instability. These two mechanisms are characterized by the ectopic invasion and annealing of single-stranded DNA between high-homology DNA bubbles and/or replication forks [18]. Coincident DNA bubbles passing through aligned *Alu* elements may expose their complementary "flipped out" bases to one another [24,25]. Complementary replication forks may also be susceptible to this type of interaction. Each pathway may result in the formation and subsequent resolution of a DNA conformation referred to as a doomsday junction. These two mechanisms are illustrated in Figures 1 and S1, respectively. Although we are unaware of other mechanisms that might also explain this imbalance, we readily acknowledge that they may exist. The ectopic DNA conformations described in these two figures are offered as possible explanations for the *Alu* pair I:D imbalance phenomenon and they are used as a platform for constructing this instability model.

Figure 1E identifies the eight potential sites where a single-strand break could occur during the resolution of a doomsday junction. These sites (illustrated by yellow lightning bolts) are created at the periphery of the doomsday junction where each single strand of DNA transitions from the original DNA double helix to the ectopic conformation of the doomsday junction. These regions of single-stranded DNA may be susceptible to attack by single strand nucleases. If only one strand at the end of each *Alu* element is cut, the doomsday junction can likely resolve itself without damage to the original sequence. However, if both strands at the same end of either of the two inverted *Alu* elements are cut, a DSB can occur (Figure 1F). This potential for a DSB at each end of an *Alu* element forms the basis for the "two-hit hypothesis" for each *Alu* element considered by this instability model.

Probability one – *Alu*-induced deletion risk

The *Alu*-induced deletion risk is the likelihood of a deletion arising from the resolution of a doomsday junction. The two-hit deletion potential of each *Alu* element results in the number of potential *Alu*-induced deletion sites within a given *Alu* landscape being twice the number of *Alu* elements. Three variables were found to significantly correlate with the *Alu* pair I:D ratio; 1) spacer size, 2) the number of *Alu* elements within the spacer and 3) the clustering state of the each *Alu* pair (discussed in more detail, below). Figures 2, 3 and S2 express the human inverted to direct *Alu* pair ratio, I:D ratio, as a function of these three variables The *Alu* pair I:D ratio was not found to significantly correlate with *Alu* length (see Methods).

Figure 2 provides a detailed view of how the I:D ratio varies with spacer size for APSNs 1–10 for Type 1 *Alu* pairs. Each of these 10 curves is plotted along 10 data points. These 10 data points represent the I:D ratio for 10 fitted spacer size groupings from Table S2. These 10 data points represent, from smallest to largest, the 2.5th, 10th, 20th, 30th, etc. through 90th percentiles of spacer sizes groupings (see Methods). The shape of the curves in these three figures illustrate that the *Alu* pair I:D ratio is not a smooth function across the full range of spacer sizes. These curves are plotted along the median of spacer size for the ten spacer size percentile groupings for each of the respective *Alu* pair sequence

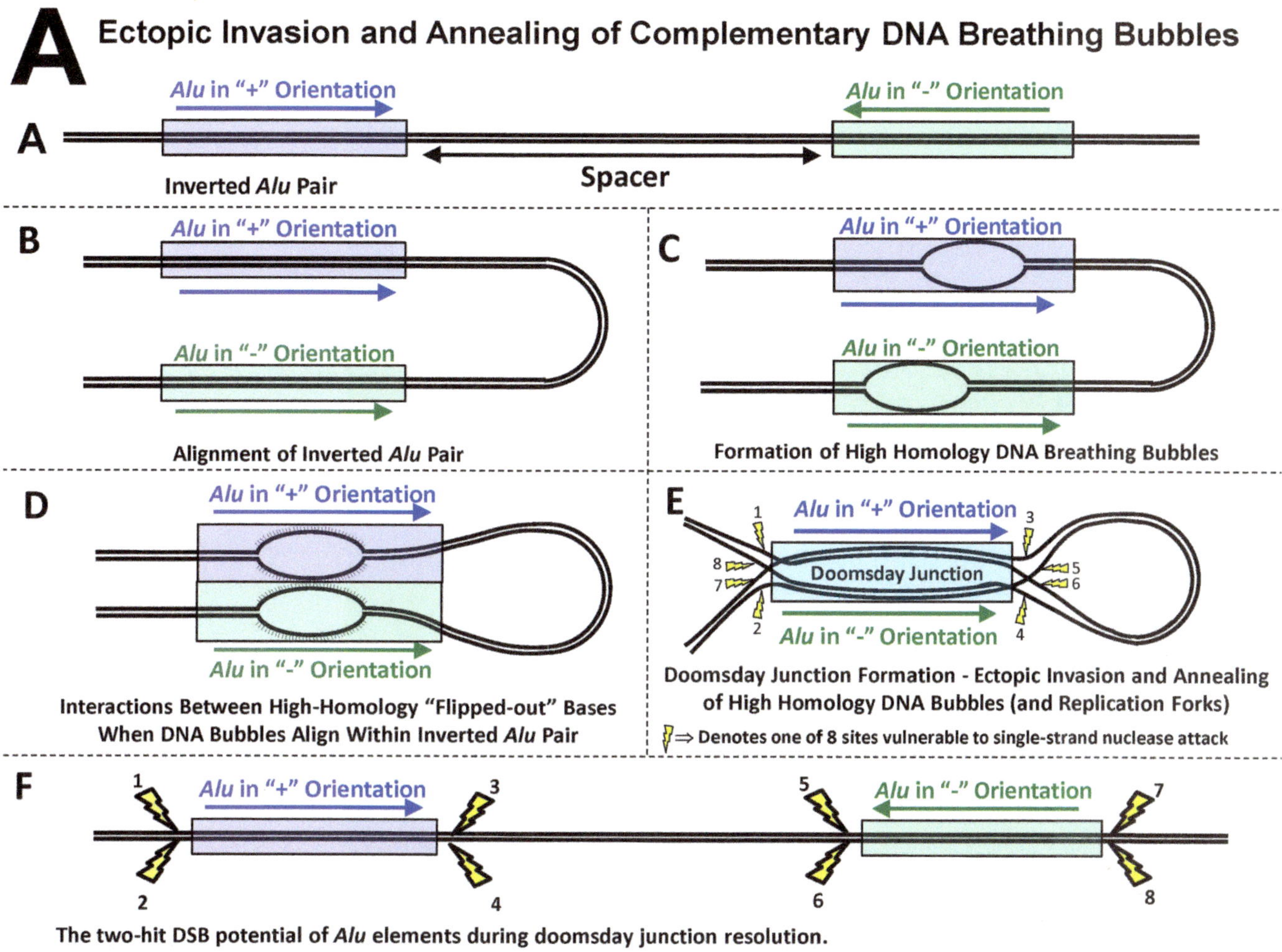

Figure 1. Proposed mechanism for formation and resolution of doomsday junction formed by ectopic invasion and annealing of complementary DNA breathing bubbles. (A) Two *Alu* elements in opposite orientations form an inverted *Alu* pair. (B) These inverted *Alu* pairs can align as high-homology regions. (C) DNA bubbles create short-lived sections of single-stranded DNA [25]. (D) The unbound bases within these bubbles are characterized by their flipping out from the centerline of the DNA strand [24]. Coincident passage of these bubbles within aligned *Alu* elements can create the opportunity for interactions between the flipped-out bases of the complementary DNA strands. (E) The ectopic invasion and annealing of single-stranded DNA associated with high-homology DNA bubbles could potentially extend to the entire length of the *Alu* elements. The hypothetical conformation created by this interaction is termed a doomsday junction. A similar interaction may also occur between high-homology replication forks and is described in Figure S1 and [18]. Eight segments of single-stranded DNA formed at the boundary of doomsday junctions create the opportunity for single-strand nuclease attack. These sites are illustrated as yellow lightning bolts. (F) As again illustrated by the yellow lightning bolts, each end of each *Alu* element involved in the doomsday junction is vulnerable to a double-strand break. This two-hit hypothesis for each *Alu* element was incorporated into the model's algorithm (see text).

numbers (APSNs). The APSN is the parameter that describes the number of *Alus* within the spacer of an *Alu* pair. The APSN for an *Alu* pair is the n+1 number of *Alu* elements residing with the spacer (see Methods).

Three possible mechanisms may explain the unusual shape of the human *Alu* pair I:D ratio versus spacer size curves. Using the APSN1 curve in Figure 2 as a reference, these three mechanisms may be as follows; 1) between the 0th and 5th spacer size percentiles (centered at ~100 bp), hairpin formation may be the predominant form of *Alu-Alu* interaction, 2) for the 10th (5th–15th), 20th (15th–25th) and 30th (25th–35th) spacer size percentiles (centered between ~100 and ~500 bp) DNA persistence (stiffness), may hinder inverted *Alu-Alu* interactions and 3) for spacer sizes between the 40th (35th–45th) and 90th (85th–95th) spacer size percentiles, DNA persistence appears to wane and the curve begins to progress toward unity.

Human *Alu*, LINE1 and SVA elements, frequently cluster together in groups where adjacent elements are separated by ≤50 bp [18]. Using this definition of clustering, four types of clustered *Alu* pairs can be described. These are identified as Types 0, Type 1, Type 2 and Type 3 *Alu* pairs. Type 0 *Alu* pairs (clustered together) have both *Alu* elements residing within the same cluster, Type 1 *Alu* pairs (clustered separately) have both *Alu* elements residing within different clusters, Type 2 *Alu* pairs (hemi-clustered) have only one of the two elements residing within a cluster and Type 3 *Alu* pairs (non-clustered) have neither element residing within a cluster (see Methods). Type 1, 2 and 3 *Alu* pairs exhibit distinctly different I:D ratios and their stabilities must therefore be estimated separately (Figure S3). Type 0 *Alu* pairs are subject to strong orientational insertion bias and their instability has been estimated via experimental studies of *Alu* elements in yeast (see Methods and [18,19]).

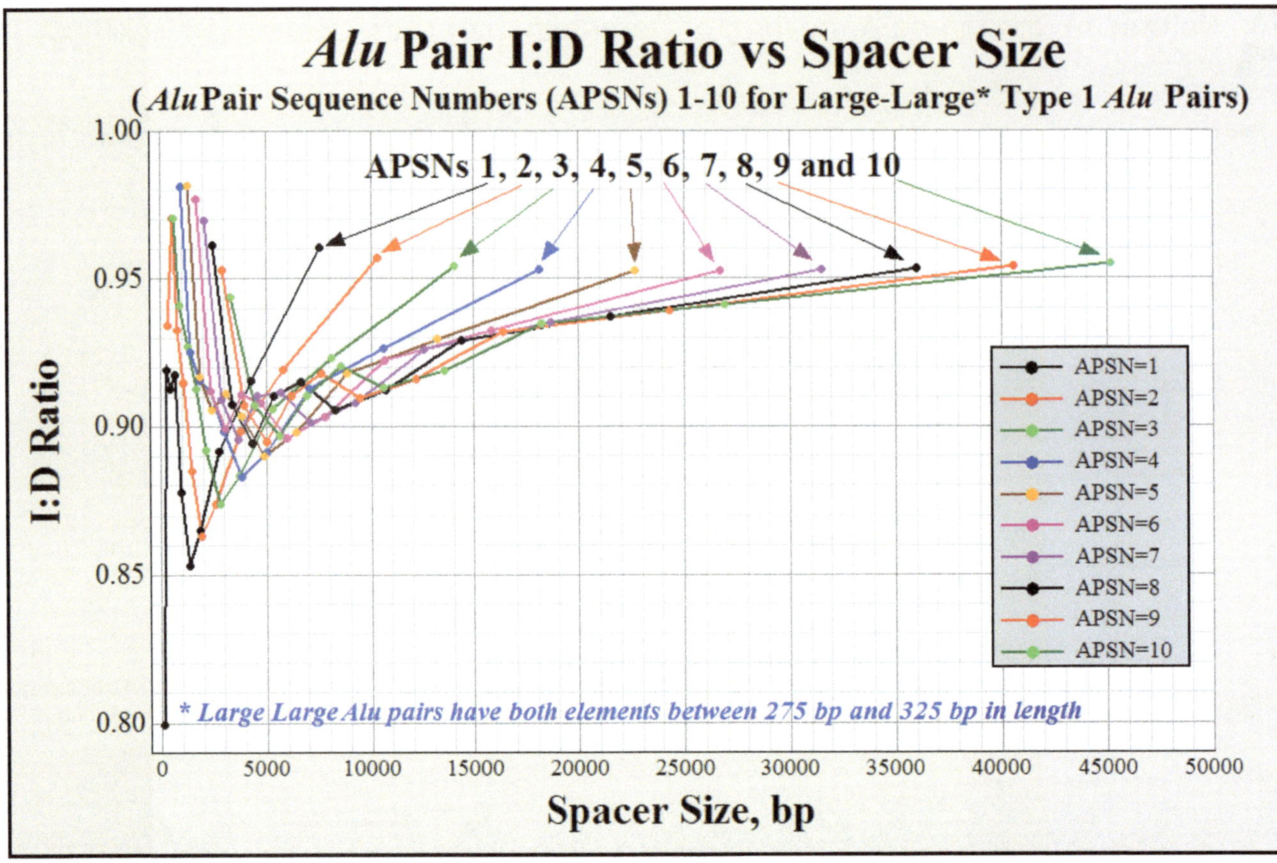

Figure 2. The *Alu* pair I:D ratio versus spacer size for Type 1 *Alu* pairs for APSNs 1–10. A total of 10 points are used to construct each of the 10 APSN curves in this figure. These points represent the respective *Alu* pair I:D ratios for 10, non-overlapping spacer size groupings. The first point (left to right) represents the I:D ratio for the smallest five percent of spacer sizes. This point is followed by nine consecutively larger spacer size groupings. Each of these nine larger sized groupings contains 10% of the *Alu* pairs found within the respective APSN family. The I:D ratio for each percentile group is plotted against its median spacer size, respectively (see Methods). This plot illustrates that the I:D ratio is not a continuous function versus spacer size and may indicate the activity of different *Alu-Alu* interaction mechanisms (see text). These curves, along with their 5′ mirror images, make up ten of the 220 (APSNs ±1–110) curves that are collectively shown in Figure 3.

Figure 2 illustrates the I:D ratio versus spacer size for Type 1 large-large (275–325 bp) *Alu* pairs for APSNs 1–10. Figure 3 is similar to Figure 2 and includes all APSNs (±110) containing at least one spacer size percentile with an I:D ratio <0.995. I:D ratios ≥0.995 do not provide statistical confidence that the I:D ratio is below unity (see Methods). Figures S2A and S2B are similar to Figure 3 and show the I:D ratio versus spacer size and APSN for Type 2 and Type 3 *Alu* pairs, respectively.

Using the I:D ratio relationships illustrated in Figures 3 and S2, the model generates a predicted stability for each *Alu* element within a gene's *Alu* landscape. The predicted I:D ratio is the predicted stability for the *Alu* pair. The contribution that an inverted *Alu* pair makes to the stability of each *Alu* element of that pair is obtained by taking the square root of that pair's predicted I:D ratio. Likewise, the stability for one end of an *Alu* element is the fourth root of that *Alu* pair's predicted I:D ratio. The overall stability of one end of an *Alu* element is the product of the fourth roots of all of the predicted I:D ratios for each of the potential 220 inverted interactions (i.e., the grand product) that an *Alu* element might form with its ±110 *Alu* neighbors (see Methods).

Figure 3 reveals an unexpected excursion of the I:D ratio above unity for the highest *Alu* density genomic regions. This excursion only exists for APSNs ≥65 and only for the most *Alu* dense regions of the genome (0–5th spacer size percentile, Table S2). This high I:D ratio may indicate that direct *Alu* pair recombination in these high *Alu* density regions of the genome may outpace the activity of inverted APE events.

Alu landscapes

Each of the genes considered in this study were evaluated using the backdrop of *Alu* elements in which they reside. These *Alu* backdrops are referred to as *Alu* landscapes. Figure 4 illustrates the *Alu* landscapes around two of the deletion-prone cancer genes evaluated in this study, *BRCA1* and *VHL*. The vertical blue lines in each figure demarcate 100,000 bp distances from the respective end of each gene and the light blue region in the center of each diagram encompasses the respective gene's coding locus.

The respective instability score (*i*Score) for each *Alu* element is plotted along the vertical axis at the locus of each *Alu* element. These *i*Score values are the inverse of the *Alu* stabilities calculated using the algorithms developed from Figures 3 and S2. Higher *i*Score values represent higher *Alu* instabilities. The red dots signify the locus versus the iScore value for each element within the *Alu* landscape.

The *Alu* landscapes illustrated in Figure 4 span ±500,000 bp from the end of each gene. Similar landscapes are shown for eight additional genes in Figure S4. The instability model only includes those *Alus* residing within ±250,000 bp from the end of each gene

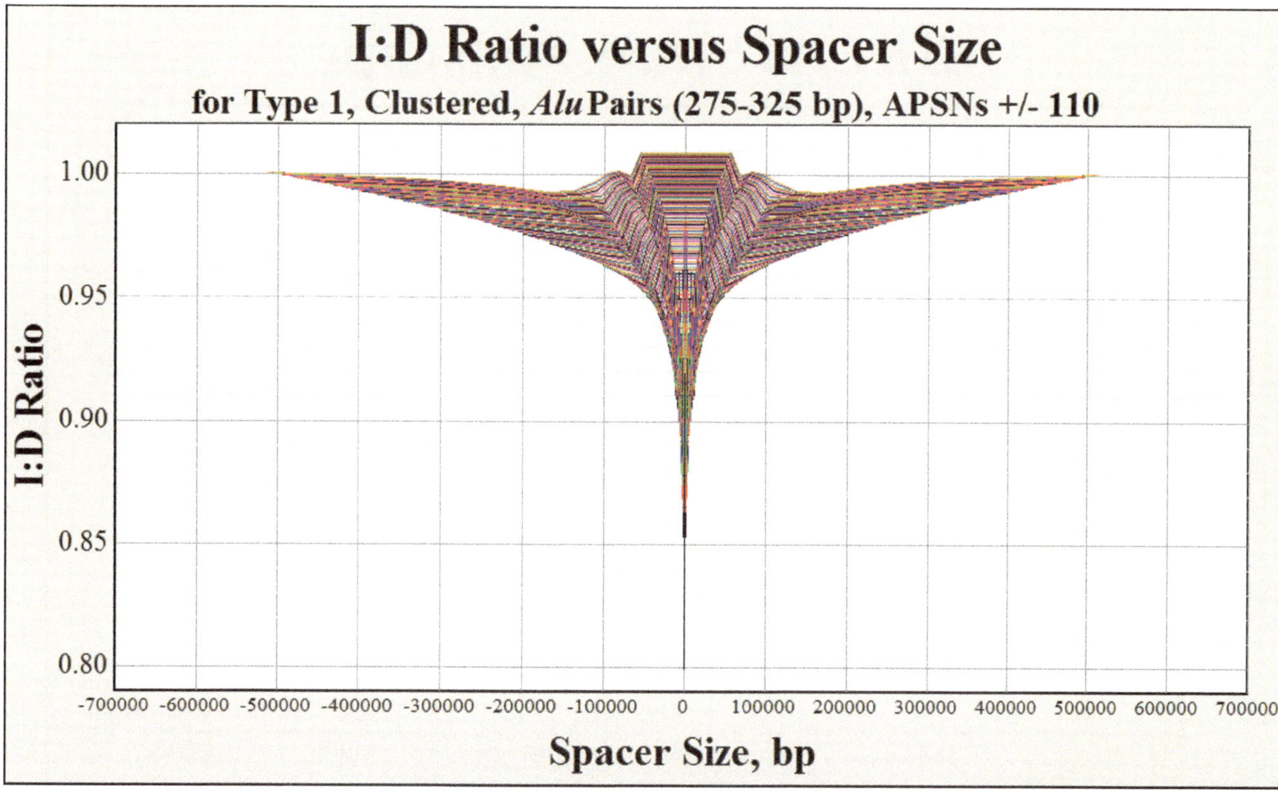

Figure 3. The *Alu* pair I:D ratio versus spacer size for Type 1 *Alu* pairs for APSNs ±1–110. This figure illustrates the ±110 APSN curves for full-length (275–325 bp) Type 1 human *Alu* pairs. The individual curves in this figure are so closely spaced that they collectively appear as a surface. An expanded view of Type 1 APSN curves 1–10 is shown in Figure 2. Similar I:D surfaces for Type 2 and Type 3 *Alu* pairs are shown in Figures S2A and S2B, respectively.

(discussed in more detail, below). The larger landscapes provided in Figures 4 and S4 are shown to illustrate the ebb and flow of *Alu* instabilities across the genome. Approximately 0.3% of the human genome is represented in the 10 *Alu* landscapes shown in these two figures.

An average of 410 *Alu* elements reside within the +/− 250,000 bp landscapes of the genes examined in this study. These gene-specific *Alu* populations are not of sufficient size to detect inverted *Alu* pair stability with statistical confidence (see Methods). However, using the genome-wide human *Alu* population, it is possible to construct a statistically relevant model to estimate relative gene instabilities based upon each gene's respective *Alu* landscape. An additional insight into this model's relevance is that the human genome-wide I:D imbalance is slightly over two percent (I:D = 0.979) [18]. While this I:D depression is statistically significant ($p<0.05$) the large majority of inverted *Alu* pairs likely remain in the human genome. Consequently, if the mechanisms that created this I:D depression remain active, the genome-wide loss of slightly over two percent of the inverted *Alu* element population would likely do little to deter the continued activity of these mechanisms. Furthermore, wet bench experimental comparisons between orthologous chimpanzee and human inverted *Alu* pair loci reveal the chimpanzee-specific loss of inverted *Alu* pairs [18]. This analysis suggests that a portion of the loss of inverted *Alu* pairs may be of recent origin.

The panes in Figure S4A–S4E illustrate the *Alu* landscapes for the five deletion-prone cancer genes, *APC, ATM, BRCA1, MLH1 and MSH2.* The panes in Figure S4F–S4H describe the *Alu* landscapes for randomly chosen genes, *GDPD2, KEAP1 and SF3B3.* Among the 100 genes examined in this study, only two of the top 10 highest *Alu* density landscapes are associated with deletion-prone cancer genes, *ARID1A* and *BRCA1*. These two genes rank 8th and 10th in this list with *Alu* landscape densities of 1,322 and 1,309 *Alus* per mega base, respectively (see Table S3). The *Alu* element density across the human genome averages 381 *Alus* per mega base. The top five most *Alu* dense landscapes (all randomly selected genes) belong to *KEAP1, NCF1, NANOS3, OPRD1,* and *SET1* with *Alu* densities of 1,916, 1,783, 1,644, 1,534 and 1,525 *Alus* per mega base, respectively (see Table S4).

Probability two – *Alu*-induced deletion size risk

Human genome indel size frequency distributions from two previous studies provide a glimpse into the shape of the overall human deletion size frequency distribution [26,27]. A hybrid deletion size frequency model was developed from these studies and is shown in Figure 5. The sum of the 500,000 deletion probabilities shown in this figure equals 1.0. This hybrid model is used to estimate the relative deletion size risks that arise from inverted *Alu*-induced DSBs (See Methods). The shape of the curve in Figure 5 reflects a deletion size frequency distribution where 95 percent of deletions are ≤50 bp. The maximum Figure 5 deletion size of 500,000 bp was chosen because this size deletion has a risk of occurrence that is less than one billionth of the risk predicted for a 1 bp deletion. This model assumes that deletions extend equidistant from an initiating DSB. Consequently, the maximum distance from which an individual *Alu* element is considered to pose a deletion risk to a coding exon is 250,000 bp (250,000 bp × 2 = 500,000 bp). In addition to considerations for

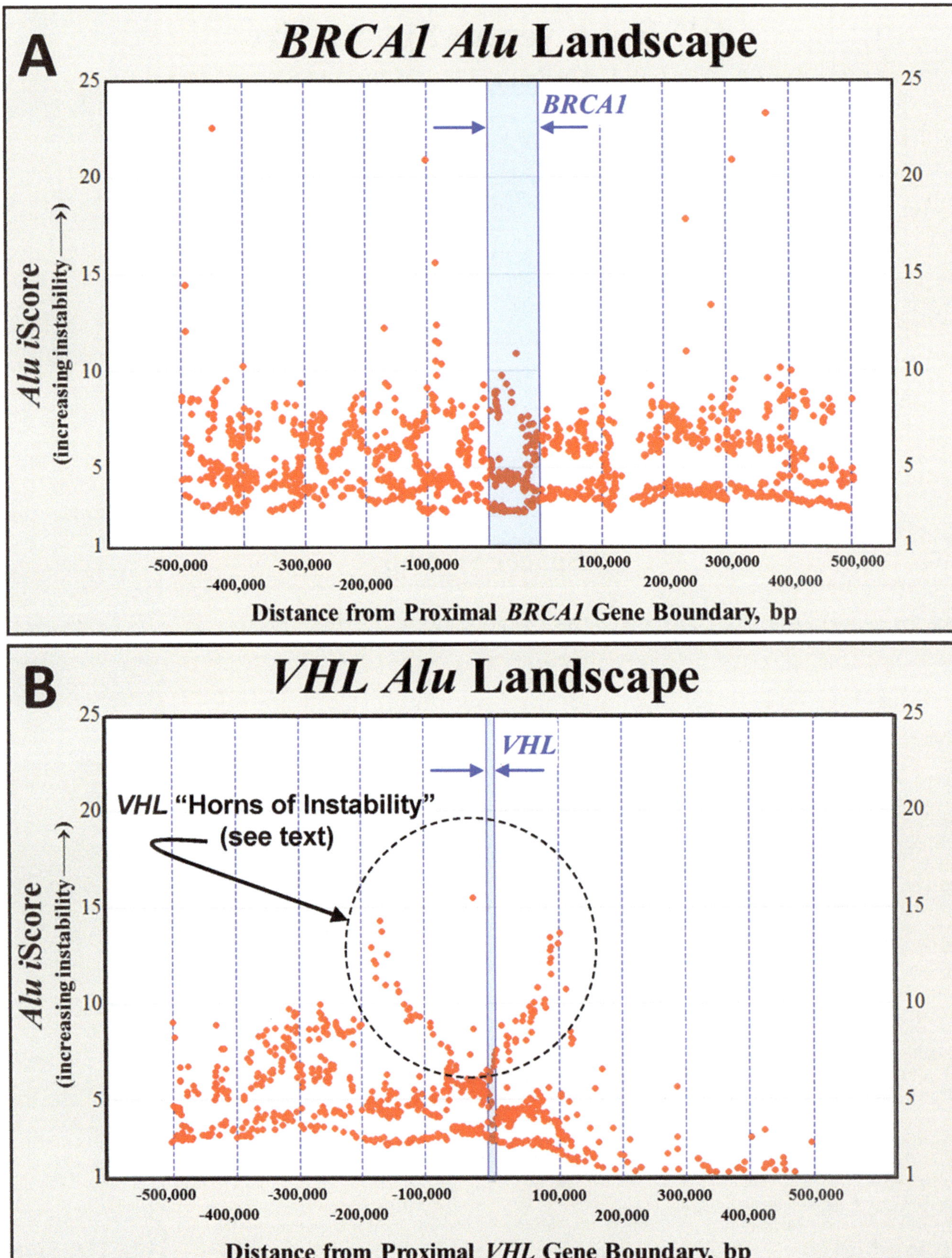

Figure 4. ***Alu*** **landscapes for** ***BRCA1*** **and** ***VHL.*** This figure characterizes the *Alu* landscapes within and 500 kbp, 5′ and 3′ of A) *BRCA1* and B) *VHL*. The midpoint for each *Alu* element is plotted against its respective instability score, *i*Score. Larger *i*Score values represent higher predicted *Alu* element instabilities (see text). Similar *Alu* landscapes for eight additional genes examined in this study are shown in figures S4A–H. These spans are

twice the size of the ±250 kbp flanking landscapes which are considered to pose a risk for an exon damaging deletion (see text). These larger spans better illustrate the ebb and flow of *Alu*-related instability around each respective gene.

maximum deletion size, additional flanking sequence must be examined within an *Alu* landscape to accommodate for the possibility that inverted *Alu* pairs can interact when separated by up to 421,000 bp. This is the spacer size (in Figure 3) that intersects with an I:D ratio of 0.995. This I:D ratio is statistically lower than unity ($p \leq 0.05$, see Methods). Therefore, an *Alu* element that is separated by as much as 671,000 bp from a coding exon could potentially threaten the coding integrity of that exon. At this distance from a coding exon, an *Alu* element could conceivably interact with a second *Alu* separated by only 250,000 bp from the same exon (spacer size between the two *Alus* = 671,000 bp - 250,000 bp = 421,000 bp). This interaction could potentially generate a DSB at the second *Alu* that could possibly extend into the coding exon.

Relative gene stabilities

The relative stability of a gene for the purpose of this study is defined as the relative likelihood that a coding exon will not be breached by a deletion. The determination of this stability must consider the collective deletion risks along with the respective deletion size risks posed by all potential DSB sites generated within a gene's *Alu* landscape. More specifically, the overall stability of a gene is the multiplied product (grand product) of the individual *Alu* element contributions to that gene's stability within its *Alu* landscape (see Methods). The required calculations to determine this stability are extensive. Estimation of the stability of *BRCA1*, because of its large *Alu* landscape, requires 171,225 consecutive calculations. As can be seen from Table S3, *BRCA1* has 761 *Alu* elements residing within its intronic regions and the 250,000 bp flanking regions, 5′ and 3′ of the gene. The majority of these calculations are associated with the 220 potential *Alu* pair interactions for each of these 761 *Alu* elements. The sheer number of required consecutive calculations raised concerns that significant adjustments would be required for proper interpretation of the raw output from the model. This concern did not materialize. The individual gene stabilities plotted in Figure 6 are the unadjusted output stability values from the model.

The uppermost histogram in Figure 6 is a distribution of the raw stabilities of the 50 deletion-prone genes taken directly from the model. The bottom histogram is a distribution of the raw stabilities of the 50 randomly selected genes. Lower values represent greater instability. Tables S3 and Table S4 list the individual gene stabilities. For reference, this instability model would generate a stability of 100 for a gene residing within an *Alu*-free landscape. The average unadjusted stabilities of the deletion-prone cancer genes and randomly chosen genes from Tables S3 and S4 are 77.7% and 85.9%, respectively. The deletion-prone cancer genes, therefore, have 58% greater likelihood of a deletion insult than that of the randomly chosen genes.

$$\frac{[(1-0.777)-(1-0.859)]}{(1-0.859)} \times 100 = 58\%$$

This likelihood increases to 78% when *GDPD2*, the randomly chosen gene with an exonized *Alu* element, is excluded from the list of random genes (discussed in more detail, below).

Only one cancer gene, *IKZF1*, was among the most stable 10% of the 100 genes analyzed, while seven deletion-prone cancer genes, *FANCA, NCOR1, BRCA1, PBRM1, ATM, FANCD2* and *MSH2* were among the most unstable 10% (10) of the 100 genes analyzed (Tables S3 and S4). The top 10% most stable genes contain an average of 4 coding exons, versus an average coding exon count of 31 for the 10% most unstable genes (Tables S3 and S4). Individual least squares correlations were subsequently performed using 1) exon number, 2) *Alu* population within each gene's *Alu* landscape and 3) gene coding length to determine the extent to which these parameters can be used to predict the estimated relative gene stabilities determined in this study. These three correlations were made using the combined set of 100 genes found in Tables S3 and S4. The best predictor of the model's estimated relative gene stability among these three variables was

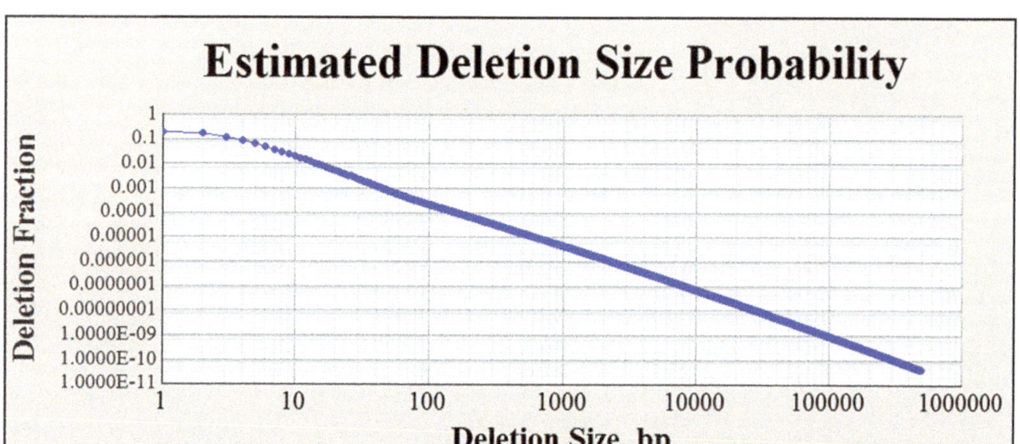

Figure 5. Estimated human deletion size frequency distribution. (A) This log-log (base 10) plot estimates the relative distribution of deletion sizes within the human genome. The curve was constructed from two different studies and predicts that 95% of deletions are ≤50 bp in size and 99% of deletions are ≤445 bp [26,27]. When combined with the two-hit hypothesis for *Alu* elements (see Figure 1F and text), this curve suggests that the two ends of an *Alu* element pose specific and different risks to an exon's coding region.

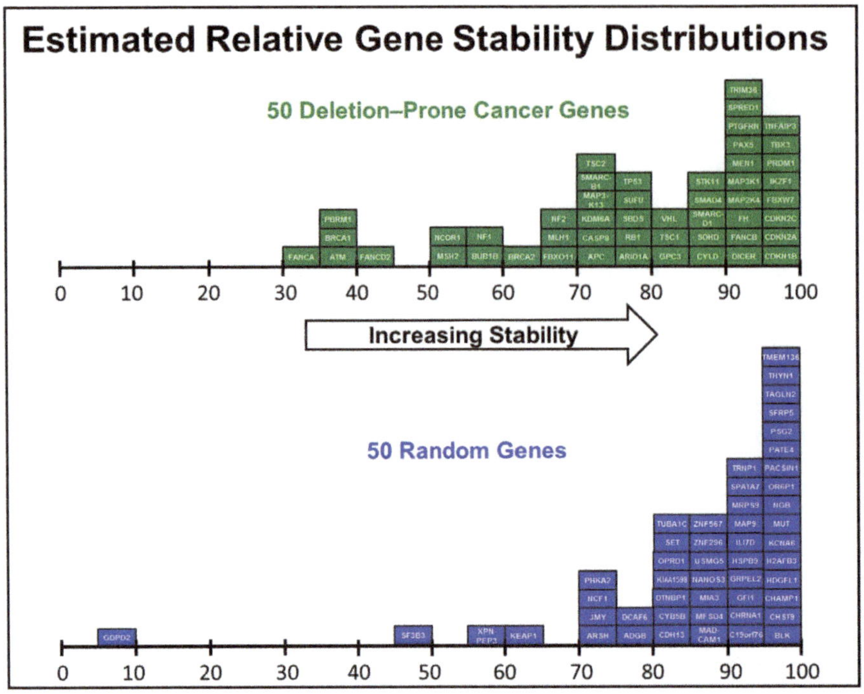

Figure 6. Distributions of estimated relative instabilities for 50 deletion-prone cancer genes and 50 randomly chosen genes. This diagram provides separate histograms that describe the relative instabilities of the 50 deletion-prone genes and the 50 randomly selected genes, respectively. The values in these histograms are the unadjusted outputs from the *Alu* element-based instability model algorithm. These stabilities are also provided in Tables S3 and Table S4, respectively. Note that the least stable of all 100 genes is the randomly selected gene, *GDPD2*. This low stability is the result of the putative exonized *Alu* that occurs in variant 1 of *GDPD2's* 12th exon (see text).

found to be gene exon number followed by the *Alu* population within the +/− 250,000 bp landscape followed by gene coding region length. Using a least squares regression for these three variables versus this model's estimated gene stabilities generated adjusted R^2 values of 41.7%, 23.2% and 1.6%, respectively. These lower adjusted R^2 values suggest that the algorithms used in this instability model provide complexity which cannot be accurately estimated with a single variable. This is consistent with the view that this study's methodology is a new approach for accessing the *Alu* element contribution to estimating relative gene instabilities.

The least stable of all 100 genes is the randomly selected gene, *GDPD2*. The low relative stability of *GDPD2* (7.1%, see Table S4) results from a putative exonized *Alu* that occurs in variant 1 of *GDPD2's* 12th exon. Four different variants of this gene are represented in the UCSC genome browser. The absence of this exon in the other three variants is consistent with this predicted instability. This *Alu* element-based instability model considers an exonized *Alu* element as the most unstable form of structural variation within a gene's coding region. Therefore, in addition to the disruption of coding sequence associated with an *Alu* insertion into an exon, subsequent disruption may also ensue because of the high potential for small deletions to occur at the ends of the *Alu* element. Both of these mechanisms may help explain the scarcity of exonized *Alus*. The potential risk of an exon-damaging deletion originating from the end of a nearby *Alu* element is consistent with the observed scarcity of *Alus* and other transposable elements within 50 bp of human exons [28,29].

An examination of the variation in relative gene instabilities with respect to variation in the deletion size frequency distribution was also conducted. This evaluation was performed by varying the ≤50 bp deletion size frequency between 90 and 99 percent in increments of one percent (Figure S5). While this analysis resulted in significant changes in absolute gene instabilities, the relative instabilities between most genes were unaltered. Exceptions to this observation occurred for *ATM* and *CASP8*. These have the two closest *Alu* elements located within 5 and 7 bp of exons 14 and 8, respectively. The next closest *Alu* to a deletion-prone cancer gene exon occurs at exon 19 of *FANCD2* with a separation of 20 bp. *ATM* and *CASP8* disproportionately increase in relative instability (compared to the other 48 genes in the deletion-prone cancer gene group) as the fraction of deletions ≤50 bp was increased (see Methods).

Relative exon stabilities

The relative stabilities of the 1,287 coding exons that make up the 100 genes evaluated in this study were also compared. Figure S6A is a boxplot of the individual exon stabilities for the 50 deletion-prone cancer genes. Figure S6B is a similar boxplot for the 50 randomly selected genes. The two figures are constructed left-to-right based upon each gene's most unstable exon. These two figures illustrate that relative exon stability values tend to cluster in a gene specific manner. Within the deletion-prone cancer gene group the two left-most genes, *ATM* and *CASP8*, have moderate mean exon stability values. However, the presence of exons with outlying high instabilities within *ATM* and *CASP8* puts these two genes first and second place of the most unstable among the deletion-prone cancer genes. These two genes have *Alu* elements that are within 5 and 7 bp of their 14th and 8th exons, respectively. When average exon instability is used as the sorting criterion (illustrated by the bold black line through each respective boxplot), *VHL*, *BRCA1*, FANCA, *TP53* and *SBDS* make up the top 10% most unstable genes among the 50 deletion-prone cancer genes. Finally, Figure S6B illustrates the very low stability value

(7.2) determined for the exon containing the putative exonized *Alu* in *GDPD2*.

Deletion sizes in VHL cancer deletion families do not recapitulate Figure 5

Figure 5 is constructed upon the premise that over 95% of deletions in the human genome are less than 50 bp in length [26,27]. In contrast, 25% of the deletions resulting in VHL cancer are greater than 10,000 bp [30]. This apparent conflict in deletion size frequency may arise from ascertainment bias as only those deletions that result in VHL cancer are detected. The *Alu* landscape flanking the VHL gene in Figure 4B reveals two regions of high *Alu* instability (*i*Scores shaped as horns) that extend in both 5′ and 3′ directions from the base of the *VHL* gene. As can be seen from the diagram, the 5′ and 3′ regions extend approximately 150,000 bp and 100,000 bp, respectively from the gene. Based on genome-wide derived deletion size frequencies in Figure 5, most of the deletions arising within these "horns of *Alu* instability" would be much shorter than the distances required to damage the VHL coding integrity and would likely go undetected.

Discussion

Evolution is a slow process. The clues to its activity reside almost exclusively in the subtle patterns that it leaves behind. Two of these patterns, chimeric *Alus* and the instability of cancer genomes are consistent with this study's model of inverted *Alu* pair instability. The implications of these two evolutionary patterns are discussed below.

Chimeric *Alus* may camouflage the instability of inverted *Alu* pairs

It is generally accepted that most chimeric *Alu* elements are formed by non-allelic homologous recombination (NAHR) between two direct oriented *Alu* elements [5]. However, chimeric *Alu* elements can also be generated by single-strand annealing repair of DSBs that occur within the spacer sequence separating a direct oriented *Alu* pair. However, single-strand annealing repair is only possible when high-homology sequences flank the DSB. Satisfying this homology requirement entails sufficient resection of the intervening spacer sequence separating the *Alu* pair [8].

The presence of a chimeric *Alu* element at the boundary, or breakpoint, of structural variation provides little evidence regarding the etiology of its formation. As a result, the mechanistic details behind this type of structural variation are difficult to ascertain. Without supporting evidence for an intervening deletion mechanism in the pre-chimeric spacer, the putative NAHR route is the most reasonable explanation for the formation of chimeric *Alu* elements.

This study's *Alu* structure-based stability algorithm was constructed upon the premise that DSBs can be generated from the interaction between inverted *Alu* pairs. It is possible that a fraction of these inverted *Alu* pair generated DSBs could be repaired through single-strand annealing repair of direct-oriented *Alu* pairs. This type of repair would generate a chimeric *Alu* element. The chimeric *Alu* element would effectively mask the inverted *Alu* pair as the source of the DSB. Further adding to this camouflage is the possibility that the chimeric *Alu* breakpoint (repair point) can be thousands of base pair removed from the initiating DSB [5,31].

Both non-allelic homologous recombination and single strand annealing repair likely contribute to the human chimeric *Alu* population. However, to our knowledge, the strongest evidence in support of either theory is the imbalance in the human *Alu* pair I:D ratio [18,20]. Chimeric *Alu* elements appear to result from repair of approximately 10 percent of inverted APE deletions [18].

Oncogenesis may be a passenger mutation to genome-wide instability

As mentioned previously in the Results section, the *Alu* element-based instability model predicts that deletion-prone cancer genes are ~58% more unstable than randomly selected genes. This 58% difference between cancer and random gene deletion rates is not sufficiently large to preclude the possibility that both rates may be common products of an insidious process that damages the genomes of somatic cells. Prior to senescence, the trillions of cells in our bodies likely provide multiple occasions for an unfortunate combination of cancer-prone genetic damage to occur [32].

Most of the mutations in a cancer cell are passenger mutations that do not appear to contribute to the cancer cell's fitness [33]. It is generally assumed that the vast majority of these passenger mutations are byproducts of oncogenesis. While passenger mutations may be more likely to occur subsequent to the oncogenic driver mutation, the assumption that somatic cell genomes are stable prior to oncogenesis has not been proven.

In final support of a model suggesting general somatic cell instability is the observation that deletion size frequencies observed in VHL cancer (see Results) do not conform to the deletion size frequency distribution that has been observed in healthy cells (Figure 5). The disproportionate number of large deletions (relative to Figure 5) observed among various VHL cancer families suggests that many smaller, non-cancerous deletions occur, but go undetected within healthy cell populations.

The human *Alu* pair I:D ratio may underrepresent inverted *Alu* pair interactions

As previously stated, a premise of this study is that the imbalance in the human *Alu* pair I:D ratio is a consequence of genomic instability. The human *Alu* pair I:D imbalances illustrated in Figures 3 and S2 may under estimate inverted *Alu* pair instability for two reasons. 1) The depression of the I:D ratio does not include inverted *Alu* pair deletions that have been lost through negative selection pressure and genetic drift. 2) The instability estimates derived from the I:D ratio assumes no instability between direct oriented pairs. Several studies have shown that both inter-chromosomal and intra-chromosomal recombination occurs between *Alu* elements [5,31,34].

The development of this genomic instability model is just one approach to finding tangible risk factors associated with mobile element-related threats to the genome. Unfortunately, we are far from a complete understanding of the entire puzzle. However, the fundamentals provided by the algorithms used in this study may lay the foundation for other computational approaches to comparing genetic risks posed by structural variations that are unique to specific individuals, families and people groups. With the advancement of genome sequencing technologies and the emergence of whole genome analyses, sophisticated modeling systems such as this *Alu*-element based instability model, will likely be essential to the future of genomics research.

Conclusions

Interactions between highly homologous *Alu* elements and their potential to result in deletions, duplications, inversions and gene conversion events has been well documented [5,10,14,15,35]. Various forms of structural variation have been shown to account for a large proportion of human genetic diversity [9,26,36]. Recent studies have suggested that common types of *Alu* induced structural variation may be just the tip of the iceberg, with far more complex mechanisms for *Alu* induced genome instability being possible [9,18–20]. The model developed in this study estimates relative human genome instability based upon the premise that inverted *Alu* pair exclusions are generated as a consequence of genomic instability.

Assuming that the basic concepts for this *Alu* element-based gene stability model are correct; five conclusions are evident from this study. 1) *Alu* landscapes create regions of genomic instability that are unique for each human gene. The majority of this instability resides within the ±250,000 bp regions flanking each gene. 2) Genes with higher exon counts are potentially more vulnerable to coding deletions. Additional exons provide more opportunities for *Alu* elements to reside in close proximity to coding regions. 3) Exonized *Alu* elements are a particularly unstable class of structural variation. This instability is inherent in exonized *Alus* because any deletion resulting from an *Alu-Alu* interaction is more likely to result in loss of coding sequence. 4) The human deletion size frequency curve predicts that large deletions detected through a cancer phenotype may be evidence that many smaller deletions also occur at the same locus, but go undetected. 5) This *Alu*-based human genome instability model may be used to evaluate the genetic risk posed by *Alu* element-based variation which is unique to specific individuals, families, and people groups.

Methods

Data acquisition and flow

The hg19, 2009 Human Genome Assembly was used for this study. Retrotransposon data was obtained from RepeatMasker [37] and downloaded from the UCSC genome BLAT Table Browser (UCSC Table Browser website. Available: http://genome.ucsc.edu/cgi-bin/hgTables?command = start. Accessed 2013 May 2). This data was imported into Excel 2010 (Microsoft Corporation; Redmond, Washington). Statistics were calculated using Excel 2010 output using Minitab 15 and Minitab 16 (Minitab Inc.; State College, Pennsylvania).

Identification of the key variables that correlate with the human *Alu* pair I:D ratio

Three variables were found to significantly correlate with the *Alu* pair I:D ratio. These three variables are 1) the spacer size separating the two members of the *Alu* pair, 2) the number of *Alu* elements within the spacer separating the two members of the *Alu* pair and 3) the clustering state (clustered or not clustered) of the each member of the *Alu* pair.

The *Alu* pair I:D ratio was not found to correlate strongly with *Alu* size. The only exception to this observation occurs between the first 10 immediate *Alu* neighbors of small-small and small-medium *Alu* pairs. Small *Alus* are between 30 and 135 bp in length and medium *Alus* are between 136 and 274 bp in length. This anomaly involves less than 0.2 percent of the *Alu* pair population. Manual inspection of several of these loci suggests that this phenomenon results from these smaller *Alu* fragments being incorporated into tandem repeats (data not shown). Incorporation of *Alu* fragments into tandem repeats lowers the I:D ratio for pairs of this size.

Description of key variables – spacer size

The spacer is the intervening sequence between the two *Alu* elements that make up an *Alu* pair. Spacer size is the number of base pairs within this intervening sequence. Additional *Alu* elements may be present within the spacer sequence.

Description of key variables – *Alu* pair sequence number (APSN)

The parameter describing the number of *Alu* elements within the spacer of an *Alu* pair is termed the *Alu* pair sequence number (APSN). The APSN would ideally be defined as the number of *Alu* elements within the spacer sequence. However, the APSN uses either a positive or negative value to discriminate between pairs formed by *Alus* located either 5′ (negative) or 3′ (positive) of each *Alu* being evaluated. As a result, mathematical confounding of 5′ and 3′ adjacent pairs precludes the use of zero to describe this parameter. The APSN is consequently defined as the "n+1" number of *Alu* elements within the spacer.

Description of key variables – clustering

The human non-LTR retrotransposons, *Alu* LINE and SVA elements, frequently cluster together in groups we previously defined as CLIQUEs, catenated LINE1 endonuclease induced queues of uninterrupted *Alu*, LINE1 and SVA elements [18]. Building on our original work, this study found that the *Alu* pair I:D ratio is a strong function of the clustering state of *Alu* pairs (Figure S3). Four types of clustered *Alu* pairs exist and are identified as types 0, 1, 2 and 3. Type 0 and Type 1 *Alu* pairs are located within CLIQUEs. Type 0 *Alu* pairs are formed when both members of the pair reside within the same CLIQUE and Type 1 *Alu* pairs are formed when both members of the pair reside within different CLIQUEs. Type 0 pairs are rare (<0.5 percent of human *Alu* pair population) and because of the inherent orientational *Alu* biases within a CLIQUE, require a different methodology than I:D ratio to determine instability [18]. This methodology is discussed separately under the heading entitled, "Determination of *Alu* pair instability within CLIQUEs", below. Type 2 *Alu* pairs are hemi-clustered. This category of *Alu* pairs occurs where only one of the two *Alus* making up the pair resides within a CLIQUE. Type 3 *Alu* pairs are non-clustered. Figures 3 and S2 illustrate the relationship of I:D ratio among different clustering conformations within the human *Alu* pair population.

Algorithm development for estimating *Alu* pair I:D ratio from key variables

Segregation of the separate contributions of spacer size, APSN and clustering to the *Alu* pair I:D ratio was accomplished using a five-step methodology.

Step one in *Alu* pair I:D ratio algorithm development was determination of the full-size *Alu* pair population (275–325 bp) with its associated I:D ratio for each APSN (from APSN = ±1 through APSN = ±110). This information is available from previously published work (for APSNs 1–107) that utilized the human genome assembly hg18 as its resource [18]. This study updated the earlier work using improved techniques and the most recent human genome assembly, hg19. The improved techniques permitted extending the number of statistically significant APSNs from 107 to 110.

Step two in I:D ratio algorithm development was accomplished by incrementally stepping through each of the populations of APSNs 1–110 in small (0.03–0.05% of each APSN population) spacer size increments. The population of *Alu* pair types 1, 2 and 3 (clustered, hemi-clustered and non-clustered) are determined within each increment. The resultant data set for each APSN and *Alu* pair type was then sorted into ten percentile groups. The first percentile accounts for the smallest five percent of the spacer sizes and the remaining nine percentiles capture sequential groupings of approximately ten percent of the APSN's *Alu* pair population. Each of these final nine percentiles is identified by its respective median point; 10th, 20th, 30th etc., through 90th percentiles. The spacer size boundaries for these final nine percentile groupings include ±5% of the *Alu* pair population for the APSN being evaluated. As examples, the 10th percentile describes the grouping that includes spacer sizes ranging between the 5th and 15th percentiles, the 20th percentile describes the spacer sizes falling between the 15th and 25th percentiles, etc. The *Alu* pair sample size for most APSN populations falls between 550,000 and 560,000. The only exceptions are the APSNs 1–4. These APSN families increase in population size from 461,054 to 548,606 because of CLIQUE (clustering) effects. An *Alu* pair population size above 507,000 is required to provide statistical confidence that an I:D value ≤0.995 is below unity ($p<0.05$).

The percentile groupings are further reduced in size by subdividing them into their respective *Alu* pair types. The median spacer sizes along with actual and fitted I:D ratios for Type 1 *Alu* pairs are shown in Table S2. As shown in Table S5, sample sizes across these spacer size percentile groupings reduce the sample size to as low as 2,611 for the 0–5th percentile grouping for Type 1 *Alu* pairs for APSN = 1. The average sample size for the larger percentiles (APSN>1) is 18,574. This sample size problem for measuring the I:D ratio for individual APSNs within percentiles and *Alu* pair types is addressed in step three of this five-step methodology.

Step three in *Alu* I:D ratio algorithm development plots each of the ten percentile groupings for APSNs 1 through 115 against its median spacer size. This approach increases the population size for each percentile grouping by approximately 115X and permits more accurate estimation of the actual I:D ratio at each APSN (see Figure S7). The smallest of these 115X sample sizes is 693,930 for the 2.5^{th} percentile of Type 1 *Alu* pairs. This sample size is larger than the 507,000 minimum sample size (see step two, above) required for I:D values of <0.995 to be statistically less than unity ($p<0.05$). Examination of these 115 groupings revealed that for APSNs >110, no percentile grouping dropped below the minimum statistically significant I:D value of 0.995 ($p\leq0.05$). Consequently, only APSNs of 1 through 110 were used in the construction of the instability model algorithm.

A total of 30 regression curves were generated; 10 for Type 1 *Alu* pairs (clustered; 13,364,142 total full-length pairs), 10 for Type 2 *Alu* pairs (hemi-clustered; 28,537,478 total full-length pairs) and 10 for Type 3 *Alu* pairs (non-clustered; 18,836,832 total full-length pairs). Each set of percentile data was then regressed versus its median spacer size (from step two). The resultant algorithm that describes the data for each respective percentile was then assembled. In several instances the best fit for the data was accomplished by using a composite of two or more regressions for one set of APSN percentile data. Examples of these curve fits are shown in Figures S7A–C for the 2.5^{th} percentile curves for Type 1, 2 and 3 *Alu* pairs.

Step four in development of the *Alu* I:D ratio prediction algorithm was the extraction of the respective I:D ratios for each of the ten regressed percentiles for each APSN for *Alu* pair types 1, 2 and 3. Each regressed I:D ratio value was plotted for each APSN against its median spacer size. This step produces 345 different I:D curves, 115 curves for each *Alu* pair type. As mentioned previously, only APSN curves 1–110 had at least one point along the spacer size percentiles with an I:D ratio that was statistically below unity ($0.995 = p<0.05$). This technique excludes *Alu* pair type zero, which was treated separately (see heading, "Determination of *Alu* pair instability within CLIQUEs", below). An example of regressed data extracted from this step for Type 1 *Alu* pairs for APSNs 1–10 is shown in Figure 2. Figures 3 and S2 show the complete set of regressed I:D data (APSNs = ± 1–110) for Type 1, 2 and 3 *Alu* pairs.

Step five in development of the *Alu* pair instability algorithm development was the regression of the ten percentile data points derived from step four (above) for each of the 345 graphs. The shape of these curves often requires more than one regression equation to accurately portray these regressed values. In addition, median spacer size values below the 2.5^{th} percentile and above the 90th percentile fall outside of the regressed region for these curves. Spacer sizes that are smaller than the median spacer size for the 2.5^{th} percentile are assigned the I:D value of the 2.5^{th} percentile. Straight lines connect the 2.5^{th} percentile midpoints for the 5′ and 3′ curves for each APSN for each of the three *Alu* pair types shown in Figures 3 and S2. Spacer sizes that are larger than the median spacer size for the 90th percentile are fit along a straight line from the I:D value at the 90th percentile to unity at the 99th percentile. The equation types and associated coefficients for the ±110 APSN curves associated with Type 1 *Alu* pairs are provided in Table S6.

Determination of *Alu* pair instability within CLIQUEs

Type 0 *Alu* pairs possess inherent *Alu* orientational insertion biases. This is reflected by the low CLIQUE I:D ratio = 0.460. These biases preclude the direct estimation of *Alu* pair instability from I:D measurements [18]. However, less than 0.5% of human *Alu* pairs reside within the same CLIQUE. Most of these Type 0 *Alu* pairs have spacer sizes of ≤50 bp [18]. Although these pairs represent a relatively small fraction of the total *Alu* pair population, their small spacer size may make a disproportionately large contribution to the total inverted *Alu* pair instability within the genome.

A solution to this stability prediction dilemma for Type 0 *Alu* pairs was resolved using data from previous work performed with a yeast experimental system. This system measured the instability of inverted *Alu* pairs when separated by 12, 20, 30 and 100 bp for homologies of 94% and 100% [19]. Typical human *Alu* pair homologies are 85% [20].

Fortunately, the median spacer size for adjacent Type 1 (clustered) *Alu* pairs in 0th–5th percentile range was 100 bp (Table S5). This data point, representing 2,611 *Alu* pairs (Table S5), is one of the four spacer sizes evaluated for determination of inverted *Alu* pair instability in the experimental yeast system. This data point was used to anchor the 85% *Alu* homology curve to the 94% and 100% homology curves used in the yeast experiments [19]. The resultant Type 0 *Alu* pair algorithm for estimating inverted *Alu* pairs with 85% homologies is as follows.

$$0.7804\text{-}(3.0271 \times e^{(-0.164251 \times \text{SpacerSize,bp})})$$

This algorithm is used to predict the I:D ratio for Type 0 *Alu* spacer sizes ≤50 bp. The algorithms developed for Type 1 *Alu* pairs were used to estimate Type 0 *Alu* pairs with spacer sizes >50 bp.

Instability estimate for individual *Alu* elements within an *Alu* pair

The I:D ratio is the stability of an *Alu* pair, not the stability of an individual *Alu* element. The instability of an individual *Alu* element within an *Alu* pair is estimated as the square root of the I:D ratio estimated for that pair. Depending upon the single-strand cleavage pattern at its eight potential cleavage sites (Figures 1 and S1; [18]), the resolution of the hypothetical doomsday junction can result in some level of gene conversion and/or from zero to four DSBs.

The I:D ratio versus spacer size relationships represented in Figures 3 and S2 are composed of the ±110 APSNs curves for Type 1, 2 and 3 *Alu* pairs. Each of these APSN curves contain at least one percentile along their spacer size interval (Figure 3) where the I:D ratio is <0.995. The I:D<0.995 cutoff represents the statistical confidence interval for full-length *Alu* pair families ($p<0.05$). These curves permit the maximum inverted *Alu* pair interaction distance to be increased from the previously reported value of APSN = ±107 to APSN = ±110 (Cook et al. 2011). Any predicted I:D ratio that is >0.995 is assigned a value of 1.0.

Alu element stability and *i*Score determination

The stability of an *Alu* element is the grand product of the square root of the I:D ratios calculated for each of the *Alu* pairs formed by its ±110 immediately flanking (5′ and 3′) *Alu* elements. This stability is expressed by the following equation.

$$\text{Stability of an } Alu \text{ element} = \prod_{APSN=-110}^{APSN=110} \sqrt{I : D(APSN)}$$

The stability of each of these 220 flanking *Alu* pairs is determined from the previously developed I:D versus spacer size versus APSN algorithms. Direct oriented *Alu* pairs are considered stable and assigned a value of 1. The *i*Score is the inverse of the estimated stability of an *Alu* element and is used only in Figures 4 and S4 to illustrate the relative stabilities of the various *Alu* elements located within a gene's *Alu* landscape.

Since each end of an *Alu* element is subject to a potential deletion, the stability of only one end of each *Alu* element is the grand product of the fourth root of the I:D ratio for all 220 potential *Alu-Alu* interactions. This stability is expressed as follows.

$$\text{Stability of either end of an } Alu \text{ element} =$$

$$Alu_{End} = \prod_{APSN=-110}^{APSN=110} \sqrt[4]{\mathrm{I : D(APSN)}}$$

These algorithms are generated from the genome-wide human *Alu* element population. The individual *Alu* landscapes of the 100 genes examined in this study are of insufficient size to reveal statistically significant imbalances within their respective I:D ratios. Using the average number of *Alus* within these landscapes (410 *Alus*), statistical relevance can only be recognized when the I:D ratio is outside of the range of 0.82–1.22 ($p<0.05$).

Estimation of deletion size probability

Two studies provided insight into the human deletion frequency distribution [26,27]. Recent cancer studies also provide similar information. However, the unique nature of cancer cells precludes the use of this data in the characterization of DNA stability in healthy cells. In this study, human indel size frequency curves are treated as having the same shape as the corresponding human deletion size frequency curve.

The deletion size frequency curve in Figure 5 was prepared from a composite of data provided in the two studies mentioned, above. The first study, Wheeler et al., 2008, provides a deletion size frequency curve that was used to estimate the deletion size frequency for deletion sizes ≤75 bp. The second study [26], is used to estimate the deletion size frequency for deletion sizes >75 bp. Modeling of the deletion/indel size frequency data from both studies excluded the *Alu* insertion perturbation present between 250 and 350 bp. This permitted smoothing of the respective regression fits.

In the first study, deletion frequency data was regressed between 1 and 400 bp and for the second study, the indel frequency data was regressed between 50 and 10,000 bp. In both studies over 95% of deletions/indels were ≤50 bp. The second study (Mills et al., 2011) used a higher number of individuals (79) and thus supplied additional data for the more rare larger deletion sizes.

The sum of the 500,000 individual deletion size probabilities illustrated in Figure 5 equal 1.0. The probability of a specific deletion size occurring is lower than the probability of that same or larger deletion size occurring. This latter probability of a "minimum required deletion size or larger" required for loss of coding sequence is used in the model's algorithm.

The model's algorithm considers each end of each *Alu* element separately in its determination of exon and gene stability. Estimation of the risk that an *Alu* end poses to an exon coding sequence first requires that the distance between the end of the *Alu* element and the proximal end of the exon be determined. This distance is defined as D_{Min}. The formula that describes the probability of a minimum deletion size is as follows.

$$D_{Min} = \text{Probability of a specific deletion size (or larger)} = P_{Deletion}$$

$$P_{Deletion} = \sum_{d=D_{Min}}^{d=500,000} \text{deletion fraction (d)}^{*}$$

* Individual deletion fractions are taken from Figure 5

Determination of relative exon instability

Individual exon instabilities are calculated through a five-step process. Step one is calculating the DSB risk posed by each end of each *Alu* element ($Risk_{End}$) within a gene's ±250,000 bp *Alu* landscape. Step two is determining the potential deletion size risk, $P_{Deletion}$, posed by each end of each *Alu* element within this landscape, to the coding exon of interest. Step three is multiplying each individual $Risk_{End}$ value by its respective $P_{Deletion}$ value. Step four calculates the grand product of these "$Risk_{End} \times P_{Deletion}$" products. This estimated relative exon stability, $Exon_{RS}$, is expressed by the following formula.

$$Exon_{RS} = \prod_{N=3'\text{end of the } 3' \text{ most } Alu \text{ in } +/-250,000 \text{ bp flanking landscape}}^{N=5'\text{end of the } 5' \text{ most } Alu \text{ in } +/-250,000 \text{ bp flanking landscape}} Alu_{End}(N) P_{Deletion}(N)$$

Step five determines the exon instability. Since exon stability plus exon instability equals 1.0, the exon instability is one minus the estimated exon stability derived from the formula above.

Determination of relative gene instability

Relative gene instability is defined as the relative likelihood of a deletion occurring at some location within a gene's coding exons. This is determined through a four-step process. The first three steps are identical to the first three steps described under the "Determination of relative exon instability" heading above. Step three in this procedure is only performed for the closest exon to each *Alu* element end. This step determines the highest risk, $Risk_{Max}$, that one end of an *Alu* element can pose to a gene. Step four multiplies each of these, $Risk_{Max}$, values determined for each *Alu* end. This grand product produces the estimate of that gene's relative stability, $Gene_{RS}$.

$$Gene_{RS} = \prod_{N=3'\text{end of the }3'\text{ most }Alu\text{ in }+/\text{-}250{,}000\text{ bp flanking landscape}}^{N=5'\text{end of the }5'\text{ most }Alu\text{ in }+/\text{-}250{,}000\text{ bp flanking landscape}} Risk_{Max}$$

Step five determines the gene instability. Since the stability of a gene plus its instability equals 1.0, gene instability is one minus the estimated gene stability derived from the formula above.

Gene selection

The 50 random human genes used in this study were selected from the list of 19,026 human protein-coding genes provided by the HUGO Gene Nomenclature Committee, HGNC. The source file containing these genes was downloaded from the HGNC website [38]. The 50 random genes were selected from this list using Minitab 16.

The 50 deletion-prone cancer genes were selected from [21,23] and the Catalogue of Somatic Mutations in Cancer [22] (Wellcome Trust Sanger Institute Cancer Genome Project website. Available: http://www.sanger.ac.uk/genetics/CGP/Census/large_deletion.shtml. Accessed 2013 May 2). Only coding exons were selected for each gene. Exon loci were obtained from the RefSeq CDS Fasta Alignment page on the UCSC genome browser (UCSC Genome Browser website. Available: http://genome.ucsc.edu/cgi-bin/hgPal. Accessed 2013 May 2). Variant 1 isoforms of all genes were selected when more than one gene was listed under RefSeq genes.

Variation in relative gene stability with variation in deletion size frequency

The deletion size frequency curve used in this study (Figure 5) illustrates that 95% of human deletions have lengths ≤50 bp. The availability of genome-wide human deletion size frequency data is limited. Consequently, the sensitivity of relative gene stabilities to the shape of this deletion frequency curve was examined. This examination was made by varying the fraction of deletions that were ≤50 bp from 0.90 to 0.99 in increments of 0.01 for the 50 deletion-prone cancer genes. The results of this examination showed that the relative stabilities of 48 of the 50 deletion-prone cancer genes remained essentially unchanged as the ≤50 bp increment was varied. Exceptions to this observation occurred with two genes, *ATM* and *CASP8*, which have the rare occurrence of *Alus* within 5 and 7 bp of their exons [28]. The results of this sensitivity analysis are shown in Figure S5.

Supporting Information

Table S1 Selected studies linking *Alu*-related deletions to disease phenotypes.

Table S2 Raw and fitted I:D ratios for the ten spacer size groupings for Type 1 *Alu* pairs.

Table S3 Characteristics of the 50 deletion-prone human cancer genes examined in this study.

Table S4 Characteristics of the 50 randomly chosen human genes examined in this study.

Table S5 Spacer size percentile samples sizes versus APSN for Type 1 *Alu* pairs.

Table S6 Coefficients for equations describing the I:D ratio versus spacer size for Type 1 *Alu* pairs.

Figure S1 A proposed mechanism for the formation of a doomsday junction that is catalyzed by the ectopic invasion and annealing of complementary replication forks. (A) Two *Alu* elements in opposite orientations form an inverted *Alu* pair. (B) Concomitant advancement of replication forks through each member of an inverted *Alu* pair. C) Bending of the DNA to permit alignment of the complementary replication forks. D) Ectopic invasion and annealing of single-stranded DNA associated between high-homology replication forks could potentially extend to the entire length of the *Alu* elements. The hypothetical conformation created by this interaction is termed a doomsday junction. As also illustrated in Figure 1, eight segments of single-stranded DNA are formed at the boundary of the doomsday junction and create the opportunity for single-strand nuclease attack. These sites are illustrated as yellow lightning bolts.

Figure S2 The *Alu* pair I:D ratio versus spacer size for Type 2 and Type 3 *Alu* pairs for APSNs ±1–110. This figure illustrates the ±110 APSN curves for full-length (275–325 bp) A) Type 2 *Alu* pairs and B) Type 3 human *Alu* pairs.

Figure S3 I:D ratios for Type 1, 2 and 3 *Alu* pair families for APSNs 1–150. Note that the departure of the I:D ratio from unity is greatest for clustered (Type 1) *Alu* pairs and closest to unity for non-clustered (Type 3) *Alu* pairs.

Figure S4 *Alu* landscapes for five deletion-prone cancer genes and three randomly chosen genes. Each *Alu* element is plotted within and 500 kbp, 5′ and 3′ flanking each gene. The locus of each *Alu* is plotted against its respective instability score, *i*Score. The *i*Score is the inverse of the model's predicted *Alu* stability and thus larger values represent higher instabilities. The five selected deletion-prone cancer genes are A) *APC*, B) *ATM*, C) *MLH1*, D) *MSH2* and E) *TP53*. The three randomly chose genes are F) *GDPD2*, G) *KEAP1* and H) *SF3B3*.

Figure S5 Sensitivity analysis of the relative stabilities of the deletion-prone cancer genes versus variation in the fraction of deletions that is ≤50 bp. The shape of the deletion size frequency curve used to determine relative gene stabilities (Figure 5) places 95% of deletions with lengths of ≤50 bp. This figure examines the variation in relative deletion-prone cancer gene stabilities as the fraction of deletions ≤50 bp is varied between 90% and 99%. As can be seen from this figure, the relative stabilities of 48 of the 50 deletion-prone cancer genes (96%) remain essentially unchanged as the ≤50 bp increment is varied. Exceptions to this observation are observed with *ATM* and *CASP8* (bolded curves), which have the rare occurrence of *Alus* within 5 and 7 bp of their exons, respectively. These two genes exhibit higher relative stabilities as the fraction of deletions ≤50 bp in length increases.

Figure S6 Estimated relative exon stability distributions for the 50 deletion-prone cancer genes and 50 randomly chosen genes. A) Boxplot of the individual exon stabilities for the 50 deletion-prone cancer genes. The genes in this figure are ordered left-to-right on the basis of each gene's least stable exon. While individual exon stabilities vary widely, they tend to cluster in a gene specific manner. Exceptions to this pattern are illustrated by the presence of a single, outlying low stability exon within *ATM* and *CASP8*. These individual exon stabilities place these two genes at first and second place of highest instability among these 50 deletion-prone cancer genes. (B) Boxplot of the individual exon stabilities for the 50 randomly selected genes. Note that a broken Y-axis scale is required to capture the low stability of the putative exonized *Alu* in the 12th exon of *GDPD2* (see text).

Figure S7 Fitted curves for 2.5th percentile spacer size groupings for type 1, 2 and 3 for APSN families 1–115. These three curves are part of thirty curves that are used to estimate the I:D ratio for the Type 1, Type 2 and Type 3 *Alu* pairs. Ten different curves are used for each *Alu* Pair Type. These ten curves are used to construct an I:D ratio curve for each APSN family versus spacer size (see Methods). The curves shown here for A) Type 1 B) Type 2 and C) Type 3 *Alu* pairs represent the I:D ratio for the smallest median spacer size percentile (2.5th percentile) of spacer size groupings.

Acknowledgments

We would like to thank LSU Chemical Engineering Professor John C. Flake for his review of the manuscript and helpful comments.

Author Contributions

Conceived and designed the experiments: GWC MKK. Performed the experiments: GWC. Analyzed the data: GWC. Contributed reagents/materials/analysis tools: GWC. Wrote the paper: GWC. Conceived and constructed math algorithm: GWC. Wrote formulas describing various math algorithms: GWC JTF. Assembled computer model: GWC HDH MLF*. Designed experiments: GWC MKK. Performed experiments: GWC. Prepared Figures and Tables: GWC MGB. Edited the manuscript: GWC MKK JAW MAB. Read and approved the final manuscript: GWC MKK JAW MGB MLF JTF HDH MAB. *MLF worked on this project in the Department of Biological Sciences, LSU-Baton Rouge as a participant in the Louisiana Biomedical Research Network (LBRN) while completing a B.S. degree in Bioengineering at Clemson University, SC.

References

1. Lander ES, Linton LM, Birren B, Nusbaum C, Zody MC, et al. (2001) Initial sequencing and analysis of the human genome. Nature 409: 860–921.
2. de Koning AP, Gu W, Castoe TA, Batzer MA, Pollock DD (2011) Repetitive elements may comprise over two-thirds of the human genome. PLoS Genet 7: e1002384.
3. Batzer MA, Deininger PL (2002) Alu repeats and human genomic diversity. Nat Rev Genet 3: 370–379.
4. Deininger PL, Batzer MA (1999) Alu repeats and human disease. Mol Genet Metab 67: 183–193.
5. Sen SK, Han K, Wang J, Lee J, Wang H, et al. (2006) Human genomic deletions mediated by recombination between Alu elements. Am J Hum Genet 79: 41–53.
6. Witherspoon DJ, Watkins WS, Zhang Y, Xing J, Tolpinrud WL, et al. (2009) Alu repeats increase local recombination rates. BMC Genomics 10: 530.
7. Konkel MK, Batzer MA (2010) A mobile threat to genome stability: The impact of non-LTR retrotransposons upon the human genome. Semin Cancer Biol 20: 211–221.
8. Hedges DJ, Deininger PL (2007) Inviting instability: Transposable elements, double-strand breaks, and the maintenance of genome integrity. Mutat Res 616: 46–59.
9. Lupski JR (2010) Retrotransposition and structural variation in the human genome. Cell 141: 1110–1112.
10. Lee J, Han K, Meyer TJ, Kim HS, Batzer MA (2008) Chromosomal inversions between human and chimpanzee lineages caused by retrotransposons. PLoS One 3: e4047.
11. Jurka J, Kohany O, Pavlicek A, Kapitonov VV, Jurka MV (2004) Duplication, coclustering, and selection of human Alu retrotransposons. Proc Natl Acad Sci U S A 101: 1268–1272.
12. de Smith AJ, Walters RG, Coin LJ, Steinfeld I, Yakhini Z, et al. (2008) Small deletion variants have stable breakpoints commonly associated with alu elements. PLoS One 3: e3104.
13. Kitada K, Aikawa S, Aida S (2012) Alu-Alu Fusion Sequences Identified at Junction Sites of Copy Number Amplified Regions in Cancer Cell Lines. Cytogenet Genome Res.
14. Kass DH, Batzer MA, Deininger PL (1995) Gene conversion as a secondary mechanism of short interspersed element (SINE) evolution. Mol Cell Biol 15: 19–25.
15. Roy AM, Carroll ML, Nguyen SV, Salem AH, Oldridge M, et al. (2000) Potential gene conversion and source genes for recently integrated Alu elements. Genome Res 10: 1485–1495.
16. Aleshin A, Zhi D (2010) Recombination-associated sequence homogenization of neighboring Alu elements: signature of nonallelic gene conversion. Mol Biol Evol 27: 2300–2311.
17. Zhi D (2007) Sequence correlation between neighboring Alu instances suggests post-retrotransposition sequence exchange due to Alu gene conversion. Gene 390: 117–121.
18. Cook GW, Konkel MK, Major JD 3rd, Walker JA, Han K, et al. (2011) Alu pair exclusions in the human genome. Mob DNA 2: 10.
19. Lobachev KS, Stenger JE, Kozyreva OG, Jurka J, Gordenin DA, et al. (2000) Inverted Alu repeats unstable in yeast are excluded from the human genome. EMBO J 19: 3822–3830.
20. Stenger JE, Lobachev KS, Gordenin D, Darden TA, Jurka J, et al. (2001) Biased distribution of inverted and direct Alus in the human genome: implications for insertion, exclusion, and genome stability. Genome Res 11: 12–27.
21. Solimini NL, Xu Q, Mermel CH, Liang AC, Schlabach MR, et al. (2012) Recurrent hemizygous deletions in cancers may optimize proliferative potential. Science 337: 104–109.
22. Forbes SA, Bindal N, Bamford S, Cole C, Kok CY, et al. (2011) COSMIC: mining complete cancer genomes in the Catalogue of Somatic Mutations in Cancer. Nucleic Acids Res 39: D945–950.
23. Stephens PJ, Tarpey PS, Davies H, Van Loo P, Greenman C, et al. (2012) The landscape of cancer genes and mutational processes in breast cancer. Nature 486: 400–404.
24. Fogedby HC, Metzler R (2007) Dynamics of DNA breathing: weak noise analysis, finite time singularity, and mapping onto the quantum Coulomb problem. Phys Rev E Stat Nonlin Soft Matter Phys 76: 061915.
25. Jeon JH, Sung W, Ree FH (2006) A semiflexible chain model of local denaturation in double-stranded DNA. J Chem Phys 124: 164905.
26. Mills RE, Pittard WS, Mullaney JM, Farooq U, Creasy TH, et al. (2011) Natural genetic variation caused by small insertions and deletions in the human genome. Genome Res 21: 830–839.
27. Wheeler DA, Srinivasan M, Egholm M, Shen Y, Chen L, et al. (2008) The complete genome of an individual by massively parallel DNA sequencing. Nature 452: 872–876.
28. Zhang Y, Romanish MT, Mager DL (2011) Distributions of transposable elements reveal hazardous zones in mammalian introns. PLoS Comput Biol 7: e1002046.
29. Lev-Maor G, Ram O, Kim E, Sela N, Goren A, et al. (2008) Intronic Alus influence alternative splicing. PLoS Genet 4: e1000204.
30. Franke G, Bausch B, Hoffmann MM, Cybulla M, Wilhelm C, et al. (2009) Alu-Alu recombination underlies the vast majority of large VHL germline deletions: Molecular characterization and genotype-phenotype correlations in VHL patients. Hum Mutat 30: 776–786.
31. Han K, Lee J, Meyer TJ, Wang J, Sen SK, et al. (2007) Alu recombination-mediated structural deletions in the chimpanzee genome. PLoS Genet 3: 1939–1949.
32. Serrano M (2010) Cancer: a lower bar for senescence. Nature 464: 363–364.
33. Greenman C, Stephens P, Smith R, Dalgliesh GL, Hunter C, et al. (2007) Patterns of somatic mutation in human cancer genomes. Nature 446: 153–158.
34. Elliott B, Richardson C, Jasin M (2005) Chromosomal translocation mechanisms at intronic alu elements in mammalian cells. Mol Cell 17: 885–894.
35. Bailey JA, Liu G, Eichler EE (2003) An Alu transposition model for the origin and expansion of human segmental duplications. Am J Hum Genet 73: 823–834.
36. Girirajan S, Campbell CD, Eichler EE (2011) Human copy number variation and complex genetic disease. Annu Rev Genet 45: 203–226.
37. Smit AFA, Hubley R, Green P (1996–2010) RepeatMasker Open-3.0. Available: http://www.repeatmaskerorg.
38. Seal RL, Gordon SM, Lush MJ, Wright MW, Bruford EA (2011) genenames.org: the HGNC resources in 2011. Nucleic Acids Res 39: D514–519.

9

Prioritization of Copy Number Variation Loci Associated with Autism from AutDB–An Integrative Multi-Study Genetic Database

Idan Menashe[1,2]*, Eric C. Larsen[1], Sharmila Banerjee-Basu[1]*

1 MindSpec, McLean, Virginia, United States of America, **2** Department of Public Health, Faculty of Health Sciences, Ben Gurion University of the Negev, Beer-Sheva, Israel

Abstract

Copy number variants (CNVs) are thought to play an important role in the predisposition to autism spectrum disorder (ASD). However, their relatively low frequency and widespread genomic distribution complicates their accurate characterization and utilization for clinical genetics purposes. Here we present a comprehensive analysis of multi-study, genome-wide CNV data from AutDB (http://mindspec.org/autdb.html), a genetic database that accommodates detailed annotations of published scientific reports of CNVs identified in ASD individuals. Overall, we evaluated 4,926 CNVs in 2,373 ASD subjects from 48 scientific reports, encompassing $\sim 2.12\times10^{9}$ bp of genomic data. Remarkable variation was seen in CNV size, with duplications being significantly larger than deletions, ($P = 3\times10^{-105}$; Wilcoxon rank sum test). Examination of the CNV burden across the genome revealed 11 loci with a significant excess of CNVs among ASD subjects ($P<7\times10^{-7}$). Altogether, these loci covered 15,610 kb of the genome and contained 166 genes. Remarkable variation was seen both in locus size (20 - 4950 kb), and gene content, with seven multigenic ($\geq$3 genes) and four monogenic loci. CNV data from control populations was used to further refine the boundaries of these ASD susceptibility loci. Interestingly, our analysis indicates that 15q11.2-13.3, a genomic region prone to chromosomal rearrangements of various sizes, contains three distinct ASD susceptibility CNV loci that vary in their genomic boundaries, CNV types, inheritance patterns, and overlap with CNVs from control populations. In summary, our analysis of AutDB CNV data provides valuable insights into the genomic characteristics of ASD susceptibility CNV loci and could therefore be utilized in various clinical settings and facilitate future genetic research of this disorder.

Editor: Yi-Hsiang Hsu, Harvard Medical School, United States of America

Funding: AutDB is funded by the Simons Foundation, which licenses it as SFARI Gene (https://gene.sfari.org/autdb/Welcome.do). The funders had no role in study design, data collection and analysis, decision to publish, or preparation of the manuscript.

Competing Interests: The authors have declared that no competing interests exist.

* E-mail: menashe.idan@gmail.com (IM); sharmila@mindspec.org (SB)

Introduction

Copy number variations (CNVs) are structural chromosomal aberrations, which are giving rise to gains or losses of certain genomic loci across the human genome [1,2]. While most CNVs have no apparent phenotypic consequences, there is increasing evidence that a number of chromosomal micro deletions or duplications at specific locations are involved in the predisposition of various human diseases [3,4]. Recent advances in high-resolution, high-throughput genomics technologies have facilitated the detection of CNVs in large-scale genetic studies. Moreover, the continuous drop in labor and cost associated with these technologies promote their inclusion in genetic screening for pre- and post- pregnancy tests.

Autism spectrum disorder (ASD) constitutes a collection of clinically heterogeneous disorders that are characterized by impairments in social interactions, deficits in language and communication, and increased repetitive or stereotypic movements [5,6]. ASD is highly heritable with estimates ranging between 40-90% heritability [7,8]. However, given the genetically heterogeneous nature of ASD, the underlying genetic mechanisms of these disorders remain vague. Recent genetic studies have indicated that rare CNVs may play an important role in ASD susceptibility [9–14], and today they are considered one of the common genetic contributors of ASD [15]. Given the evidence that CNVs are a significant genetic risk factor not only for ASD, but for other developmental deficits and congenital anomalies, it has recently been proposed that chromosomal microarray (CMA) screening replace conventional cytogenetic techniques as a first-tier clinical diagnostic test for individuals with these disorders [16,17]. However, the relatively low frequency and widespread genomic distribution of these variants in ASD cases complicates the clinical utilization of CMA screening as a potential diagnostic tool.

Given the important role of CNVs in ASD genetics and the increasing usage of CMA screening for the genetic evaluation of ASD individuals, there is a tremendous need to consolidate the large amounts of CNV data that were generated from ASD subjects and subsequently prioritize the most consistent genomic loci associated with ASD susceptibility. To this end, we explored the CNV data available at the online autism genetic database AutDB (http://www.mindspec.org/autdb.html) which, to the best of our knowledge, is the most comprehensive online resource of curated genetic data of ASD from published scientific reports. Using a range of statistical and bioinformatics analyses we performed a comprehensive and rigorous assessment of CNVs that were observed in ASD cohorts across multiple published reports consisting of both large-scale whole-genome studies and smaller-scale case studies. We subsequently determined the genomic boundaries and genetic characteristics of 11 loci demonstrating significant CNV burden among ASD subjects.

Materials and Methods

CNV data

For this study, we used CNV data available at the CNV module of the AutDB database (data freeze of October 2011)[18]. As with the other modules of AutDB, content of the CNV module originates entirely from published, peer-reviewed scientific literature and is rigorously annotated by scientists. Preliminary screening of reports for inclusion in the database resulted from a search of the scientific literature using PubMed (http://www.ncbi.nlm.nih.gov/pubmed/) with the following keywords: "autism/autistic/ASD" and "copy number/CNV/deletion/duplication/chromosome/structural variant". Furthermore, CNV reports listed in ASD review articles that were not identified in the initial PubMed search were included for consideration. Next, the initial candidate CNV reports were filtered to remove those reports that did not contain at least one ASD individual in which one or more CNVS were identified. This restriction has since been relaxed to include reports describing patients with other neurodevelopmental or neuropsychiatric disorders, such as mental retardation/intellectual disability and developmental delay; in some cases, these patients also display ASD traits, but no formal diagnosis of ASD. Detailed information on ASD and control subjects from the studied cohorts was extracted from each selected CNV report for inclusion in the CNV module database.

For the purpose of CNV prioritization we aimed at analyzing a homogeneous subset of the CNV data set from AutDB by using several filtering criteria (number of CNVs removed from the data are in square brackets): only ASD cases were used [12,416 CNVs]; patients with a disease other than ASD, such as schizophrenia, developmental delay, etc. were excluded [74 CNVs]; a genome-wide CNV discovery method (array-CGH, SNP array, or solid phase hybridization) was required for inclusion; CNVs discovered by a targeted discovery method such as FISH or qPCR were excluded [215 CNVs]; we manually screened individuals to ensure minimal overlap of patients; in some cases a given CNV report may include individuals that had previously been described in another report [104 CNVs]. In such cases, we used patient ID information included in the scientific report to identify duplicate patient entries in the CNV module dataset and subsequently use the CNV data for a given individual that contained the largest number of CNV loci [104 CNVs]. Furthermore, we restricted our analysis to CNVs with defined start and end points [218 CNVs]. Finally, to maintain a uniform CNV/individual ratio in our data, we used CNVs identified in a control cohort of unaffected matched siblings as a filter to generate a set of "case-specific" CNVs in the accompanying case cohort of the Sanders *et. al* study [14] [13,036 CNVs]. This filtering process resulted in a homogeneous data set of 4,926 CNVs in 2,373 ASD subjects gathered from 48 scientific reports.

CNV burden

To search for genomic loci demonstrating excess of CNVs among ASD subjects, we divided the genome into consecutively distributed regions of 10 kb, and evaluated the CNV burden in each region as follows:

$$CNV\ burden\ (\theta) = \sum_{i=1}^{N} C_{i,j} \times W_j \quad (1)$$

where $C_{i,j} = 1$ if a CNV (i) from a particular study (j) of all CNVs (N) in the data, overlaps with the 10 kb genomic region, and W_j is the weight associated with the study (j) calculated as the Loci/CNVs ratio in the study.

To assess the statistical significance of θ, we first randomly distributed the CNVs in our dataset in the human genome and then calculated their corresponding 10 kb θs. We repeated this procedure 10,000 times to generate a null genome-wide distribution of θs that fitted a Poisson distribution with $\lambda = 0.8363$. We then used this Poisson distribution to calculate the statistical significance of θ associated with each 10 kb region among individual with ASD in our data. Given the median CNV length in our data was ~ 43 kb, the maximal number of non-overlapping CNV loci in the human genome is $\sim 7\times10^4$. Hence, we used this number to set a Bonferroni corrected cutoff for genome-wide significance of $P<7\times10^{-7}$ ($0.05/7\times10^4$). All analyses were performed using a commercial software package (MATLAB R2011b, The MathWorks Inc., Natick, MA, 2000).

CNV loci characterization

We used the RefSeq genes track in the UCSC Genome Browser on Human Mar. 2006 (NCBI36/hg18) assembly, (http://genome.ucsc.edu) to locate genes overlapping with ASD susceptibility CNV loci. Further, we used the December 2012 release of the Human Gene module of AutDB [18] to identify genes that have been reported as containing potential susceptibility variants for ASD.

Analysis of control CNV data

The majority of CNV loci curated in the CNV module of AutDB and used in our analysis were curated from published scientific reports in which one or more filtering steps were used to remove variants previously identified in unaffected individuals within the general population, resulting in a population of case-specific or case-enriched CNVs that were subsequently annotated. However, we concluded that an independent analysis of control CNV datasets using our prioritization strategy could be useful both in more accurately defining the boundaries of the eleven susceptibility loci identified in our initial analysis, as well as in allowing us to differentiate between potential false-positives or polymorphic CNV regions that would likely confer decreased risk of ASD susceptibility than loci with little or no control CNV overlap. Therefore, we examined the overlap of these 11 ASD susceptibility loci with CNVs among 4400 individuals with no diagnosis of ASD (controls) using data from three genome-wide CNV analyses varying in their sample sizes, studied cohorts, and CNV detection method/platform [19–21]. This control CNV data were collected from the Database of Genomic Variants (DGV) [22], and were analyzed using the same 10 kb regions described above.

Results

The CNV module of AutDB

The CNV module of AutDB as of October 2011 consisted of CNV data from 72 annotated publications, encompassing 30,989 CNVs from 4,359 individuals (3,099 ASD cases, 66 cases of other neurodevelopmental or neuropsychiatric disorders, and 1,194 control individuals). These have been summarized into 2,429 unique CNV reports classified as 'major' (i.e. independently validated) or 'minor' (1,047 and 1,382 CNV reports respectively), and distributed across 1035 CNV loci genome-wide (Figure S1). The median number of reports per CNV locus was two (one 'major' and one 'minor'), with the highest number of reports per locus reaching 16 (14 and 2 'major' and 'minor' respectively) for the 16p11.2 locus (Figure S1).

CNV dataset

For the purpose of CNV prioritization, we applied a set of stringent filtering criteria (see methods) to the AutDB data to generate a uniform subset of CNVs. This resulted in a dataset containing 4,926 CNVs in 2,373 ASD subjects (Mean = 2.08, STD = 1.79 CNVs per individual) collected from 48 scientific reports (Table S1) that encompassed $\sim 2.13\times10^{9}$ bp of genomic data. The maximal number of CNVs per individuals seen in our data was 18. Of the 2,373 ASD subjects in our data, 1,532 were males and 330 females, which is consistent with the 4:1 reported male-to-female ratio of ASD prevalence in the general population [23]. Notably, the gender of 511 subjects in our data was not reported. Of the 4,926 CNVs in our data, 1,923 (39%) were duplications and 3,003 (61%) were deletions (Table 1). In addition, 3,377 (68.6%) of the CNVs were inherited, 239 (4.9%) were *de novo*, and 1,310 (26.6%) had no indicated inheritance (Table 1). We observed remarkable variation in CNV size and inheritance patterns, with duplications being significantly larger than deletions ($P = 3\times10^{-105}$; Wilcoxon rank sum test; Figure S2), and *de novo* CNVs tended to be more prevalent among females than males ($X^2 = 18.6$; $P<0.0001$).

CNV characterization

To identify genomic loci with an excess of CNVs among ASD subjects, we divided the genome into consecutively distributed regions of 10 kb and assessed the burden of CNVs within them (See Materials and Methods). The genomic distribution of CNV burden among individuals with ASD is depicted in Figure 1. Overall, there were eleven genomic loci displaying significant burden score ($P<7\times10^{-7}$) distributing along eight chromosomes (Figure 2; Table 2), and containing 166 RefSeq genes [24]. Of these, four loci contained only one gene (*BCL9, NLGN1, DOCK8,* and *KPNA3* in 1q21.1, 3q26.31, 9p24.3, and 13q14.3, respectively), one locus contained three genes (*TRIML1, TRIML2,* and *LOC401164*, on 4q35.2) and six loci contained ≥ seven genes (Table 2). Seven of these loci included a relatively equal number of duplications and deletions, whereas four loci contained a majority (≥75%) of duplications.

The highest CNV burden was seen in a locus on human chromosome 16p11.2 ($\theta = 30.96$; $P < 1\times10^{-20}$), with 27 duplications and 36 deletions identified in ASD subjects from 10 different studies (Figure 3). No CNVs within this region were observed in 4400 controls. This locus spanned 750 kb and contained 31 RefSeq genes, three of which (*SEZ6L2*, MAPK3, and *KCTD13*) have been reported as containing susceptibility genetic variants in ASD individuals [25–27]. Interestingly, while duplications in the 16p11.2 locus tended to be inherited, deletions were overwhelmingly *de novo* in origin (Table 2).

The next top-ranked CNV loci were located in three regions on human chromosome 15q11.2-13.3 (Figure 4). The largest of the three loci (15q11.2-q13.1) spanned 4.95 Mb between chromosomal breakpoints BP2 and BP3 and consisted predominantly of duplications among ASD subjects (31 duplications vs. 2 deletions). Two additional susceptibility loci on chromosome 15 (15q11.2 and 15q13.2-q13.3, within BP1-BP2 and BP4-BP5, respectively) were approximately half the size (~2.4 Mb) and demonstrated equivalent prevalence of duplications and deletions in ASD individuals. A total of 44 RefSeq genes reside within these three regions, 10 of which have already been associated with ASD: *CYFIP1, NIPA1, NIPA2,* and *TUBGCP5* within the 15q11.2 locus; *UBE3A, ATP10A, GABRB3, SNRPN,* and *HERC2* within the 15q11.2-q13.1 locus; and *CHRNA7* within the 15q13.2-q13.3 locus. A closer inspection of the inheritance patterns of CNVs in these three genomic regions revealed that duplications tended to be *de novo* in origin, whereas deletions tended to be inherited (Table 2).

Two multigenic ASD susceptibility CNV loci were identified on human chromosome 22 (Figure 5). The 22q11.21 and 22q13.32-q13.33 loci spanned 2.5 Mb and 1.5 Mb, and overlapped with 51 and 33 genes, respectively. The CNV locus on 22q11.21 was predominantly enriched in duplications in ASD cases (21 duplications vs. 1 deletion), with a sharp increase in the number of duplications within a 20 kb region (chr22:19,345,000-19,365,000) that was implicated as a CNV enriched region in ASD cases (Glessner et al) (Figure 5A). Of the 51 genes within the 22q11.21 locus, two (*TBX1* and *GNBL1*) have already been associated with ASD. The locus on 22q13.32-q13.33 was primarily

Table 1. CNV characteristics.

	Male			Female		
	Duplications	**Deletions**	**sum**	**Duplications**	**Deletions**	**sum**
De novo	62 (1.7%)	109 (3.1%)	171 (4.8%)	27 (3.5%)	40 (5.1%)	67 (8.9%)
Inherited	1046 (29.5%)	1663 (46.9%)	2709 (76.4%)	200 (26.5%)	343 (45.7%)	543 (71.0%)
NR*	188 (5.3%)	479 (13.5%)	667 (18.8%)	32 (4.0%)	122 (15.2%)	154 (20.1%)
Sum	1296 (36.5%)	2251 (63.5)	3547 (100.0%)	259 (34.0%)	505 (66.0%)	764(100.0%)
	NR*			**All**		
	Duplications	**Deletions**	**sum**	**Duplications**	**Deletions**	**sum**
De novo	1 (0.2%)	0 (0.0%)	1 (0.2%)	90 (1.8%)	149 (3.0%)	239 (4.9%)
Inherited	50 (8.1%)	75 (12.2%)	125 (20.3%)	1296 (26.3%)	2081 (42.3%)	3377 (68.6%)
NR*	317 (51.5%)	172 (28.0%)	489 (79.5%)	537 (10.9%)	773 (15.7%)	1310 (26.6%)
Sum	368 (59.8%)	247 (40.2%)	615 (100.0%)	1923 (39.0%)	3003 (61.0%)	4926 (100.0%)

*NR = Not reported.

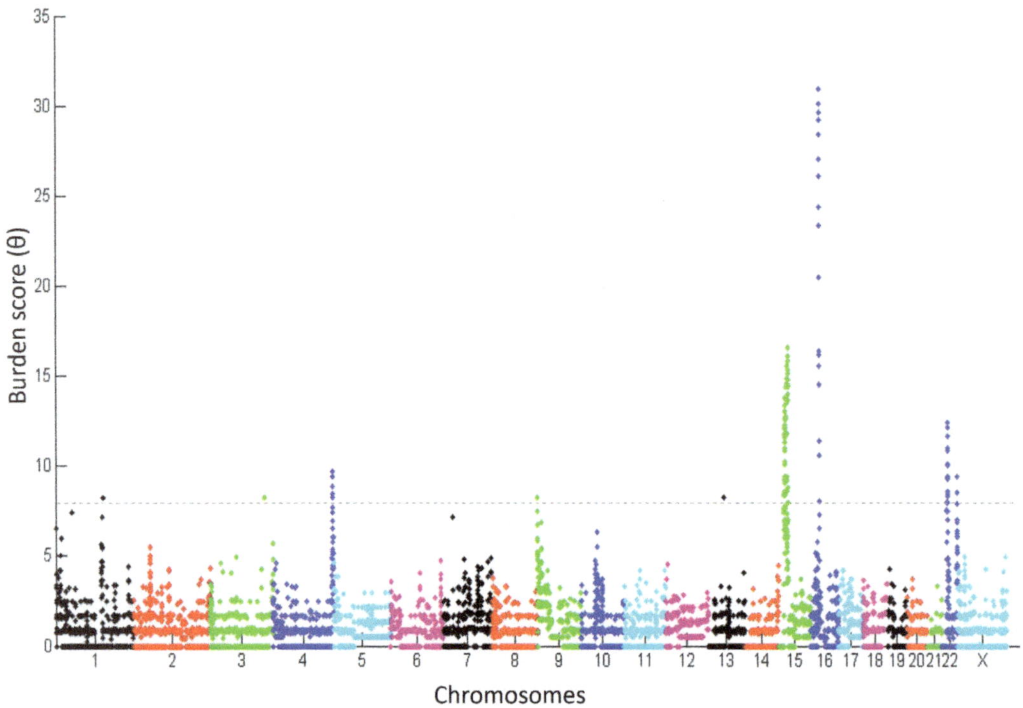

Figure 1. Whole-genome distribution of CNV burden. A Manhattan plot showing CNV burden among ASD subjects in 10 kb regions continuously distributed across the human genome. A dashed horizontal line indicate the burden score of 6.5 (0.995 quantile) that was used a threshold to determine the top ranked ASD susceptibility CNV loci.

enriched in deletions among individuals with ASD (nine deletions vs. three duplications), and did not include CNVs in controls (Figure 5B). No previously characterized ASD-associated genes resided within the boundaries of this genomic locus; however, the ASD-associated gene SHANK3 is directly adjacent to the telomeric end of this region.

Finally, we examined the overlap of these eleven ASD susceptibility loci with CNV data from 4400 control individuals.

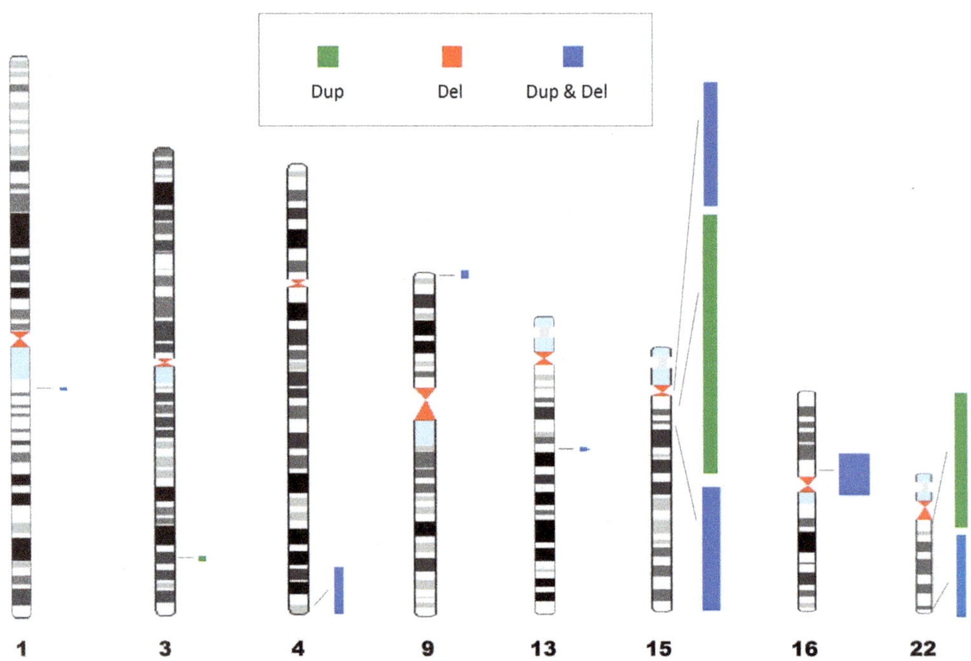

Figure 2. Physical locations of the top ranked 11 ASD susceptibility CNV loci on Human G-banded ideogram. CNV loci length and width are proportional to their genomic size and burden score respectively. Green, red, and blue are for CNV loci containing primarily copy number gains (Duplications), copy number losses (Deletions), or both Duplications and Deletions, respectively.

Table 2. Top ASD susceptibility CNV loci.

Genomic locus					CNVs										
Locus	Start[a] (kb) (hg18)	End[a] (kb) (hg18)	Size (kb)	RefSeq Genes[b]	Duplications[c]			Deletions[c]			Total	# Studies	Θ[d]	P-value[e]	# CNVs among controls[f]
					Inh	DN	NR	Inh	DN	NR					
1q21.1	145555	145575	20	BCL9	3	3	3	-	-	4	13	4	8.15	2.61×10^{-7}	0
3q26.31	174745	174775	30	NLGN1	1	-	114	-	-	-	115	2	8.23	2.61×10^{-7}	31
4q35.2	189295	190195	900	TRIML1 TRIML2 LOC401164	1	-	8	1	1	5	15	5	9.38	2.16×10^{-8}	4
9p24.3	235	375	140	DOCK8	4	1	4	-	2	1	12	5	8.23	2.61×10^{-7}	0
13q14.3	49265	49275	10	KPNA3	4	-	-	6	-	-	10	1	8.22	2.61×10^{-7}	0
15q11.2	18525	20845	2320	9	5	6	1	6	1	-	19	7	13.38	4.31×10^{-13}	384
15q11.2-q13.1	21185	26135	4950	28	3	7	20	1	-	1	32	8	13.01	4.31×10^{-13}	159
15q13.2-q13.3	27985	30425	2440	7	4	8	2	8	1	3	26	9	16.60	1.11×10^{-16}	272
16p11.2	29495	30245	750	31	13	5	9	2	21	13	62	10	30.96	$< 1\times10^{-20}$	0
22q11.21	17265	19805	2540	51	8	3	10	-	1	-	22	7	12.15	7.24×10^{-12}	46
22q13.32-q13.33	47925	49435	1510	33	3	-	-	4	3	2	12	6	9.41	2.16×10^{-8}	0
Sum			15610	167	59	32	173	45	31	34	374				
					264			110							

[a]The boundaries of the ASD susceptibility CNV loci were determined as the midpoint of the 10kb region with a $\theta>6.5$.
[b]RefSeq genes including both protein-coding and non-coding RNA genes, and excluding Pseudogenes. For loci with >3 genes, only the number of RefSeq genes is given.
[c]CNV type. Inh = Inherited, DN = De-Novo, NR = Not Reported.
[d]CNV burden score.
[e]P-value calculated based on a Poisson distribution of CNV burden scores.
[f]4400 individuals with no ASD diagnosis from three large genome-wide studies.

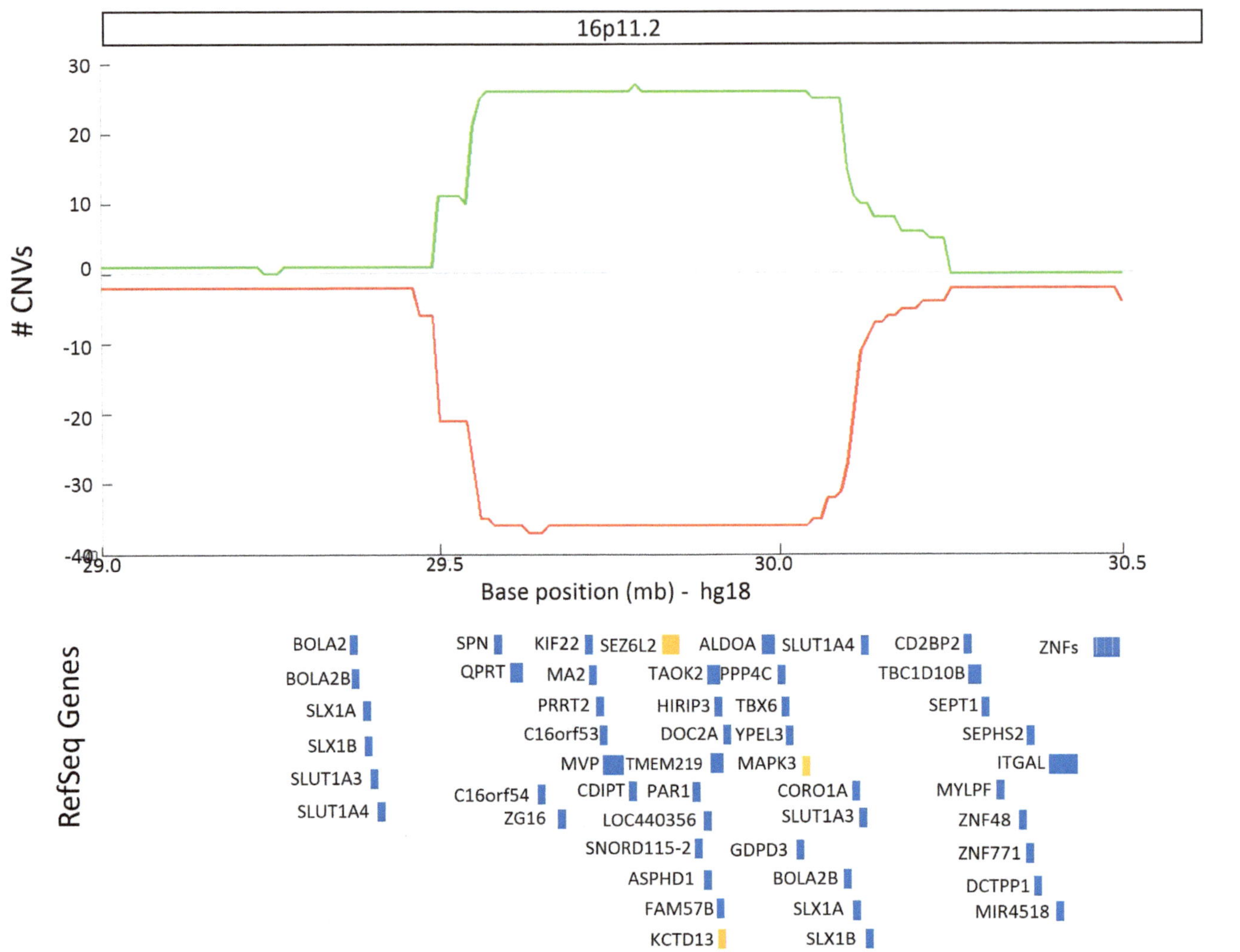

Figure 3. ASD susceptibility CNV locus on human chromosome 16. The number of individuals with duplications (green) and deletions (red) are plotted along human chromosome 16p11.2. RefSeq Genes overlapping with this region are depicted in blue rectangles. Genes that have been associated with ASD according to AutDB [18] are colored in orange.

Consequently, three of these loci (3q26.31, 4q35.2; Figure S3, and 15q11.2; Figure 4) showed a complete overlap with CNVs among controls, suggesting potential false positives, or polymorphic loci that confer lesser risk than initially estimated. In addition, three other ASD susceptibility loci (15q11.2-q13.1, 15q13.2-q13.3, and 22q11.21; Figures 4 and 5) demonstrated partial overlap with CNVs among controls, allowing us to refine the boundaries of this susceptibility loci.

Discussion

The major goals of this study were to: (A) prioritize ASD susceptibility loci based on their CNV burden among ASD individuals, and (B) determine the genomic and genetic characteristics of these loci. The CNV dataset used for our analysis was curated from published scientific reports and primarily included CNVs that have been exclusively seen among individual with ASD in each of these studies. Yet, we compared these data with a control CNV dataset derived from three genome-wide CNV reports [19–21] to further refine the boundaries of our susceptibility loci and identify potential false-positives or lower-risk susceptibility loci. In addition, we contrasted our results with those of two other similar studies [28,29], and a comprehensive review of the scientific literature describing ASD-associated CNVs [30]. We found that five (45.45%) of the susceptibility loci reported in this study were also identified in all of these other reports, and two other loci (18.2%) overlapped with one large population study of CNVs in human genetic disease [28] (Table S2). These overlapping findings support the validity of our analysis, especially given the relatively low frequency and widespread genomic distribution of ASD susceptibility CNV loci, as well as the differences in population makeup in the other studies.

The highest CNV burden in our study was seen in the 16p11.2 locus. This genomic locus have long been known as, a genetic risk factor for ASD [31], as well as other disorders including schizophrenia [32], developmental delay and cognitive impairment [33], major depressive disorder [34], and obesity [35]. Three genes within this locus (*SEZ6L2*, *MAPK3*, and *KCTD13*) have been independently identified as ASD genetic risk factors [25–27], but the genetic mechanisms by which deletions or duplications within the 16p11.2 locus contribute to ASD susceptibility remain unknown. Initial hints for potential functional mechanism might be found in a recent study in zebrafish [27] demonstrated that over-expression of the ASD-associated gene, *KCTD13*, led to

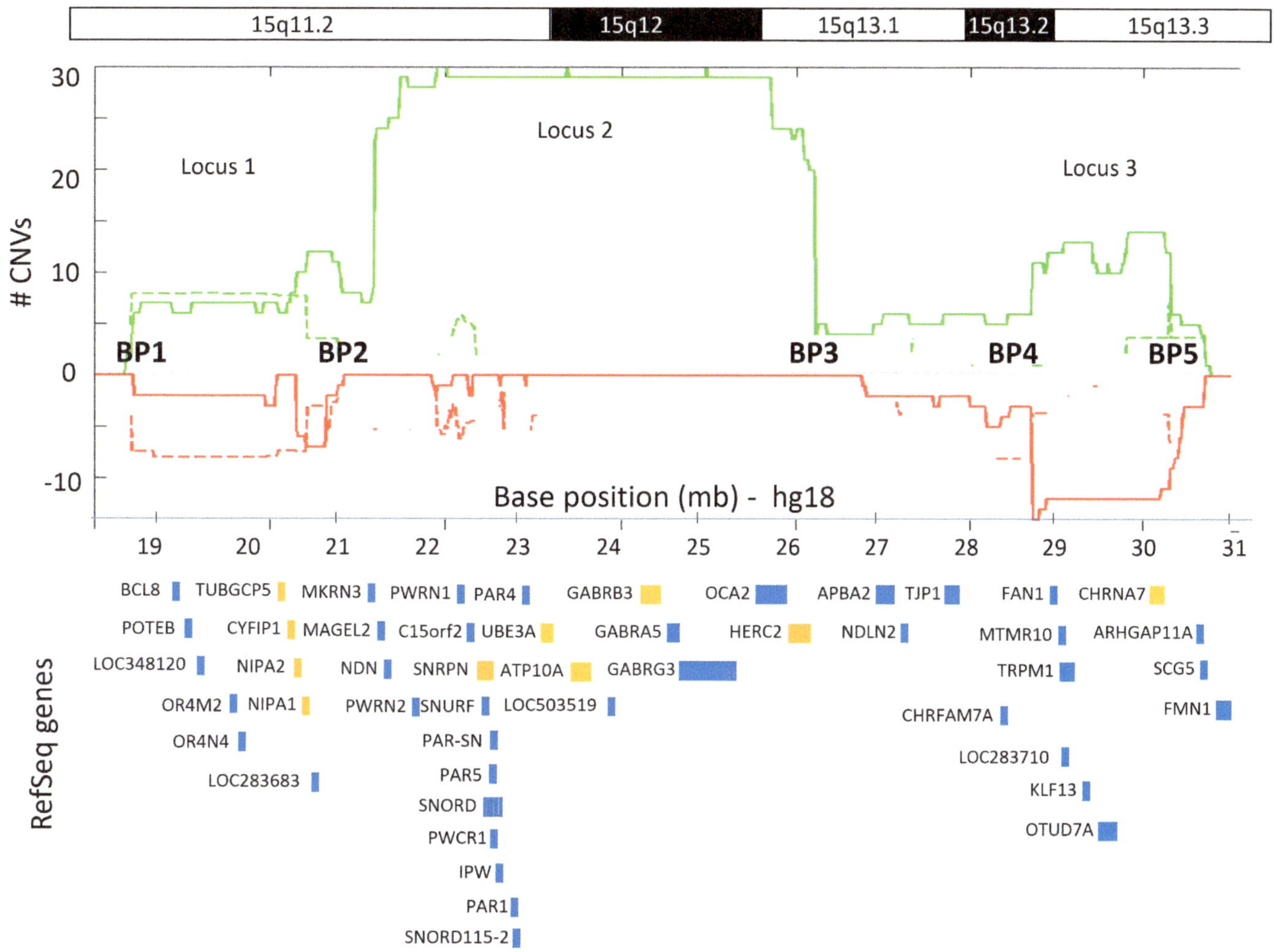

Figure 4. ASD susceptibility CNV loci on human chromosome 15 The number of individuals with duplications (green) and deletions (red) are plotted for both ASD cases (continuous lines) and controls (broken lines, on a $\log_2$ scale) along human chromosome 15q11.2-13.3. RefSeq genes overlapping with these regions are depicted in blue rectangles. Genes that have been associated with ASD according to AutDB [18] are colored in orange. The variation in CNV burden along the 15q11.2 – 13.3 region, suggests three distinct ASD susceptibility loci: Locus 1 within breakpoints (BP) 1-2, Locus 2 within BPs 2-3, and Locus 3 within BPs 4-5.

decreased proliferation of neural progenitor cells and reduced head size, mirroring the microcephaly phenotype commonly seen in cases with 16p11.2 duplication. Alternatively, suppression of *KCTD13* led to increased neural progenitor cell proliferation and increased head size, mirroring the macrocephaly phenotype observed in many cases with 16p11.2 deletion. Likewise, another gene within the 16p11.2 locus, *TAOK2*, was recently shown to influence the formation of basal dendrites in the developing cortex [36]. Therefore, it is likely that these two genes, as well as other genes within the 16p11.2 susceptibility loci, act in concert and contribute to ASD susceptibility.

Our analysis implicated three distinct ASD loci within the 15q11-q13 genomic locus. Duplications within the 15q11.2-q13.1 region, located between chromosomal breakpoints (BPs) BP2-BP3, have long been strongly implicated in ASD pathogenesis [37], whereas deletions of this region are a primary cause of Angelman and Prader-Willi syndromes [38]. The 15q11.2-q13.1 region is flanked by two genomic loci, 15q11.2 and 15q13.2-q13.3 that have demonstrated association with not only autism but also other neurodevelopmental and neuropsychiatric disorders. CNVs within the 15q11.2 genomic loci (BP1-BP2) have been shown to confer risk to epilepsy and developmental delay [39,40], while deletions of the 15q13 region (BP4-BP5) were associated with increased risk for intellectual disability and epilepsy [41]. Notably, the complete overlap of CNVs among controls with the 15q11.2 locus, as seen in other two others CNV loci in this study (3q26.31, and 4q35.2) support the two-hit premise of ASD etiology [42].

Four monogenic genomic loci were identified in our analysis that merit further investigation. The 9p24.3 locus, which contains the *DOCK8* gene overlaps with linkage regions identified in large autism extended pedigrees [43,44], and disruption of *DOCK8* has previously been implicated in two cases of intellectual disability and developmental delay [45]. Therefore, while there is no direct evidence implicating *DOCK8* in ASD, these previous results and our findings argue for a potentially critical role for this gene in ASD susceptibility. The *BCL9* gene resides within a ~1.5–Mb genomic region of the 1q21.1 locus in which both deletions and duplications can result in syndromes associated with numerous phenotypes, including autism [46]. Furthermore, common variants in the *BCL9* gene are associated with schizophrenia, bipolar disorder, and major depressive disorder [47]. *BCL9* is a component of the canonical Wnt signaling pathway, which has

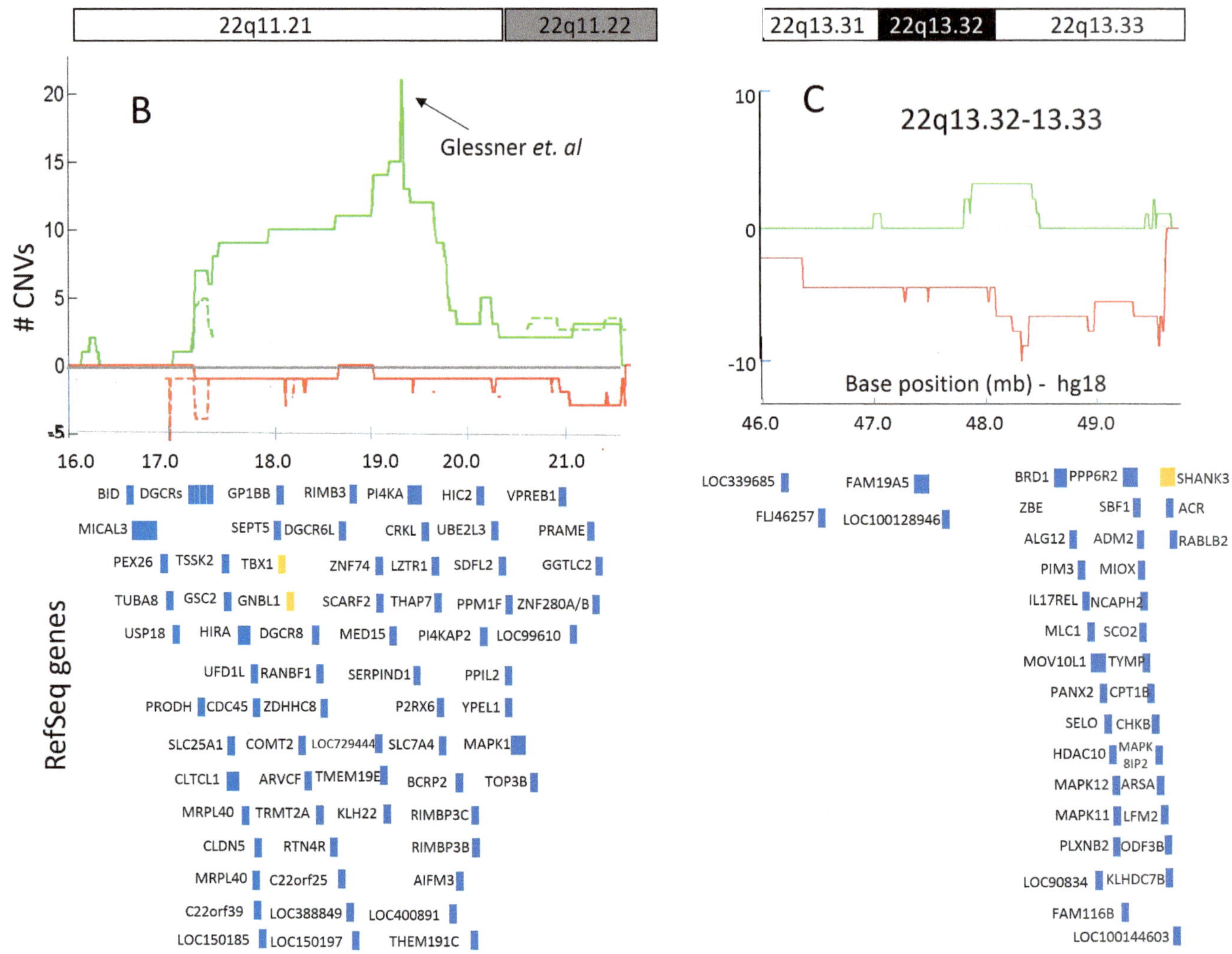

Figure 5. ASD susceptibility CNV loci on human chromosome 22. The number of individuals with duplications (green) and deletions (red) are plotted for both ASD cases (continuous lines) and controls (broken lines, on a $\log_2$ scale) along human chromosome 22q11.21-13.33. RefSeq genes overlapping with these regions are depicted in blue rectangles. Genes that have been associated with ASD according to AutDB [18] are colored in orange. (**A**) The CNV locus on 22q11.21 contains primarily copy number gains (duplications). A black arrow indicates the peak in duplications count due to the CNV data from Glessner *et. al*. (**B**) The CNV locus on 22q13.32-13.33 contains primarily copy number losses (deletions).

been proposed to be affected in ASD [48]. While there is no direct genetic evidence demonstrating that the *BCL9* gene confers genetic risk to ASD susceptibility, our analysis in combination with previous findings strongly implies a potential pathogenic role for this gene. Duplications of the 3q26.31 locus within the *NLGN1* gene were also seen among control individuals, thus questioning the association of this locus with ASD susceptibility [10]. Yet, the functional relevance of *NLGN1* to ASD is supported by a recent report identified a duplication of the *NLGN1* gene in an autistic patient with mild intellectual disability [49], as well as by functional studies demonstrating a role for this gene in neurite outgrowth via interactions with the ASD-associated gene NRXN1 [50]. Finally, the *KPNA3* gene at the 13q14.3 locus, has also been implicated as a potential schizophrenia susceptibility gene [51]. Further investigation will be required to ascertain the relevance of *KPNA3* to these two related psychiatric disorders.

The incorporation of CNV data from both large-scale whole-genome studies and smaller-scale case studies as a framework for ASD-related CNV loci characterization is the major strength of our study. However, this approach has some limitations. First, some of the studies in our dataset included genome-wide CNV data, while other studies focused on the identification of CNVs within specific genomic loci or with a specific mechanism of CNV inheritance. To account for the potential bias arising from this variation, we assigned different weights to studies based on the CNVs/Loci ratio reported in them, and incorporated this weight in the calculation of the CNV burden score (See methods). Another potential source for variation in our data is the differing CNV detection technologies used by the different studies. Accordingly, one might suspect that the higher burden scores observed among larger genomic regions (>500 kb) would be due to the greater likelihood of CNVs within these regions to be detected by all CNV detection methods, whereas smaller CNVs would not be detected by many of the older lower-resolution microarray platforms. While we cannot rule out this possibility, Sanders et. al [14], which employed a high-resolution CNV detection method, reported that the burden of CNVs in their study was remarkably similar to previously published results using lower-resolution CNV detection platforms.

The results of this analysis could have broad clinical and scientific implications. For example, one could use these loci as a guideline for the evaluation of chromosomal microarray (CMA) screening, a procedure that is being increasingly used in genetic evaluation of ASD subjects. Alternatively, the detailed characteristics provided for each of the CNV loci highlighted in the study may be used for further, in-depth exploration of the biological mechanism underlying their role in ASD susceptibility. An intriguing premise is whether the different genetic mechanisms trigger ASD susceptibility in subjects containing CNVs at distinct genomic loci, and whether these genetically diverse individuals present different ASD related phenotypes. We anticipate to increase the resolution of these analyses with continued updates to AutDB and additional control CNV data, which will subsequently provide both scientists and clinicians with a valuable resource for genetics research and clinical diagnostic efforts of ASD.

Supporting Information

Figure S1 Distribution of CNV loci reports. Distribution of 'major' and 'minor' reports across CNV loci in AutDB[18], (A) Venn Diagram, (B) cumulative distribution function (cdf). (C) Top 1% reported CNV loci.

Figure S2 Distribution of CNV sizes. (A) Histogram of the log10 (CNV size) indicate that CNV sizes in our data have a lognormal distribution with a mean = 42.8 kb. (B) CDF plots for the sizes of copy number gains (green), and copy number losses (red).

Figure S3 ASD susceptibility CNV loci on human chromosomes 3 & 4. The number of individuals with duplications (green) and deletions (red) are plotted for both ASD cases (continuous lines) and controls (broken lines) along human chromosomes 3q26.31 & 4q35.2. RefSeq genes overlapping with these regions are depicted in blue rectangles. Genes that have been associated with ASD according to AutDB [18] are colored in orange.

Table S1 Scientific reports of CNVs in ASD individuals. Details of the 48 scientific reports and their CNV data used in this study are listed in the table.

Table S2 CNV loci associated with ASD across different studies. A list of CNV loci associated with ASD from four different studies are depicted in the table. CNV loci highlighted in multiple studies are highlighted in bold.

Acknowledgments

We thank Catherine C. Swanwick for her review and valuable editorial comments and the IT team of MindSpec for the development of AutDB.

Author Contributions

Conceived and designed the experiments: IM ECL SB. Performed the experiments: ECL. Analyzed the data: IM. Contributed reagents/materials/analysis tools: SB. Wrote the paper: IM ECL SB.

References

1. Iafrate AJ, Feuk L, Rivera MN, Listewnik ML, Donahoe PK, et al. (2004) Detection of large-scale variation in the human genome. Nat Genet 36: 949–951.
2. Sebat J, Lakshmi B, Troge J, Alexander J, Young J, et al. (2004) Large-scale copy number polymorphism in the human genome. Science 305: 525–528.
3. Lee JA, Lupski JR (2006) Genomic rearrangements and gene copy-number alterations as a cause of nervous system disorders. Neuron 52: 103–121.
4. Stankiewicz P, Lupski JR (2010) Structural variation in the human genome and its role in disease. Annu Rev Med 61: 437–455.
5. Lord C, Cook EH, Leventhal BL, Amaral DG (2000) Autism spectrum disorders. Neuron 28: 355–363.
6. Nazeer A, Ghaziuddin M (2012) Autism spectrum disorders: clinical features and diagnosis. Pediatr Clin North Am 59: 19–25, ix.
7. Muhle R, Trentacoste SV, Rapin I (2004) The genetics of autism. Pediatrics 113: e472–486.
8. Hallmayer J, Cleveland S, Torres A, Phillips J, Cohen B, et al. (2011) Genetic Heritability and Shared Environmental Factors Among Twin Pairs With Autism. Arch Gen Psychiatry.
9. Anney R, Klei L, Pinto D, Regan R, Conroy J, et al. (2010) A genome-wide scan for common alleles affecting risk for autism. Hum Mol Genet 19: 4072–4082.
10. Glessner JT, Wang K, Cai G, Korvatska O, Kim CE, et al. (2009) Autism genome-wide copy number variation reveals ubiquitin and neuronal genes. Nature 459: 569–573.
11. Pinto D, Pagnamenta AT, Klei L, Anney R, Merico D, et al. (2010) Functional impact of global rare copy number variation in autism spectrum disorders. Nature 466: 368–372.
12. Weiss LA (2009) Autism genetics: emerging data from genome-wide copy-number and single nucleotide polymorphism scans. Expert Rev Mol Diagn 9: 795–803.
13. Gilman SR, Iossifov I, Levy D, Ronemus M, Wigler M, et al. (2011) Rare de novo variants associated with autism implicate a large functional network of genes involved in formation and function of synapses. Neuron 70: 898–907.
14. Sanders SJ, Ercan-Sencicek AG, Hus V, Luo R, Murtha MT, et al. (2011) Multiple Recurrent De Novo CNVs, Including Duplications of the 7q11.23 Williams Syndrome Region, Are Strongly Associated with Autism. Neuron 70: 863–885.
15. Swanwick CC, Larsen EC, Banerjee-Basu S (2011) Genetic Heterogeneity of Autism Spectrum Disorders. In: Deutsch SI, Urbano MR, editors. Autism Spectrum Disorders: The Role of Genetics in Diagnosis and Treatment: InTech. pp. 65.
16. Miller DT, Adam MP, Aradhya S, Biesecker LG, Brothman AR, et al. (2010) Consensus statement: chromosomal microarray is a first-tier clinical diagnostic test for individuals with developmental disabilities or congenital anomalies. Am J Hum Genet 86: 749–764.
17. Shen Y, Dies KA, Holm IA, Bridgemohan C, Sobeih MM, et al. (2010) Clinical genetic testing for patients with autism spectrum disorders. Pediatrics 125: e727–735.
18. Basu SN, Kollu R, Banerjee-Basu S (2009) AutDB: a gene reference resource for autism research. Nucleic Acids Res 37: D832–836.
19. Zogopoulos G, Ha KC, Naqib F, Moore S, Kim H, et al. (2007) Germ-line DNA copy number variation frequencies in a large North American population. Hum Genet 122: 345–353.
20. Shaikh TH, Gai X, Perin JC, Glessner JT, Xie H, et al. (2009) High-resolution mapping and analysis of copy number variations in the human genome: a data resource for clinical and research applications. Genome Res 19: 1682–1690.
21. International HapMap C, Altshuler DM, Gibbs RA, Peltonen L, Altshuler DM, et al. (2010) Integrating common and rare genetic variation in diverse human populations. Nature 467: 52–58.
22. Zhang J, Feuk L, Duggan GE, Khaja R, Scherer SW (2006) Development of bioinformatics resources for display and analysis of copy number and other structural variants in the human genome. Cytogenet Genome Res 115: 205–214.
23. Newschaffer CJ, Croen LA, Daniels J, Giarelli E, Grether JK, et al. (2007) The epidemiology of autism spectrum disorders. Annu Rev Public Health 28: 235–258.
24. Pruitt KD, Tatusova T, Brown GR, Maglott DR (2012) NCBI Reference Sequences (RefSeq): current status, new features and genome annotation policy. Nucleic Acids Res 40: D130–135.
25. Kumar RA, Marshall CR, Badner JA, Babatz TD, Mukamel Z, et al. (2009) Association and mutation analyses of 16p11.2 autism candidate genes. PLoS One 4: e4582.
26. Schaaf CP, Sabo A, Sakai Y, Crosby J, Muzny D, et al. (2011) Oligogenic heterozygosity in individuals with high-functioning autism spectrum disorders. Hum Mol Genet 20: 3366–3375.
27. Golzio C, Willer J, Talkowski ME, Oh EC, Taniguchi Y, et al. (2012) KCTD13 is a major driver of mirrored neuroanatomical phenotypes of the 16p11.2 copy number variant. Nature 485: 363–367.
28. Itsara A, Cooper GM, Baker C, Girirajan S, Li J, et al. (2009) Population analysis of large copy number variants and hotspots of human genetic disease. Am J Hum Genet 84: 148–161.
29. Kaminsky EB, Kaul V, Paschall J, Church DM, Bunke B, et al. (2011) An evidence-based approach to establish the functional and clinical significance of copy number variants in intellectual and developmental disabilities. Genet Med 13: 777–784.

30. Marshall CR, Scherer SW (2012) Detection and characterization of copy number variation in autism spectrum disorder. Methods Mol Biol 838: 115–135.
31. Walsh KM, Bracken MB (2011) Copy number variation in the dosage-sensitive 16p11.2 interval accounts for only a small proportion of autism incidence: a systematic review and meta-analysis. Genet Med 13: 377–384.
32. McCarthy SE, Makarov V, Kirov G, Addington AM, McClellan J, et al. (2009) Microduplications of 16p11.2 are associated with schizophrenia. Nat Genet 41: 1223–1227.
33. Shinawi M, Liu P, Kang SH, Shen J, Belmont JW, et al. (2010) Recurrent reciprocal 16p11.2 rearrangements associated with global developmental delay, behavioural problems, dysmorphism, epilepsy, and abnormal head size. J Med Genet 47: 332–341.
34. Degenhardt F, Priebe L, Herms S, Mattheisen M, Muhleisen TW, et al. (2012) Association between copy number variants in 16p11.2 and major depressive disorder in a German case-control sample. Am J Med Genet B Neuropsychiatr Genet 159B: 263–273.
35. Jacquemont S, Reymond A, Zufferey F, Harewood L, Walters RG, et al. (2011) Mirror extreme BMI phenotypes associated with gene dosage at the chromosome 16p11.2 locus. Nature 478: 97–102.
36. de Anda FC, Rosario AL, Durak O, Tran T, Graff J, et al. (2012) Autism spectrum disorder susceptibility gene TAOK2 affects basal dendrite formation in the neocortex. Nat Neurosci 15: 1022–1031.
37. Hogart A, Wu D, LaSalle JM, Schanen NC (2010) The comorbidity of autism with the genomic disorders of chromosome 15q11.2-q13. Neurobiol Dis 38: 181–191.
38. Buiting K (2010) Prader-Willi syndrome and Angelman syndrome. Am J Med Genet C Semin Med Genet 154C: 365–376.
39. Burnside RD, Pasion R, Mikhail FM, Carroll AJ, Robin NH, et al. (2011) Microdeletion/microduplication of proximal 15q11.2 between BP1 and BP2: a susceptibility region for neurological dysfunction including developmental and language delay. Hum Genet 130: 517–528.
40. de Kovel CG, Trucks H, Helbig I, Mefford HC, Baker C, et al. (2010) Recurrent microdeletions at 15q11.2 and 16p13.11 predispose to idiopathic generalized epilepsies. Brain 133: 23–32.
41. van Bon BWM, Mefford HC, de Vries BBA (1993) 15q13.3 Microdeletion.
42. Gau SS, Liao HM, Hong CC, Chien WH, Chen CH (2012) Identification of two inherited copy number variants in a male with autism supports two-hit and compound heterozygosity models of autism. Am J Med Genet B Neuropsychiatr Genet 159B: 710–717.
43. Allen-Brady K, Miller J, Matsunami N, Stevens J, Block H, et al. (2009) A high-density SNP genome-wide linkage scan in a large autism extended pedigree. Mol Psychiatry 14: 590–600.
44. Coon H, Villalobos ME, Robison RJ, Camp NJ, Cannon DS, et al. (2010) Genome-wide linkage using the Social Responsiveness Scale in Utah autism pedigrees. Mol Autism 1: 8.
45. Griggs BL, Ladd S, Saul RA, DuPont BR, Srivastava AK (2008) Dedicator of cytokinesis 8 is disrupted in two patients with mental retardation and developmental disabilities. Genomics 91: 195–202.
46. Brunetti-Pierri N, Berg JS, Scaglia F, Belmont J, Bacino CA, et al. (2008) Recurrent reciprocal 1q21.1 deletions and duplications associated with microcephaly or macrocephaly and developmental and behavioral abnormalities. Nat Genet 40: 1466–1471.
47. Li J, Zhou G, Ji W, Feng G, Zhao Q, et al. (2011) Common variants in the BCL9 gene conferring risk of schizophrenia. Arch Gen Psychiatry 68: 232–240.
48. Kalkman HO (2012) A review of the evidence for the canonical Wnt pathway in autism spectrum disorders. Mol Autism 3: 10.
49. Leblond CS, Heinrich J, Delorme R, Proepper C, Betancur C, et al. (2012) Genetic and functional analyses of SHANK2 mutations suggest a multiple hit model of autism spectrum disorders. PLoS Genet 8: e1002521.
50. Gjorlund MD, Nielsen J, Pankratova S, Li S, Korshunova I, et al. (2012) Neuroligin-1 induces neurite outgrowth through interaction with neurexin-1beta and activation of fibroblast growth factor receptor-1. FASEB J 26: 4174–4186.
51. Morris CP, Baune BT, Domschke K, Arolt V, Swagell CD, et al. (2012) KPNA3 variation is associated with schizophrenia, major depression, opiate dependence and alcohol dependence. Dis Markers 33: 163–170.

10

Quantitative Detection of *Borrelia burgdorferi* sensu lato in Erythema Migrans Skin Lesions Using Internally Controlled Duplex Real Time PCR

Maria O'Rourke[1*9], Andreas Traweger[1,4 9], Lara Lusa[2], Dasa Stupica[3], Vera Maraspin[3], P. Noel Barrett[1], Franc Strle[3], Ian Livey[1]

1 Vaccine R&D, Baxter Bioscience, Orth/Donau, Austria, **2** Faculty of Medicine, Institute for Biostatistics and Medical Informatics, Ljubljana, Slovenia, **3** Department of Infectious Diseases, University Medical Centre Ljubljana, Ljubljana, Slovenia, **4** Paracelsus Medical University, Spinal Cord Injury and Tissue Regeneration Center Salzburg, Institute of Tendon and Bone Regeneration, Salzburg, Austria

Abstract

B. burgdorferi sensu stricto, *B. afzelii*, *B. garinii* and *B. bavariensis* are the principal species which account for Lyme borreliosis (LB) globally. We have developed an internally controlled duplex quantitative real time PCR assay targeting the *Borrelia* 16S rRNA and the human RNAseP genes. This assay is well-suited for laboratory confirmation of suspected cases of LB and will be used to assess the efficacy of a vaccine against LB in clinical trials. The assay is highly specific, successfully detecting DNA extracted from 83 diverse *B. burgdorferi* sensu lato strains representing all major species causing LB, while 21 unrelated microbial species and human genomic DNA tested negative. The assay was highly reproducible and sensitive, with a lower limit of detection of 6 copies per PCR reaction. Together with culture, the assay was used to evaluate paired 3 mm skin biopsy samples taken from 121 patients presenting with solitary erythema migrans (EM) lesion. PCR testing identified more positive biopsy samples than culture (77.7% PCR positive versus 55.1% culture positive) and correctly identified all specimens scored as culture positive. OspA-based typing identified the majority of isolates as *B. afzelii* (96.8%) and the bacterial load was significantly higher in culture positive biopsies than in culture negative biopsies ($P<0.001$). The quantitative data also enabled relationships between *Borrelia* burden and patient symptoms to be evaluated. The bacterial load was significantly higher among patients with systemic symptoms than without ($P=0.02$) and was significantly higher for biopsies retrieved from patients with EM lesions with central clearing ($P<0.001$). 16S copy numbers were moderately lower in samples from patients reporting a history of LB ($P=0.10$). This is the first quantitative PCR study of human skin biopsies predominantly infected with *B. afzelii* and the first study to demonstrate a clear relationship between clinical symptoms in *B. afzelii*-infected patients and *Borrelia* burden.

Editor: J. Stephen Dumler, The Johns Hopkins University School of Medicine, United States of America

Funding: This study was funded by Baxter and supported in part by the Slovenian Research Agency (Grant Numbers P3-0296 and J3-3636). Baxter employees were involved in study design, data collection and analysis, the decision to publish, and preparation of the manuscript. The Slovenian Research Agency had no role in study design, data collection and analysis, decision to publish, or preparation of the manuscript.

Competing Interests: A. Traweger is a former employee of Baxter. M. O'Rourke, P.N. Barrett and I. Livey are salaried employees of Baxter, holding stocks/shares. M. O'Rourke and I. Livey hold a patent for a vaccine against LB. L. Lusa, D. Stupica, V. Maraspin, and F. Strle report no conflicts of interest.

* E-mail: maria_o'rourke@baxter.com

9 These authors contributed equally to this work.

Introduction

Borrelia burgdorferi sensu lato (s.l.) is transmitted through the bite of infected ticks in the *Ixodes* family and is the causative agent of Lyme borreliosis (LB), the most common vector borne disease in Europe and North America. *Borrelia burgdorferi* s.l. currently comprises a clade of 17 named species, the most recent addition being *B. finlandensis sp. nov.* [1]. *B. burgdorferi* sensu stricto (s.s) is the sole species known to cause human infection in the United States, whereas in Europe *B. afzelii*, *B. garinii* (including recently designated *B. bavariensis sp. nov* [2]) and *B. burgdorferi* s.s are responsible for most cases of LB [3]. However, *B. spielmanii* [4,5], *B. valaisiana* [6,7], *B. bissettii* [8,9] and *B. lusitaniae* [10,11] have been detected in samples of human skin and cerebrospinal fluid, suggesting that these species can also give rise to LB.

Although it is not routine practice to request laboratory testing to confirm LB presenting as erythema migrans (EM) [3], laboratory testing is highly valuable when assessing vaccine efficacy in clinical trials [12]. Moreover, demonstration of *Borrelia* in tissues is also valuable for confirming a diagnosis based on less clear manifestations of LB, particularly in regions endemic for the disease, where positive serology may be due to past exposure to *B. burgdorferi* s.l. Direct detection of *Borrelia* in clinical specimens is typically accomplished by culture or PCR [13]. Culture is slow and labor intensive, while PCR has the advantage of sensitivity, simplicity and speed. In addition, real time PCR allows the quantification of organisms present in infected samples, such as animal tissues [14–16], ticks [17], human skin [18–20] and CSF [21]. In animal studies, it has been shown that the number of *B. burgdorferi* s.s. in tissues is closely correlated with the degree of joint swelling and inflammation [14,16,22]. *B. burgdorferi* s.s has been quantified in human skin biopsies [19], however to our knowledge, quantification of *B. afzelii* or *B. garinii* in European human skin biopsies has not yet been reported. We have developed an internally controlled, duplex quantitative real time PCR assay targeting the *Borrelia* single copy 16S rRNA gene and the single copy human RNAseP gene. RNAseP has been extensively used in molecular diagnostic assays to monitor the efficiency of DNA extraction [23,24]. In addition, by co-amplification of the human RNAseP gene, the bacterial load can be expressed in relation to human genomic DNA, thereby compensating for variations arising from the DNA extraction procedure and/or the size and quality of the biopsy (e.g. variations in tissue composition due to differences in the site from which the biopsy was obtained). We used this assay together with culture, to evaluate skin biopsies from 121 patients presenting with a solitary EM at the University Medical Centre Ljubljana in Slovenia. These data enabled us to compare the ability of both methods to confirm a clinical diagnosis of EM as well as providing information on quantification of *B. burgdorferi* s.l. in skin biopsies. The results of this proof-of-concept study also serve to demonstrate the benefit of this PCR assay to determine the efficacy of a LB vaccine in clinical trials.

Materials and Methods

Borrelia 16S rRNA Real Time PCR

A TaqMan® assay amplifying and targeting a conserved 139-bp fragment of the gene encoding the *Borrelia* 16S rRNA was designed encompassing the following primers: p16Swt-fwd (5′-GGATA-TAGTTAGAGATAATTATTCCCCGTTTG-3′) and p16Swt-rev (5′-CATTACATGCTGGTAACAGATAACAAGG-3). The corresponding probe pro16Swt (5′-ACAGGTGCTGCATGGT–3′) was labeled at the 5′-end with the fluorescent reporter 6-carboxy flourescein (6FAM) and at the 3′-end with a Minor-Groove-Binder Non-Fluorescent Quencher (MGBNFQ). The real-time PCR reactions (20 µl final volume) contained 900 nM of each primer, 200 nM probe, 10 µl 2x environmental master mix 2.0 (Applied Biosystems), and the respective indicated amount of template DNA was added in a volume of 5 µl. For assay development, serial dilutions of genomic DNA extracted from the *Borrelia burgdorferi* s.s. strain ZS7 or a synthetic DNA oligonucleotide harboring the 16S target region (Sigma Proligo; 5′-AGGATATAGTTAGAGATAATTA TTCCCCGTTTGGGGTCTATATACAGGTGCTG-CATGGTTGTACCCTTGTTATCTGTTACCAGCATG-TAATGG-3′) were used. Amplification was carried out on an Applied Biosystems 7900Ht Fast Real-Time PCR System using a cycling protocol of 50°C for 2 minutes, 95°C for 10 minutes followed by 40 cycles of 95°C for 15 seconds and 60°C for 60 seconds. Fluorescent signals were recorded using SDS Software v2.3 (Applied Biosystems) and the automatic baseline option was chosen to have the software calculate the baseline for the detectors, whereas the Cq threshold was set manually.

In order to monitor the recovery of human genomic DNA from clinical skin biopsy samples, the 16S rRNA assay was duplexed with a real-time PCR assay targeting the human single-copy gene encoding the RNA moiety of the human RNAseP enzyme. The RNAseP target region was co-amplified and detected by adding 1 µl of 20x TaqMan® RNAseP control reagent (labeled at the 5′-end with VIC; Applied Biosystems) to the 16S rRNA real-time PCR assay and amplification was carried out as described above. The respective standard curves generated for the singleplex 16S qPCR and duplex 16S/RNAseP assays were used to determine PCR efficiency, slope, R^2 and Y intercept and this data was used to compare the 16S singleplex and duplex assays. For absolute quantification of the target gene/s in clinical test samples, serial dilutions of a template combining human genomic DNA isolated from blood (Roche) and the synthetic 16S oligonucleotide were used.

All skin biopsy specimens, including the negative controls, were tested in triplicate and every PCR run included a minimum of 3 non template controls (NTC), an extraction control (EXC) and a positive control consisting of *Borrelia*-negative human genomic DNA spiked with 250 genome equivalents of ZS7. For a run to be valid, the positive control had to be 16S/RNAseP positive and NTCs and EXC had to test negative for the presence of 16S. Real-time PCR reactions yielding a Cq value of less than 40 were scored as positive. The threshold for Cq determination was set in the early exponential phase of the amplification curve, using a similar setting for all samples. All samples were tested in triplicate and specimens for which only one 16S rRNA PCR reaction was positive were repeated. The sample was considered *Borrelia*-positive if at least one of the repeat 16S PCR reactions was positive. In addition, at least two out of three RNAseP reactions had to yield a Cq value below 32 and every sample had to test negative for PCR inhibition (SPUD assay; see below).

SPUD PCR-Inhibition Assay

A TaqMan®-based SPUD assay, modified from that reported previously [25], was performed in parallel to control for the presence of PCR inhibitors (or enhancers) in nucleic acids extracted from the skin biopsy samples. The assay, originally developed as a SYBR Green-based assay, was adapted using a TaqMan® probe for detection (SPUD-SPTM2; 5′-6FAM-CAT-AGC TTGTGCACACTCA-MGBNFQ-3′) and the primers SPUD-SPTF (5′-ACTTGGCT TTAATGGACCTCCAATT-3′) and SPUD-SPTR 5′-CCGTTTTATGTCTTAC GTGGTGTTC-3′ for amplification. Test samples were spiked with 2,000 copies of the synthetic SPUD oligonucelotide (5′AACTTGGCTTTAATGGACCTCCAATTT TGAGTGTG-CACAAGCTATGGAACACCACGTAAGACATAAAACGGC-CACATATGGTGCCATGTAAGGATGAATGT-3′; MWG Eurofins) and the results of all test samples were compared to a reference control sample. A difference in Cq of +/−0.5 was considered acceptable evidence for the lack of PCR inhibition/ enhancement.

Nucleic Acid Isolation

Nucleic acids were extracted from skin biopsy specimens and control skin portions using a QIAamp DNA Mini kit (Qiagen) and eluted in 80 µl of AE buffer. Extraction controls were set up in parallel.

DNA from *Borrelia* strains and unrelated microorganisms (see below), was extracted from stock cultures which were thawed and vortexed for 15 seconds. Subsequently, 30 µl of 0.1% Tween 20 in 1x Tris EDTA (pH 7.4) was added to a 20 µl culture aliquot and the samples were heat-denatured for 15 minutes at 95°C. After centrifugation for 5 min at 14,000 rpm and DNA quantification, the indicated amount of total DNA was added in 5 µl for subsequent real time PCR analysis. DNA quantification was performed using a Trinean DropSense96 UV/VIS droplet reader (Trinean nv, Belgium).

Analytical Specificity

Analytical specificity was determined using 1. A panel of 83 Borrelia burgdorferi sensu lato strains comprising B. burgdorferi s.s., B. afzelii, B. garinii, B. bavariensis sp.nov., B. garinii, B. andersonii, B. bissettii, B. finlandensis sp. nov, B. japonica, B.

lusitaniae, B. spielmanii and B. valaisiana. 25 ng of total DNA was used as template for qPCR testing of B.burgdorferi s.l. strains. 2. DNA isolated from 21 unrelated microbial species including Bacillus subtilis subspecies spizizenii, Bacteroides vulgatus, Clostridium sporogenes, Escherichia coli, Pseudomonas aeruginosa, Streptococcus faecalis, Staphylococcus aureus, Staphylococcus epidermidis, Micrococcus luteus, Corynebacterium jeikeium, Bacillus cereus, Brevundimonas vesicularis, Ralstonia pickettii, Sphingobacterium multivorum, Staphylococcus gallinarum, Streptococcus equi zooepidemicus, Corynebacterium renale, Paenibacillus polymyxa, Lactobacillus paracasei spp. Paracasei, Klebsiella oxytoca, Bacillus licheniformis. The amount of template used per reaction was greater than 1.3 µg, thus excluding false negative results. 3. DNA isolated from human skin from 6 healthy control subjects was evaluated. All samples were tested in duplicate.

Determination of Lower Limit of Detection (95% LOD)

Quantitative sensitivity of the 16S/RNAseP assay was determined using limiting dilutions of genomic DNA extracted from the *B. burgdorferi* s.s. strain ZS7 spiked into a constant amount of human genomic DNA isolated from "*Borrelia*-negative" skin (background matrix). For each dilution (10, 8, 6, 4, 2, 1, 0.1 copies) 12 replicates were analyzed and in total 3 runs were performed on 3 consecutive days (i.e. n = 36 for each dilution). Based on these results the positive cut-off point (LOD) was determined, which is defined as the number of target sequences per volume of sample which can be detected in 95% of test runs (95% detection limit).

Culture

Culture medium for *B. burgdorferi* s.l. (a modified BSK medium termed BSK-B) was prepared using the following components; 900 ml distilled water, 100 ml CMRL 1066 (10x), 5 g neopeptone, 6 g HEPES, 0.54 g tri- Sodium citrate-Dihydrate, 3 g D(+)-Glucose-Monohydrate, 0.8 g Sodium pyruvate, 0.4 g N-acetyl-D-glucosamine, 0.5 g Yeast extract, 2.2 g Sodium hydrogen carbonate, 200 ml Gelatin solution (3.5%), 81.7 ml Bovine Serum Albumin, (30%), 80 ml rabbit serum (heat inactivated), pH adjusted with 7.6 ml 1 N Sodium hydroxide and final volume adjusted to 1350 ml. Medium was sterilized by filtration and a sterility control was performed for each medium lot. The quality control of each batch of medium was performed to ensure that growth of an inoculum of 10 cells of *B. afzelii* (ACA1), *B. garinii* (VSBM) and *B. burgdorferi* s.s. (ZS7) strains could be supported. Medium was shipped to the clinical site (University Medical Centre Ljubljana) in tubes with 6 ml volumes. Un-used medium was also returned to the test laboratory and quality control was performed to ensure that transport to and from the clinical site did not impair the ability of the medium to support growth of small *Borrelia* inocula. Upon receipt at the test laboratory (Baxter), a further 4 ml modified BSK-B medium was added and the culture tubes were capped tightly and incubated at 33°C ±2°C and 5% ±1% CO_2 for up to 6 weeks. Samples were inspected by dark field microscopy for potential growth of *B. burgdorferi* s.l. If no spirochetes were detected by dark field microscopy, in 40 viewing fields at a magnification of 400x, after 6 weeks of cultivation, the samples were subcultured (5 ml fresh medium was added to 5 ml of culture) for a further 6 weeks. Samples were deemed culture negative if no spirochetes were detected on sub-culture, thus the maximum time in culture was 12 weeks. Stocks were made from positive cultures. Blood was also cultured from the 121 patients as described previously [26].

Typing of *Borrelia* Isolates

Isolates were typed according to their *ospA* sequences as outlined previously [27].

Patients and Controls

This study included 121 untreated patients (aged 16 to 84) from the central part of Slovenia with clinically diagnosed LB, who presented at University Medical Centre Ljubljana between June and December, 2010. All patients had a typical single EM lesion defined according to recently published European criteria [28]. A clinical history was obtained from each patient and relevant clinical details recorded including patient age, patient sex, size of EM lesion, time from appearance of EM to biopsy, date of tick bite, location of EM, presence of central clearing, presence/absence of systemic symptoms (fatigue, headache, myalgia, arthralgia, malaise, fever, dizziness, rigors and nausea), presence/absence of local symptoms (itching, burning and/or pain), history of previous LB and incidence and details of other underlying illness and chronic illness requiring treatment (such as arterial hypertension, hyperlipidemia, osteoporosis, diabetes mellitus, thyroid disease, cardiac rhythm abnormality, psychiatric illness, ischemic heart disease, osteoarthritis, asthma, malignancy or Crohns disease).

Skin taken from control subjects without active LB was tested in order to assess the specificity of the culture and PCR methods. Two pairs of skin biopsies from healthy volunteers were tested blinded by PCR and culture. Skin taken from patients who had undergone cosmetic surgery was also tested by PCR (in total 50 specimens taken from 4 patients).

Skin Biopsy

Local disinfectant (70% ethanol) and anaesthetic (1% xylocaine, injected) were applied to skin before biopsy. Paired punch skin biopsies (3 mm diameter) were taken from the leading edge of the EM lesion from each patient, one of which was transferred directly to culture tubes containing 6 ml modified BSK-B medium, which were held at room temperature for up to one week prior to shipment to the test laboratory (Baxter). The second biopsy was placed in 70% ethanol for PCR and was stored at +2 to +8°C for up to one week, prior to shipment. Biopsy samples for culture and PCR were transported at weekly intervals with temperature monitoring during shipment, samples for culture were transported at room temperature and samples for PCR transported at +2° to +8°C. Patients, from whom a first biopsy tested culture positive, were invited to attend for re-biopsy 2–3 months after antibiotic treatment, to verify the efficacy of the treatment. Re-biopsies were obtained from 93.8% (61/65) of the patients and these samples also served as post-treatment controls.

Ethics Statement

The study was approved by the National Medical Ethics Committee of the Republic of Slovenia (No 127/06/10) and conducted according to the principles expressed in the Declaration of Helsinki. Signed written informed consent was obtained from all study participants.

Statistical Analysis of Clinical Data

Data were summarized as means with standard deviation (SD) or medians with range or interquartile range (IQR) for numerical variables; categorical variables were summarized as frequencies and percentages (%). The number of *Borrelia* targets observed for the different groups was compared using the Mann-Whitney test. Chi-squared test with Yates' correction was used to compare the proportion of positive biopsies in different groups. A logistic regression model was used to examine the association between PCR biopsy positivity and time from EM to biopsy using restricted cubic splines (RCS) [29] to flexibly model the association, the estimated association was displayed graphically. A similar model was fitted using culture positivity as outcome. Data were analyzed using R statistical language [30].

Results

Duplex Quantitative Real Time PCR using Human RNAseP as an Internal Positive Control

A quantitative real-time PCR assay was developed to detect a highly conserved region of the *Borrelia* 16S rRNA gene and the human RNAseP gene simultaneously. The latter target was used for data normalization and as an internal extraction control. A successful multiplexed quantitative real-time PCR assay only yields reliable results if multiplex and singleplex reactions produce similar Cq values for the amplification of a particular target sequence. To assess the quantitative performance of the duplex assay in comparison to the 16S singleplex assay, serial dilutions of the template (either synthetic 16S oligonucleotide or total *B. burgdorferi* s.s. DNA) were amplified. All PCR reactions were carried out in the presence of a constant amount (approx. 250 ng) of human genomic DNA as background matrix. Nine independent PCR runs were set up in each case, with 3 PCR runs performed on 3 consecutive days by 3 different operators.

As shown in Table 1, co-amplification of the human RNAseP did not affect the PCR efficiency, sensitivity or correlation coefficients, since almost identical Cq values were obtained using the 16S singleplex and 16S/RNAseP duplex assay. In addition, when superimposing the recorded fluorescent traces, the exponential phase of the singleplex and duplex real-time PCR reactions were similar (data not shown). Finally, no differences were observed for serial dilutions prepared using either a synthetic 16S oligonucleotide or genomic DNA isolated from the *B. burgdorferi* s.s. strain ZS7 (Table 1), demonstrating that a synthetic template can be used for generating reliable standard curves. Therefore, for quantifying *Borrelia* in clinical specimens, a synthetic 16S oligonucleotide combined with human genomic DNA was used as template for the generation of standard curves.

As a constant amount of human genomic DNA was present in the samples used to generate the standard curves, the impact of increasing amounts of 16S copies on the RNAseP assay could also be assessed. The distribution of the Cq values generated using a singleplex RNAseP (median: 21.90; IQR: 21.82 to 22.23; n = 126) or duplex 16S/RNAseP assay (Median: 21.86; IQR: 21.78 to 22.13; n = 126) was virtually identical. In addition, RNAseP Cq values generated in the presence of either 10,000 copies or 1 copy of 16S template DNA (oligonucleotide or total ZS7 DNA) were also similar (data not shown), indicating that a high copy number of *Borrelia* 16S rRNA does not negatively influence the performance of the RNAseP assay.

Taken together, these data clearly demonstrate that the 16S rRNA and human RNAseP assays can be run in a duplex format without negatively impacting the performance of either real-time PCR assay.

Assay Specificity

The 16S rRNA gene was selected as the target for PCR, since it is an essential, chromosomally encoded gene and the sequence is highly conserved in *B. burgdorferi* s.l. The assay successfully amplified the conserved 139-bp 16S target using 25 ng of total DNA extracted from a panel of 83 diverse *B. burgdorferi* sensu lato strains selected on the basis of unique *ospA* sequences (data not shown) which included representatives of all major *Borrelia* species causing Lyme disease, notably *B. burgdorferi* s.s. *(n = 15)*, *B. afzelii* (*n* = 21), *B. garinii (n = 30)*, *B. bavariensis sp.nov. (n = 4)*, *B. andersonii* (*n* = 2), *B. bissettii* (*n* = 1), *B. finlandensis sp.nov. (n = 2)*, *B. japonica* (*n* = 2), *B. lusitaniae* (*n* = 1), *B. spielmanii (n = 3)* and *B. valaisiana* (*n* = 2). As shown in Figure 1, the distribution of the determined Cq values was comparable, with a median Cq of 27.4 and an IQR from 26.28 to 29.36. Importantly, the assay was highly specific as it did not amplify DNA from 21 unrelated microorganisms, nor was amplification evident from *Borrelia*-negative human DNA (data not shown).

Table 1. Comparison of Cq values (Mean ±1 SD; n = 9) from single and duplex real-time PCR assays obtained with dilution series of *Borrelia* DNA or a synthetic *Borrelia* 16S oligonucleotide in a background matrix of human genomic DNA.

Copies 16S	Synthetic *Borrelia* 16S		Total *Borrelia* DNA	
	16S	16S/RNAseP	16S	16S/RNAseP
10,000	26.37±0.28	26.36±0.20	26.44±0.11	26.46±0.11
1,000	29.78±0.22	29.74±0.19	29.89±0.12	29.87±0.06
100	33.17±0.34	33.32±0.24	33.42±0.32	33.12±0.20
10	36.58±0.46	37.03±1.23	37.00±0.77	36.58±0.55
5	37.30±0.72	37.83±1.16	38.33±1.17	37.74±0.85
2.5	38.84±1.25	38.37±0.84	39.02±1.01	38.45±0.84
1	39.49±0.80	39.10±0.98	39.66±0.58	39.52±0.63
PCR Efficiency	104% ±4%	104% ±6%	98% ±1%	103% ±3%
R^2	0.99±0.004	0.98±0.01	0.98±0.15	0.99±0.004
Y-Intercept	39.45±0.24	39.48±0.27	40.03±0.19	39.06±1.14
Slope	−3.23±0.08	−3.22±0.13	−3.37±0.12	−3.25±0.06

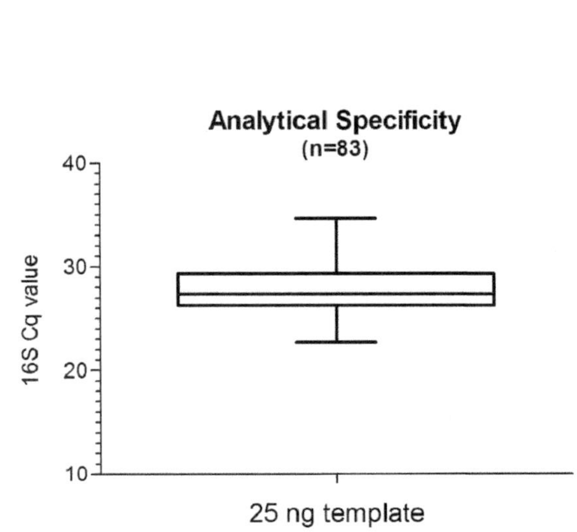

Figure 1. Distribution of Cq values determined for 83 different *B. burgdorferi* s.l. strains. Boxplot analysis of 16S Cq values using the 16S/RNaseP real-time assay probing 25 ng total DNA extracted from a panel of *B. burgdorferi* s.l. including all species known to cause LB. Whiskers indicate the minimal and maximal 16S Cq value determined.

Quantitative Sensitivity and Determination of Detection Limit using the Duplex Assay

Quantitative sensitivity of the duplex PCR assay was evaluated using a background matrix of human genomic DNA spiked with limiting dilutions of *B. burgdorferi* sensu stricto DNA (ZS7). In this case, 12 replicates of each dilution (10, 8, 6, 4, 2, 1 and 0.1 *Borrelia* genome equivalents) were tested in 3 runs performed on 3 consecutive days (i.e. n = 36 for each dilution, Table 2). The limit of detection was determined as 6 target copies per PCR reaction (20 µl) and the PCR reaction was positive on 97.2% occasions (Table 2).

Clinical Data

Paired skin biopsies were taken from 121 untreated adult patients with solitary EM lesions between June and December 2010. Their median age was 54 years (IQR: 43 to 61), and seventy (57.9%) were female. Fifty three (44%) patients had EM skin lesion with central clearing, 29 (24%) reported systemic symptoms in addition to EM, and 48 (40%) complained of symptoms at the site of the skin lesion (mild itching, burning and/or pain). Forty eight (39.7%) patients had an underlying chronic illness requiring treatment, while 7 (5.8%) had cancer. Twenty nine (24%) patients presenting with EM lesion had a history of previous LB, 4 (13.8%) of whom were diagnosed with extracutaneous manifestations of LB with/without EM at first presentation.

Evaluation of Clinical Samples and Controls by PCR and Culture

The skin biopsy samples collected were evaluated by real time PCR and culture (Table 3). In total, 3 cultures became contaminated and hence could not be evaluated. PCR identified more positive biopsy samples, with 94/121 positive (77.7%) compared to 65/118 (55.1%) biopsies positive by culture. None of the samples showed signs of PCR inhibition. Importantly, all specimens that were positive by culture were also positive by PCR. DNA isolated from biopsy samples was tested in triplicate by real-time PCR and for 8 of the 121 samples (6.6%), seven of which were culture negative, only one of three technical replicates tested *Borrelia* positive. However, as repeat testing yielded a minimum of one positive result, the samples were considered positive for *Borrelia* infection, albeit with a low copy number (0.2 to 7 *Borrelia* 16S copies per 10,000 human genome equivalents). RNAseP values for these samples indicated successful DNA extraction and there were no signs of PCR inhibition. Control biopsies which were obtained from 2 healthy volunteers and from 4 subjects who had undergone plastic surgery, were correctly identified as PCR negative, which was confirmed by repeat testing. Finally, all re-biopsies taken from the patients after antibiotic treatment (n = 61) were culture negative, although one re-biopsy tested PCR positive.

Biopsy samples were collected over a one week period (stored at room temperature for culture and at 4°C for PCR) and shipped to the test lab weekly. The effect of the length of the holding time, between taking biopsies to their incubation at 33°C, on subsequent culture isolation rates was investigated in the subset of patients who had PCR confirmed EM (n = 92, 2 PCR positive samples which were culture contaminated were excluded from analysis).

For PCR positive samples processed rapidly (incubation at 33°C ≤2 days after biopsy) 65% (17/26) of the samples were culture positive, while for samples processed less rapidly (incubation at 33°C >2, ≤8 days after biopsy) 73% (48/66) were culture positive; the difference was not statistically significant ($P = 0.66$, Chi-squared test with Yates' correction). Similar results were obtained for the complete data set [46% (17/37)] and for culture positives [59% (48/81)], for rapidly and less rapidly processed samples, respectively, ($P = 0.25$, Chi-squared test with Yates' correction).

There was also no difference in how rapidly the cultures became positive, 11.7% (2/17) of those processed more rapidly (incubation at 33°C ≤2 days after biopsy) were culture positive in less than 6 weeks, versus 12.5% (6/48) of those processed less rapidly (incubation at 33°C >2, ≤8 days after biopsy, $P > 0.99$). Thus, the length of holding time at room temperature prior to incubation (≤8 days) appeared to have no detrimental effect on the isolation rate of *Borrelia* in culture.

Table 3. Comparison of PCR and culture for the detection of *B. burgdorferi* s.l. in EM lesions and comparison of median copy number in culture positive and culture negative biopsies.

Detection by:		Number (%)	Median copies (IQR)
PCR	Culture		Per 10,000 human genome equivalents
+	+	65 (55.1%)[a]	22 (6–53)*
+	–	27 (22.9%)	6 (2–31.5)
–	–	26 (22.0%)	0
–	+	0	0

*No of spirochetes is significantly higher (P<0.001, Mann-Whitney test all patients, P = 0.04 for PCR+ patients) in culture+versus culture – biopsies.
[a]3 cultures became contaminated, (2 PCR positive and 1 PCR negative), hence 118 samples could be evaluated by culture.

Quantitative PCR Results in Culture Positive and Culture Negative Specimens

The number of *Borrelia* 16S rRNA targets detected by qPCR for PCR positive patients varied considerably, ranging from 0.2 to 321 *Borrelia* 16S rRNA target copies per 10,000 human genome equivalents with a median of 15 (IQR: 5 to 39.75). The median number of *Borrelia* 16S rRNA targets detected in culture-positive

Table 2. Limit of detection of the 16S rRNA qRT-PCR assay in the presence of human genomic DNA.

	Copies ZS7 genome equivalents						
No. target copies*	10	8	6	4	2	1	0.1
Percent positive (No) **	100% (35/35)	97.2% (35/36)	97.2% (35/36)	86.1% (31/36)	58.3% (21/36)	33.3% (12/36)	5.6% (2/36)

*Dilution series of *Borrelia* DNA from strain ZS7 in a matrix of human genomic DNA.
**Positive samples have Cq <40. Probability of detecting various copy numbers of ZS7 genomes demonstrated. The number of positive samples per tested dilution (typically 36) is indicated in brackets.

specimens was significantly higher than in culture-negative specimens (Figure 2; P=<0.001 considering all patients and P=0.04 for PCR positive patients, Mann-Whitney test, Table 3). RNAseP was detected for all samples (n=121) and the copy number per reaction ranged from 5,364 to 131,027, with a median of 29,640 (IQR: 21,080 to 44,220). For *Borrelia* PCR positive biopsies (n=94), the RNAseP per PCR reaction ranged from 5,364 to 131,027 (median 30,849, IQR: 22,170 to 38,800). While for *Borrelia* PCR negative biopsies (n=27), the RNAseP value per PCR reaction ranged from 6,471 to 62,531 (median 26,122, IQR: 17,900 to 39,690).

Typing of *Borrelia* Strains Isolated in Culture

Borrelia isolated in culture were typed (62/65 cultures) by sequencing a large fragment of the *ospA* gene as outlined previously

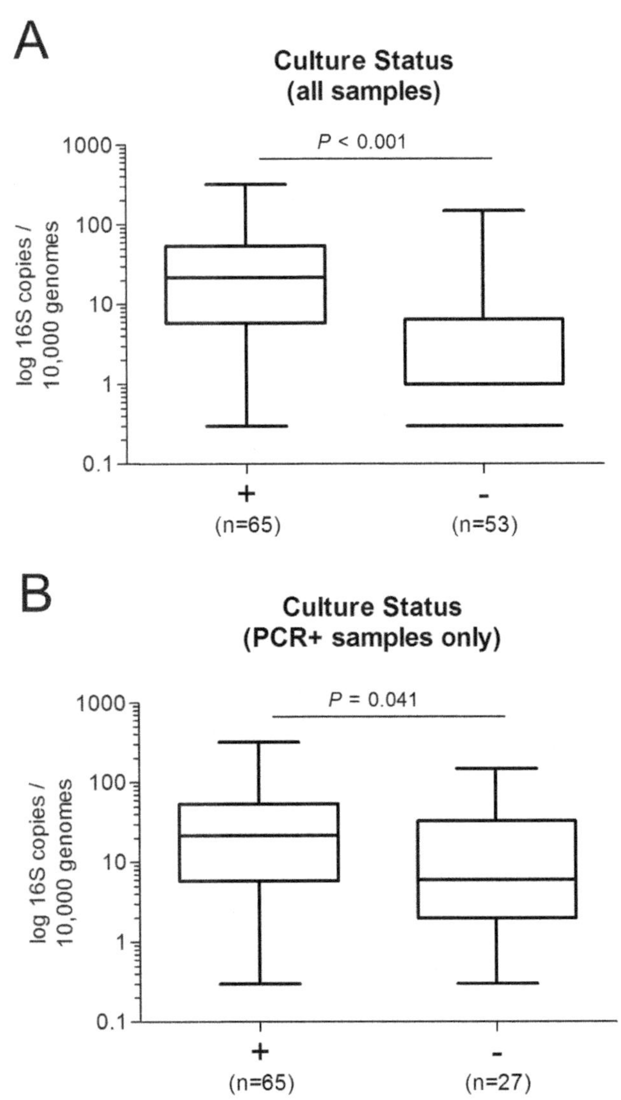

Figure 2. 16S copy number of *Borrelia* positive and negative cultures. Boxplot showing the 16S copies ($\log_{10}$) detected in skin biopsy specimens for which the paired biopsies were either positive or negative for *Borrelia* in culture. **(A)** Analysis for all samples and **(B)** only samples which were positive by PCR. Whiskers indicate the minimal and maximal copy number determined. Data were analyzed using the Mann–Whitney test.

[27]. Sixty of the isolates (96.8%) had an OspA sequence which was identical (82.3%, 51/62) or highly similar (99% OspA sequence identity) to the OspA sequence for *B. afzelii* strain PKo (GenBank Acc. No. S48322 [31]). The remaining two isolates had an OspA sequence which was identical to the OspA sequence for *B. garinii* strain Tlsl (GenBank Acc. No. X85440 [32]).

Relationship between Clinical Picture and PCR Quantification

Systemic symptoms such as fatigue, headache, myalgia, arthralgia, malaise, rigors, dizziness, nausea and fever (Table 4) were reported by 29 patients (24%). The median number of *Borrelia* per human 10,000 genome equivalents was significantly higher (P=0.02) in patients with systemic symptoms than without (Figure 3A and Table 4). This association was not statistically significant for any individual systemic symptom (data not shown). Of the 29 patients with systemic symptoms 26 (89.7%) were PCR positive and 17 (58.6%) were culture positive for *B. afzelii*. Two of these patients also had a positive blood culture (*B. afzelii*), only one other patient, who did not report systemic symptoms, had a *B. afzelii* positive blood culture. All three patients from whom blood cultures were positive also had positive skin biopsy cultures.

Local symptoms at the biopsy site were reported by 48 patients (40%) and comprised itching, burning and pain. Thirty five of the patients with local symptoms (72.9%) were PCR positive and 23 (47.9%) were culture positive, but there was no significant association between these symptoms and the number of spirochaetes ($P=0.72$, Table 4).

Of the twenty nine patients who had a previous history of LB, 23 (79.3%) and 16 (55.1%) were PCR and culture positive respectively. The median *Borrelia* 16S rRNA target copy number was 2-fold lower in biopsies from patients who reported a previous history of LB ($P=0.10$, Figure 3B and Table 4).

Fifty five of the biopsies (45%) were from patients who had an underlying illness, but there was no significant association between underlying illness and the number of 16S rRNA *Borrelia* targets or the probability of a positive result by PCR or culture ($P=0.92$). Six of the patients were immunocompromised with reported malignancy (patients diagnosed with malignancy with current chemo- or/and radiotherapy, or chemo−/radiotherapy within the last year). Copy number in these patients was comparable to copy number in non-immunocompromised patients.

Central clearing of EM lesions was observed significantly more often among patients with positive PCR (50% vs 22%, $P=0.02$), also the number of spirochetes was significantly higher among patients with central clearing ($P<0.001$, Figure 3C and Table 4). There was no association between central clearing of EM lesions and systemic symptoms ($P=0.93$).

Time from the onset of EM (as appreciated by patients) to biopsy ranged from 0 to 296 days (median 10.5 days). The association between time from EM to biopsy and PCR and culture positivity was also investigated (Figure 4). The probability of a positive result markedly increased with time from biopsy for the biopsies taken within about 20 days ($P=0.07$), while it decreased afterwards. The association between culture positivity and time from EM to biopsy was not statistically significant ($P=0.45$), but the shape of the estimated association closely mirrored the results obtained for PCR positivity (Figure 4).

Discussion

Conventional diagnosis of LB is often based solely on the evaluation of an EM lesion. However, due to the high degree of variability of skin lesions and the subjective nature of the

Table 4. Summary of clinical characteristics for 121 patients with EM and association of the significance of spirochete number with clinical characteristic.

	Present		Absent		
Characteristic	Number patients (%)	Median copy number per 10,000 human genomes (IQR)	Number patients (%)	Median copy number per 10,000 human genomes (IQR)	P[a]
Systemic symptoms[b]	29 (24%)	22 (6 to 39)	92 (76%)	5 (0 to 31.5)	0.02
Symptoms at EM site[c]	48 (40%)	12.5 (0 to 36.25)	73 (60%)	6 (1 to 25)	0.72
History of LB[d]	29 (24%)	5 (0.8 to 9)	91 (76%)	10 (0.65 to 39.5)	0.10
Underlying illness[e]	55 (45%)	5 (1 to 35)	66 (55%)	7.5 (0.3 to 33)	0.92
Immunocompromised[f]	6 (5%)	5 (0.25 to 33.75)	115 (95%)	6 (0.9 to 33.5)	0.81
Central clearing	53 (44%)	18 (4 to 56)	68 (56%)	4 (0 to 21.25)	<0.001

[a]As determined by Mann-Whitney test.
[b]Systemic symptoms comprised fatigue, headache, myalgia, arthralgia, malaise, rigors, dizziness, nausea and fever.
[c]Symptoms at site comprised itching, burning and pain.
[d]History of extracutaneous LB and/or EM, previously diagnosed by a physician; data not available for 1 patient (i.e.) n = 120.
[e]Chronic underlying illness; arterial hypertension, hyperlipidemia, osteoporosis, diabetes mellitus, thyroid disease, cardiac rhythm abnormality, psychiatric illness, ischemic heart disease, osteoarthritis or asthma.
[f]Patients had malignancy and were treated with chemo−/radiotherapy within the last year.

interpretation, misdiagnosis is possible. Isolating *Borrelia* from skin biopsy specimens is considered the gold-standard, however it is time consuming and does not contribute to early diagnosis [13]. Quantitative real-time PCR for direct and rapid molecular detection and quantification of pathogens has become a widely used technology for clinical applications and can be valuable for confirming a diagnosis based on less clear manifestations of LB or for investigating controversial disease syndromes attributed to infection with *B. burgdorferi* s.l. In addition, PCR testing can complement serology testing in endemic areas for LB, where positive serology may be due to past exposure to *B. burgdorferi* s.l.

Numerous real time PCR assays for the detection of *B. burgdorferi* s.l. have been reported previously, however few have included an internal control [21,22] nor has a quantitative PCR study of *Borrelia* in European clinical skin biopsy samples been published to date.

We have developed a *Borrelia* species-specific TaqMan® probe–based qPCR assay targeting the 16S rRNA gene of all *B. burgdorferi* s.l. strains known to cause LB. The inclusion of an assay simultaneously targeting the human RNAseP gene as an internal control enabled sample-to-sample normalization of the bacterial burden (i.e. copies of *Borrelia* 16S rRNA per 10,000 human genome equivalents) and allowed successful DNA extraction to be monitored. The assay described is highly sensitive, detecting 6 copies of 16S rRNA gene per PCR reaction, which is similar to values reported for *B. burgdorferi* s.l. from previous studies [7,21]. Furthermore, the assay is highly specific, as it failed to detect the 16S rRNA gene of 21 unrelated microbial species or any bacterial DNA that may have been co-extracted from skin commensals present in control skin biopsies. Based on these observations, we are confident that the assay can detect all known pathogenic *B. burgdorferi* s.l. in clinical skin biopsy samples.

To evaluate the diagnostic performance of the real-time qPCR assay in comparison to culture for detecting *B. burgdorferi* s.l. from skin biopsies, we undertook a study of paired 3 mm skin biopsy samples taken from 121 patients with typical solitary EM [28].

PCR testing identified more positive samples than culture (77.7% PCR positive versus 55.1% culture positive) and correctly

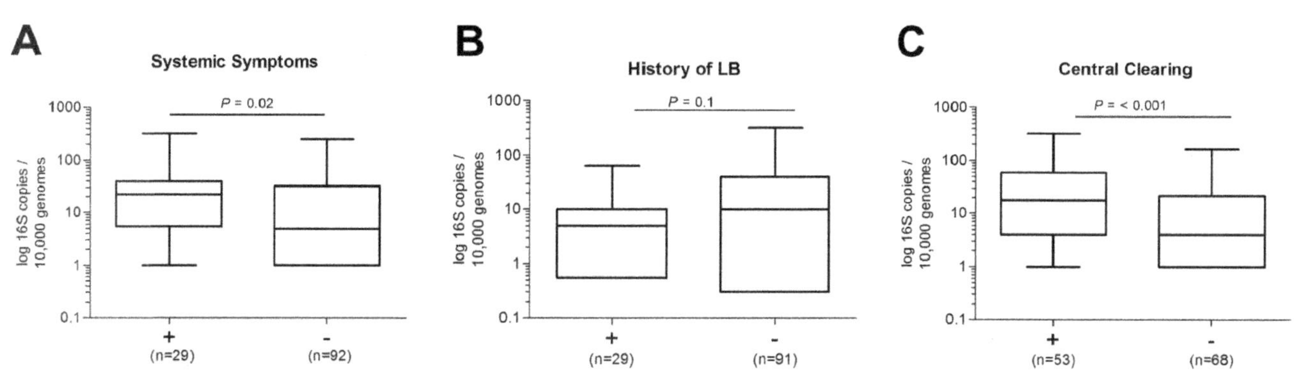

Figure 3. Association of *Borrelia* 16S rRNA targets per 10,000 genome equivalents and clinical symptoms. (**A**) Boxplot showing $\log_{10}$ 16S copies determined for biopsy specimens from patients presenting with (+) or without (−) systemic symptoms. (**B**) Distribution of *Borrelia* 16S copies quantified from specimens taken from patients reporting a previous diagnosis of LD. (**C**) 16S copies quantified for biopsies taken from EM lesions with or without central clearing. Whiskers indicate the minimal and maximal copy number determined. Data were analyzed using the Mann–Whitney test.

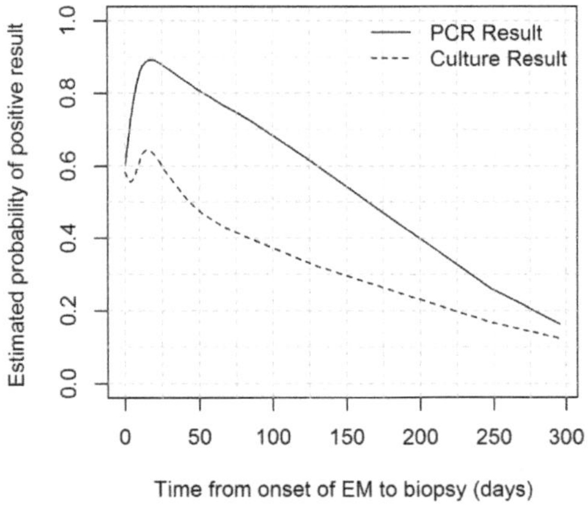

Figure 4. Estimated probability of positive detection of *B. burgdorferi* s.l. by culture and PCR as a function of time from onset of EM to biopsy.

identified all specimens scored as positive by culture. Importantly, all control samples tested negative for the presence of *B. burgdorferi* s.l. nucleic acids. In comparison, PCR assays for detection of *Borrelia* from skin biopsies described previously have reported sensitivity ranging from 25% [33] to 88% [34]. Finally, 96.8% of the isolates were typed as *B. afzelii*, which is consistent with previous studies conducted in the central part of Slovenia where *B. afzelii* accounted for approximately 86% of the strains isolated [35–37].

The median number of 16S rRNA targets per 10,000 human genome equivalents was significantly higher in culture positive biopsies than in culture negative biopsies, which is consistent with findings reported from quantitative PCR studies of *B. burgdorferi* s.s. in skin biopsies [19]. However, although we observed a clear relationship between a high *Borrelia* DNA load and culture positivity, individual samples failed to grow in culture despite having a high *Borrelia* DNA load. Possibly the distribution of *Borrelia* may have differed between the 2 separate biopsy samples used for culture and PCR [33] or particular strains may grow less well in culture medium or may not have been viable in the retrieved biopsies.

There was a significant association between the estimated probability of PCR and culture positivity and time from the onset of EM to biopsy. The probability of a positive result increased within the first 20 days after the appearance of EM and decreased afterwards, possibly reflecting dissemination of spirochetes from the EM lesion, and/or a decrease in spirochete numbers due to the host immune response. Our results are in line with a previous study reporting an inverse relationship between the recovery of *B. burgdorferi* s.s. from a skin specimen and EM duration [38]. Furthermore, significantly fewer spirochetes have been reported in older EM lesions [19]. Therefore, the possibility of confirming a clinical diagnosis of EM by either PCR or culture is potentially lower for long standing lesions.

Biopsies were also taken after completion of antibiotic treatment (post-treatment biopsies) in subjects from whom the first biopsy was culture positive. All post-treatment biopsies were culture negative and, all but one, were PCR negative. The single PCR positive (and culture negative) post-treatment biopsy was taken from a subject who responded well and whose EM disappeared after initial treatment, with no further antibiotic intervention deemed necessary. These results suggest that the infection had recently resolved and that no viable *Borrelia* were present, highlighting a potential limitation of PCR-based assays in terms of assessing on-going infection (especially after treatment), where the detection of DNA may indicate a recent infection and does not necessarily mean that live *Borrelia* are present.

A clear relationship between clinical symptoms and *Borrelia* burden was evident from this study. The median number of 16S targets detected per 10,000 human genomes was significantly higher among patients with systemic symptoms than patients who did not report systemic symptoms. A positive correlation between clinical symptoms and spirochete burden has been reported in the mouse model [14,16,22], however this is the first study demonstrating an association between spirochete burden and clinical disease symptoms in humans. The median number of 16S targets was not different in patients with another underlying disease. Also no association between local symptoms at the site of EM and 16S rRNA target copy number was found, although host specific factors may likely contribute to local symptoms.

EM is described as an expanding erythematous skin lesion with or without central clearing [28]. In our study which comprised predominantly *B. afzelii* isolates, central clearing of EM lesions was observed more often among patients with positive PCR and the number of spirochetes was significantly higher among patients with central clearing. These findings may pertain only to EM caused by *B. afzelii* and may not be valid for EM lesions caused by other *Borrelia* species such as *B. burgdorferi* s.s. and *B. garinii*, since different *Borrelia* species cause distinct clinical presentations [39,40]. Possibly the strains isolated from cases of EM with central clearing evoked a less intense host response, which is less likely to result in spirochaetal killing.

To our knowledge, this is the first quantitative PCR study to have evaluated patients with a history of LB. The median number of 16S rRNA targets was 2 fold lower in patients with a history of LB. Re-infection in patients with LB has been reported in Europe and in the US [41,42]. In a study of patients with EM, [43] *B. burgdorferi* s.s. was significantly less likely to be recovered from blood samples of patients with a prior history of LB, implying that partial immunity may protect against disseminated infection. Moreover, studies in dogs have shown that the number of *B. burgdorferi* s.s organisms detected in skin biopsies was inversely correlated with the antibody levels measured by ELISA [16]. Knowledge of the immune status of the individuals with a reported history of LB at the time of re-infection was not determined during this study, however it is possible that prior immunity may have limited the bacterial burden during a subsequent infection.

In conclusion, we have developed an internally controlled, sensitive duplex 16S rRNA/RNaseP qPCR assay specific for all *B. burgdorferi* s.l. strains known to cause LB. The results presented demonstrate that the assay is a sensitive, rapid and reliable approach for detection of *Borrelia* in skin biopsy samples retrieved from EM lesions and support its use for laboratory confirmation of clinically diagnosed LB, e.g. to assess the efficacy of a LB vaccine during clinical trials. Furthermore, this is the first quantitative PCR study of human skin biopsies predominantly infected with *B. afzelii* and the first study to report a clear relationship between clinical systemic symptoms and *Borrelia* target copy number.

Acknowledgments

We gratefully acknowledge Prof. Uroš Ahčan, M.D for providing control skin biopsies from subjects who had undergone cosmetic surgery and Marie Luise Zips, Doris Koelch, Hann Wei Chen, Daniela Nowak and Helmut Schmidt for excellent technical assistance.

Author Contributions

Conceived and designed the experiments: MO AT FS IL. Performed the experiments: MO AT DS VM. Analyzed the data: LL MO AT PNB FS IL. Contributed reagents/materials/analysis tools: DS VM PNB FS. Wrote the paper: MO. Critically revised the manuscript and approved the final version to be published: MO AT LL DS VM PNB FS IL.

References

1. Casjens SR, Fraser-Liggett CM, Mongodin EF, Qiu WG, Dunn JJ, et al. (2011) Whole genome sequence of an unusual Borrelia burgdorferi sensu lato isolate. J Bacteriol 193: 1489–1490.
2. Margos G, Vollmer SA, Cornet M, Garnier M, Fingerle V, et al. (2009) A new Borrelia species defined by multilocus sequence analysis of housekeeping genes. Appl Environ Microbiol 75: 5410–5416.
3. Stanek G, Wormser GP, Gray J, Strle F (2011) Lyme borreliosis. Lancet 379: 461–473.
4. Fingerle V, Schulte-Spechtel UC, Ruzic-Sabljic E, Leonhard S, Hofmann H, et al. (2008) Epidemiological aspects and molecular characterization of Borrelia burgdorferi s.l. from southern Germany with special respect to the new species Borrelia spielmanii sp. nov. Int J Med Microbiol 298: 279–290.
5. Wang G, van Dam AP, Dankert J (1999) Phenotypic and genetic characterization of a novel Borrelia burgdorferi sensu lato isolate from a patient with Lyme borreliosis. J Clin Microbiol 37: 3025–3028.
6. Diza E, Papa A, Vezyri E, Tsounis S, Milonas I, et al. (2004) Borrelia valaisiana in cerebrospinal fluid. Emerg Infect Dis 10: 1692–1693.
7. Rijpkema SG, Tazelaar DJ, Molkenboer MJ, Noordhoek GT, Plantinga G, et al. (1997) Detection of Borrelia afzelii, Borrelia burgdorferi sensu stricto, Borrelia garinii and group VS116 by PCR in skin biopsies of patients with erythema migrans and acrodermatitis chronica atrophicans. Clin Microbiol Infect 3: 109–116.
8. Rudenko N, Golovchenko M, Mokracek A, Piskunova N, Ruzek D, et al. (2008) Detection of Borrelia bissettii in cardiac valve tissue of a patient with endocarditis and aortic valve stenosis in the Czech Republic. J Clin Microbiol 46: 3540–3543.
9. Rudenko N, Golovchenko M, Ruzek D, Piskunova N, Mallatova N, et al. (2009) Molecular detection of Borrelia bissettii DNA in serum samples from patients in the Czech Republic with suspected borreliosis. FEMS Microbiol Lett 292: 274–281.
10. Collares-Pereira M, Couceiro S, Franca I, Kurtenbach K, Schafer SM, et al. (2004) First isolation of Borrelia lusitaniae from a human patient. J Clin Microbiol 42: 1316–1318.
11. De Carvalho I, Fonseca JE, Marques JG, Ullmann A, Hojgaard A, et al. (2008) Vasculitis-like syndrome associated with Borrelia lusitaniae infection. Clin Rheumatol 27: 1587–1591.
12. Smith RP, Schoen RT, Rahn DW, Sikand VK, Nowakowski J, et al. (2002) Clinical characteristics and treatment outcome of early Lyme disease in patients with microbiologically confirmed erythema migrans. Ann Intern Med 136: 421–428.
13. Aguero-Rosenfeld ME, Wang G, Schwartz I, Wormser GP (2005) Diagnosis of Lyme borreliosis. Clin Microbiol Rev 18: 484–509.
14. Pahl A, Kuhlbrandt U, Brune K, Rollinghoff M, Gessner A (1999) Quantitative detection of Borrelia burgdorferi by real-time PCR. J Clin Microbiol 37: 1958–1963.
15. Saidac DS, Marras SA, Parveen N (2009) Detection and quantification of Lyme spirochetes using sensitive and specific molecular beacon probes. BMC Microbiol 9: 43.
16. Straubinger RK (2000) PCR-Based quantification of Borrelia burgdorferi organisms in canine tissues over a 500-Day postinfection period. J Clin Microbiol 38: 2191–2199.
17. Wilhelmsson P, Fryland L, Borjesson S, Nordgren J, Bergstrom S, et al. (2010) Prevalence and diversity of Borrelia species in ticks that have bitten humans in Sweden. J Clin Microbiol 48: 4169–4176.
18. Li X, McHugh GA, Damle N, Sikand VK, Glickstein L, et al. (2011) Burden and viability of Borrelia burgdorferi in skin and joints of patients with erythema migrans or lyme arthritis. Arthritis Rheum 63: 2238–2247.
19. Liveris D, Wang G, Girao G, Byrne DW, Nowakowski J, et al. (2002) Quantitative detection of Borrelia burgdorferi in 2-millimeter skin samples of erythema migrans lesions: correlation of results with clinical and laboratory findings. J Clin Microbiol 40: 1249–1253.
20. Nowakowski J, Schwartz I, Liveris D, Wang G, Aguero-Rosenfeld ME, et al. (2001) Laboratory diagnostic techniques for patients with early Lyme disease associated with erythema migrans: a comparison of different techniques. Clin Infect Dis 33: 2023–2027.
21. Gooskens J, Templeton KE, Claas EC, van Dam AP (2006) Evaluation of an internally controlled real-time PCR targeting the ospA gene for detection of Borrelia burgdorferi sensu lato DNA in cerebrospinal fluid. Clin Microbiol Infect 12: 894–900.
22. Germer J, Ryckmann B, Moro M, Hofmeister E, Barthold SW, et al. (1999) Quantitative detection of Borrelia burgdorferi with a microtiter-based competitive polymerase chain reaction assay. Mol Diagn 4: 185–193.
23. Chen CY, Chi KH, Alexander S, Ison CA, Ballard RC (2008) A real-time quadriplex PCR assay for the diagnosis of rectal lymphogranuloma venereum and non-lymphogranuloma venereum Chlamydia trachomatis infections. Sex Transm Infect 84: 273–276.
24. Pau CP, Wells SK, Rudolph DL, Owen SM, Granade TC (2010) A rapid real-time PCR assay for the detection of HIV-1 proviral DNA using double-stranded primer. J Virol Methods 164: 55–62.
25. Nolan T, Hands RE, Ogunkolade W, Bustin SA (2006) SPUD: a quantitative PCR assay for the detection of inhibitors in nucleic acid preparations. Anal Biochem 351: 308–310.
26. Ruzic-Sabljic E, Arnez M, Lotric-Furlan S, Maraspin V, Cimperman J, et al. (2001) Genotypic and phenotypic characterisation of Borrelia burgdorferi sensu lato strains isolated from human blood. J Med Microbiol 50: 896–901.
27. Livey I, O'Rourke M, Traweger A, Savidis-Dacho H, Crowe BA, et al. (2011) A new approach to a Lyme disease vaccine. Clin Infect Dis 52 Suppl 3: s266-s270.
28. Stanek G, Fingerle V, Hunfeld KP, Jaulhac B, Kaiser R, et al. (2011) Lyme borreliosis: clinical case definitions for diagnosis and management in Europe. Clin Microbiol Infect 17: 69–79.
29. Harrell FE Jr, Lee KL, Pollock BG (1988) Regression models in clinical studies: determining relationships between predictors and response. J Natl Cancer Inst 80: 1198–1202.
30. R Development Core Team (2009) R: A language and environment for statistical computing. R Foundation for Statistical Computing, Vienna, Austria.
31. Zumstein G, Fuchs R, Hofmann A, Preac-Mursic V, Soutschek E, et al. (1992) Genetic polymorphism of the gene encoding the outer surface protein A (OspA) of Borrelia burgdorferi. Med Microbiol Immunol 181: 57–70.
32. Will G, Jauris-Heipke S, Schwab E, Busch U, Rossler D, et al. (1995) Sequence analysis of ospA genes shows homogeneity within Borrelia burgdorferi sensu stricto and Borrelia afzelii strains but reveals major subgroups within the Borrelia garinii species. Med Microbiol Immunol 184: 73–80.
33. Picken MM, Picken RN, Han D, Cheng Y, Ruzic-Sabljic E, et al. (1997) A two year prospective study to compare culture and polymerase chain reaction amplification for the detection and diagnosis of Lyme borreliosis. Mol Pathol 50: 186–193.
34. Brettschneider S, Bruckbauer H, Klugbauer N, Hofmann H (1998) Diagnostic value of PCR for detection of Borrelia burgdorferi in skin biopsy and urine samples from patients with skin borreliosis. J Clin Microbiol 36: 2658–2665.
35. Stupica D, Lusa L, Cerar T, Ruzic-Sabljic E, Strle F (2011) Comparison of post-Lyme borreliosis symptoms in erythema migrans patients with positive and negative Borrelia burgdorferi sensu lato skin culture. Vector Borne Zoonotic Dis 11: 883–889.
36. Stupica D, Lusa L, Ruzic-Sabljic E, Cerar T, Strle F (2012) Treatment of erythema migrans with doxycycline for 10 days versus 15 days. Clin Infect Dis 55: 343–350.
37. Ruzic-Sabljic E, Maraspin V, Lotric-Furlan S, Jurca T, Logar M, et al. (2002) Characterization of Borrelia burgdorferi sensu lato strains isolated from human material in Slovenia. Wien Klin Wochenschr 114: 544–550.
38. Nadelman RB, Nowakowski J, Forseter G, Bittker S, Cooper D, et al. (1993) Failure to isolate Borrelia burgdorferi after antimicrobial therapy in culture-documented Lyme borreliosis associated with erythema migrans: report of a prospective study. Am J Med 94: 583–588.
39. Strle F, Nadelman RB, Cimperman J, Nowakowski J, Picken RN, et al. (1999) Comparison of culture-confirmed erythema migrans caused by Borrelia burgdorferi sensu stricto in New York State and by Borrelia afzelii in Slovenia. Ann Intern Med 130: 32–36.
40. Strle F, Ruzic-Sabljic E, Logar M, Maraspin V, Lotric-Furlan S, et al. (2011) Comparison of erythema migrans caused by Borrelia burgdorferi and Borrelia garinii. Vector Borne Zoonotic Dis 11: 1253–1258.
41. Pfister HW, Neubert U, Wilske B, Preac-Mursic V, Einhaupl KM, et al. (1986) Reinfection with Borrelia burgdorferi. Lancet 2: 984–985.
42. Nadelman RB, Wormser GP (2007) Reinfection in patients with Lyme disease. Clin Infect Dis 45: 1032–1038.
43. Wormser GP (2006) Hematogenous dissemination in early Lyme disease. Wien Klin Wochenschr 118: 634–637.

Finding Protein-Coding Genes through Human Polymorphisms

Edward Wijaya[1,2]*, Martin C. Frith[2], Paul Horton[2], Kiyoshi Asai[1,2]

1 Graduate School of Frontier Sciences, University of Tokyo, Kashiwa, Japan, **2** Computational Biology Research Center, National Institute of Advanced Industrial Science and Technology (AIST), Tokyo, Japan

Abstract

Human gene catalogs are fundamental to the study of human biology and medicine. But they are all based on open reading frames (ORFs) in a reference genome sequence (with allowance for introns). Individual genomes, however, are polymorphic: their sequences are not identical. There has been much research on how polymorphism affects previously-identified genes, but no research has been done on how it affects gene identification itself. We computationally predict protein-coding genes in a straightforward manner, by finding long ORFs in mRNA sequences aligned to the reference genome. We systematically test the effect of known polymorphisms with this procedure. Polymorphisms can not only disrupt ORFs, they can also create long ORFs that do not exist in the reference sequence. We found 5,737 putative protein-coding genes that do not exist in the reference, whose protein-coding status is supported by homology to known proteins. On average 10% of these genes are located in the genomic regions devoid of annotated genes in 12 other catalogs. Our statistical analysis showed that these ORFs are unlikely to occur by chance.

Editor: Denis Dupuy, Inserm U869, France

Funding: Financial support for this study was provided by a Grant-in-Aid for Scientific Research on Innovative Areas (221S002) from Japanese Ministry of Education, Sport, Science and Technology (MEXT). The funders had no role in study design, data collection and analysis, decision to publish, or preparation of the manuscript.

Competing Interests: The authors have declared that no competing interests exist.

* E-mail: e-wijaya@cb.k.u-tokyo.ac.jp

Introduction

Compilation of an accurate catalog of protein-coding genes encoded in human genomes is a critical step to fully understand the functional elements in human genomes. Many annotations of protein-coding genes have been published [1] and a plethora of gene finding software has been introduced [2]. Nevertheless, the task has remained as a great challenge [3]. As stated by Brent [4] the difficulty lies in the limitations of sequencing protocols, ways to combine the predicted genes and finally the limitations of human curators.

In response to these findings we also believe that there is an increasing amount of genomic evidence that may affect protein-coding gene detection, which has not been taken into account. Human polymorphism is one example of such evidence [5]. It has been suggested that such polymorphism affects protein-coding genes, and that they are responsible for various human diseases [6–8].

Several studies have tried to account for damaging polymorphism in known protein-coding genes. Such polymorphism could affect the amino acid sequence, alter protein function and contribute to disease. For example Ng and Henikoff [8] discovered that two known genes were mistakenly classified as pseudogenes because of mutations. Others analyzed predicted protein-coding genes and found known SNPs and insertions and deletions in them [9,10]. However, to our knowledge no work has been done on the influence of polymorphism on the finding of novel genes.

In this article we provide a framework for finding new unannotated genes based on human polymorphism. We examine several types of polymorphism: SNPs, insertions, deletions, multiple nucleotide polymorphisms, and microsatellites. With this information we reconstruct the mRNAs and finally define the new open reading frames (ORFs). Figure 1a depicts the standard model for general transfers in biological sequence information. Figure 1b further illustrates the effect of polymorphism on the mature mRNA and its effect in defining the new ORF.

Demonstrating definitively that a transcript encodes a protein is a difficult, time-consuming and expensive task. One approach is to synthesize the protein artificially, raise an antibody against it and use the antibody to test whether it is expressed in vivo. Even this does not discriminate functional proteins from translation noise caused by stochastic nonfunctional translation by ribosomes [11,12]. On the other hand, proving that a transcript does not encode protein is an impossible task because the protein might only be expressed in very rare circumstances [13]. Thus we apply several bioinformatics criteria to improve the reliability of our findings: the ORF length, absence of repeats, E-value of homology to Swiss-Prot and shortness of 5′ untranslated region (UTR). Figure 2 illustrates two pipelines in parallel, those with and without polymorphism.

We identified 5,737 putative protein-coding genes that result from mRNA modified by human polymorphisms and have significant homology to known proteins. On average 10% of these genes are located in genomic regions unannotated by 12 other gene catalogs. A genomic coordinate list of these protein-coding genes is available as Table S1.

Among our other findings, we also discover that some of these novel genes are orthologous to genomic regions of other species

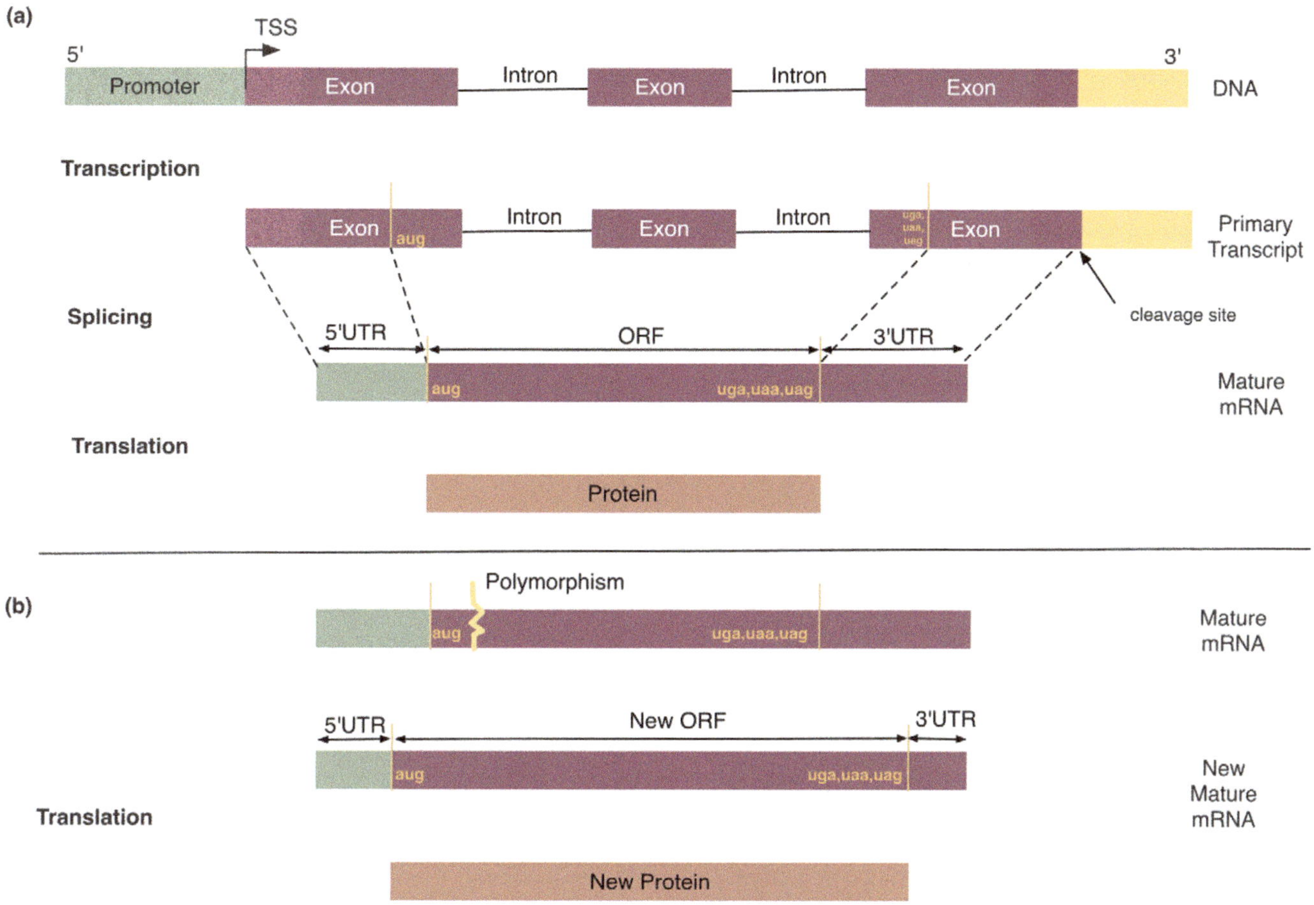

Figure 1. The standard model of genetic information transfer in molecular biology. Panel (a) shows the transfer begins with the DNA being transcribed into mRNA, and continues with protein being synthesized using information in mRNA as a template (translation). We investigate the effect of polymorphic modification of the mRNA. Panel (b) depicts how the new longer ORF was formed. The starting position of the new ORF in the mRNA is before that of original ORF. The new ORF may or may not overlap the original ORF, and if it does overlap, it is in different reading frame, so that the proteins are completely different.

where genes are annotated. Furthermore, we exhibit three examples of longest ORFs found after modification of the mRNA. They show significant homology to Swiss-Prot proteins and were not predicted as coding regions by any other gene finders. We cannot rule out that these mRNAs are fragments of longer mRNAs (i.e. artifacts), however the presence of supporting ESTs suggests that they are not.

Results

We will first highlight the results by three examples. The genomic positions of the mRNA in these examples are provided in Table 1. There we also indicate the genomic positions of the longest ORF, the polymorphism that causes the modification, the length of 5′ UTR after modification, the E-value of homology to known proteins and the sequence orientation.

Figure 3 shows a modification by polymorphism of mRNA AK124706. The new ORF is located upstream of the initial ORF. A repeating element overlaps the initial ORF, but none is found for the new ORF. There is also supporting evidence for transcription with the presence of EST (DA187884) at the 5′ end of the mRNA. It is aligned to Swiss-Prot *Integrin beta-5* protein (Acc: P18084) [14].

Figure 4 shows modification of mRNA AK127273. The mRNA and ORFs are located on the forward strand of the chromosome. We note that the predicted longest ORF after modification does overlap a pseudogene (HIT00004716) predicted by H-Inv, however this pseudogene is on the reverse strand of the chromosome. The new ORF is 5′ of the initial ORF. We verified that an EST (DA317450) is present at the 5′ end of the mRNA. None of the existing gene finders predict a coding region in the same location as our new ORF, which aligns to Swiss-Prot protein *capicua homolog* (Acc: Q96RK0) [15].

Figure 5 shows modification of mRNA AY129028. Although there is an existing coding region annotation in the mRNA, this just overlaps the initial ORF. The mRNA and ORFs are located on the reverse strand of the chromosome. Moreover, the initial ORF is located 3′ of the new ORF, and it overlaps a repeat element. This suggests that the ribosome would have to scan over the new ORF without translating it in order to reach the known ORF. Even though AY129028 looks like a fragment of pre-spliced mRNA, we found two ESTs (A1290869 and GD144663) at its 5′ end, evidence for a real transcript starting here. It appears in this figure that GD144663 starts at chr14:93,406,483–93,406,489 instead of chr14:93,406,109–93,406,489, however we believe the latter is correct. The alignment in the former position lacks a standard splice acceptor sequence, and alignment in the latter

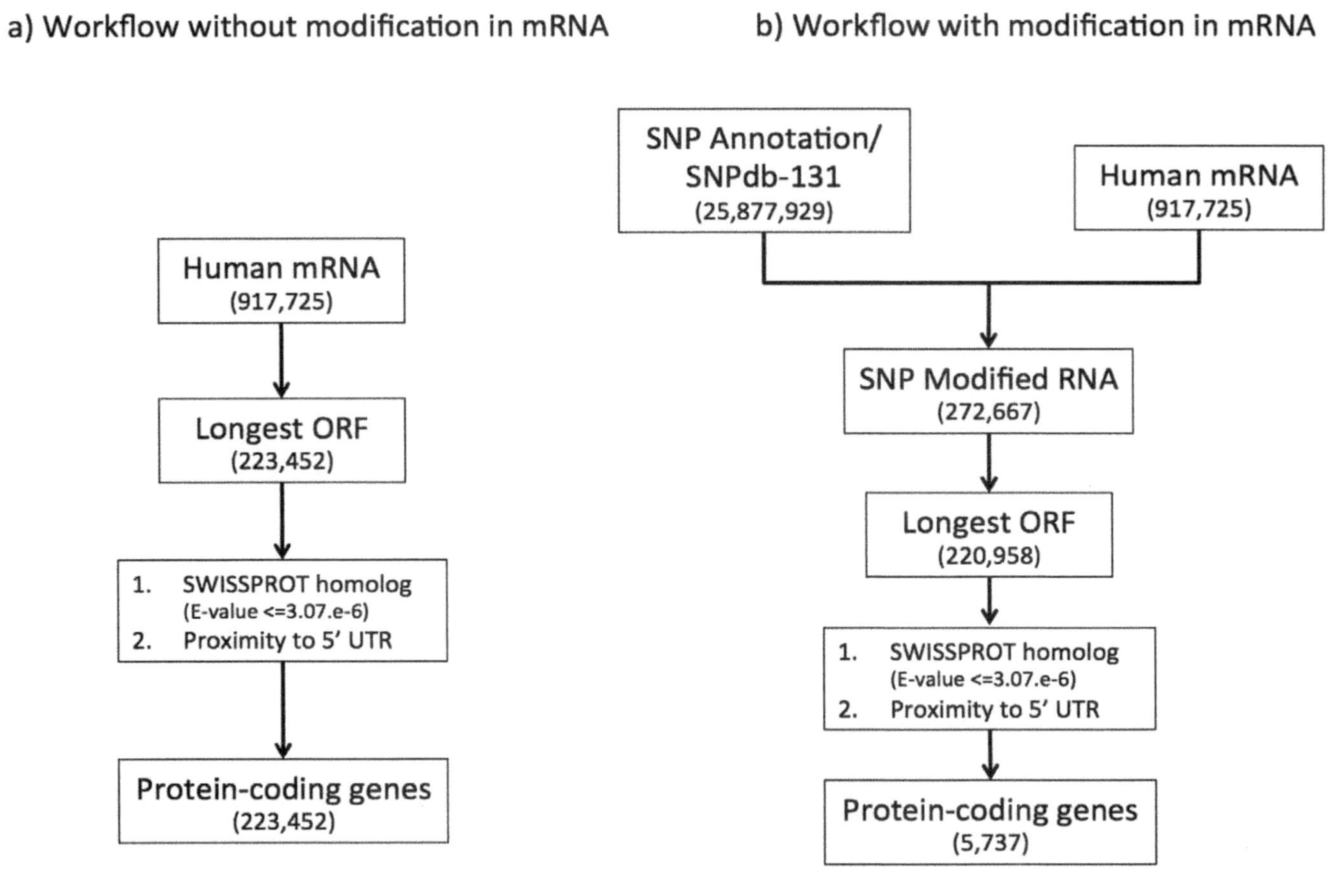

Figure 2. Workflows for finding protein-coding genes. Panel (a) describes the workflow of gene-finding without applying human polymorphism and (b) with human polymorphism. The values inside the brackets refer to the number mRNAs, ORFs and genes respectively. The final number of genes in workflow (b) refers to the genes where the ORFs change after modification, but in workflow (a) such change does not apply. For the second workflow (b) two main sources of data are used: human mRNA sequences and polymorphism data (dbSNP 131). Based on the polymorphism information we redefine the mRNA sequences. Out of the modified mRNA sequences we derived the longest ORFs. These ORFs are further refined by filtering them based on significant homology to Swiss-Prot and proximity to 5′ UTR. Finally we construct the genes from the refined ORFs.

position would have 5 matches out of 7 (see Figure S1). The new ORF also aligns to Swiss-Prot protein FLJ4630 (Acc: Q6ZRH3) [14].

Population Differences in ORF changes

Human populations from different parts of the world are surprisingly similar genetically. When averaged over all genes, the majority of genetic variation present in the human species can be found in any human ethnic group [16]. However the small level of differentiation among these groups is large enough for geneticists to make estimates on human diseases related to their ethnicity [17]. Results from large scale undertakings such as 1000 Genome [5] or HapMap Projects [18] for identifying human polymor-

Table 1. Genomic positions of mRNA for the examples shown in Figure 3, 4, 5.

	AK124706	AK127273	AY129028
Chr. name	chrUn_gl000222	chr7	chr14
Strand	−	+	-
mRNA pos. in chr.	25,008–28,821	128,295,697–128,299,178	93,403,259–93,406,150
dbSNP ref.	rs66651466	rs71162510	rs8011546
ORF pos. after modification	28,249–28,585	128,296,043–128,296,637	93,404,946–93,405,303
ORF pos. before modification	27,864–28,166	128,298,398–128,298,944	93,404,502–93,404,811
5′ UTR	236	347	847
E-value of new ORF alignment to Swiss-Prot	3.21E-10	4.74E-50	1.66E-08

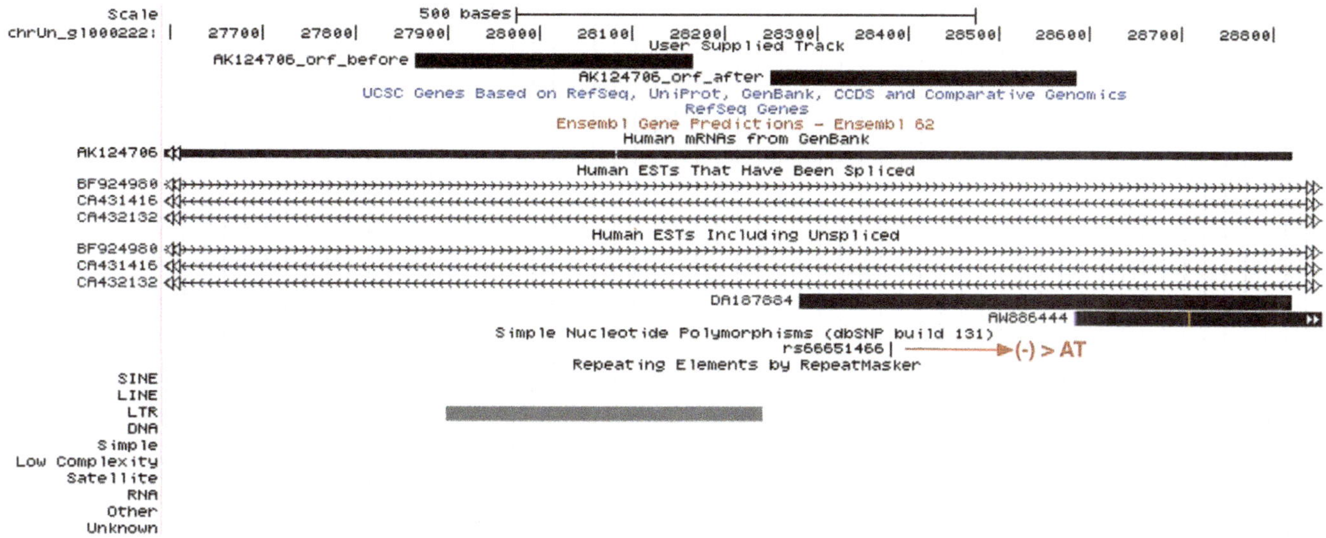

Figure 3. Modification by polymorphism of mRNA AK124706 and its ORFs. In the reference genome the modification is caused by an insertion (rs66651466) with 'AT' as the allele. The initial longest ORF before modification has length 302 bp. The new longest ORF has length 336 bp, and it aligns to Swiss-Prot Integrin beta-5 protein (Acc:P18084). Annotation of start/stop codon in the translation process and alleles that cause the change can be found in Figure 2.

phisms from diverse populations enable us to examine its significance in great detail.

We investigate the percentage of ORF-changing alleles for eleven different populations. These populations include: European-derived individuals from Utah pedigrees (CEU), Han Chinese in Beijing (CHB), Japanese in Tokyo (JPT) and Yoruba in Ibadan Nigeria (YRI). In addition to that we also included the expanded populations: African Ancestry in SouthWestern United States (ASW), Chinese Ancestry in Metropolitan Denver, CO, US (CHD), Gujarati Indians in Houston, TX (GIH), Luhya in Webuye, Kenya (LWK), Mexican Ancestry in Los Angeles, CA, US (MEX), Masai in Kinyawa, Kenya (MKK) and Toscani in Italia (TSI).

The normalized cumulative frequency of allele percentage can be seen in Figure 6a. We observed that there are slight differences in the cumulative frequency on seven populations: ASW, CHD, GIH, LWK, MEX, MKK, TSI. It is interesting to note that some populations are closely grouped together, e.g. ASW-MKK and CHB-JPT.

Four of the populations (CEU, CHB, JPT, and YRI) stand out in Figure 6a. In these populations around 40% of the new ORFs are extremely rare. In order to understand this, we plot the

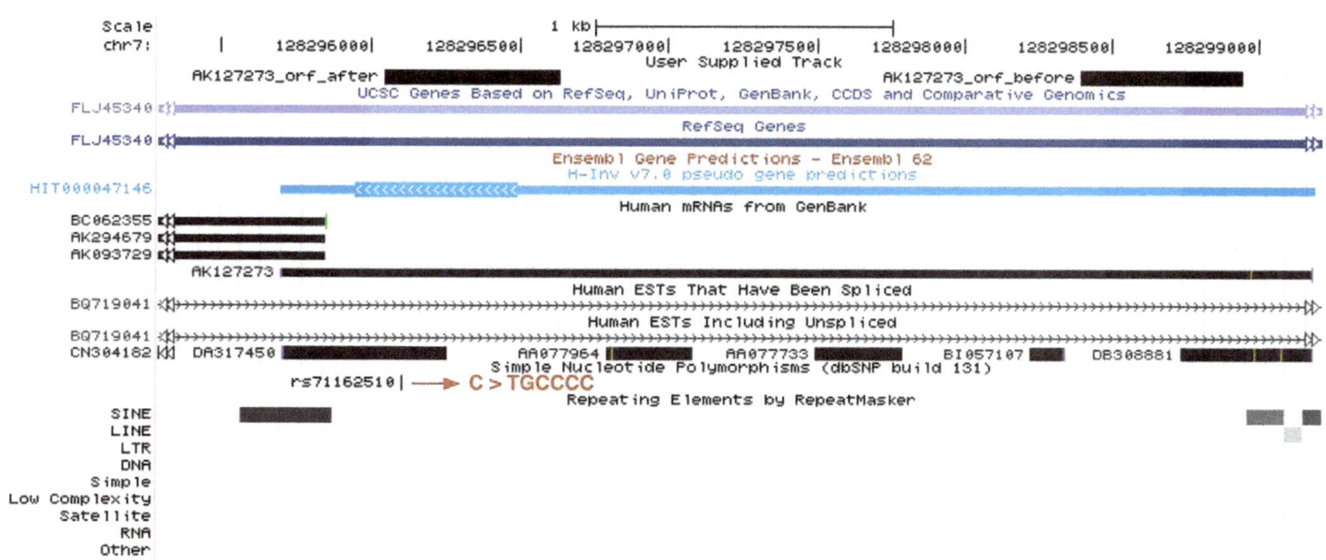

Figure 4. Modification by polymorphism of mRNA AK127273 and its ORFs. The initial longest ORF before modification has length 546 bp. The longest ORF after modification has length 594 bp. The polymorphism responsible for the modification is an in-del (rs71162510) which replaces the reference genome allele 'C' with 'TGCCCC'.

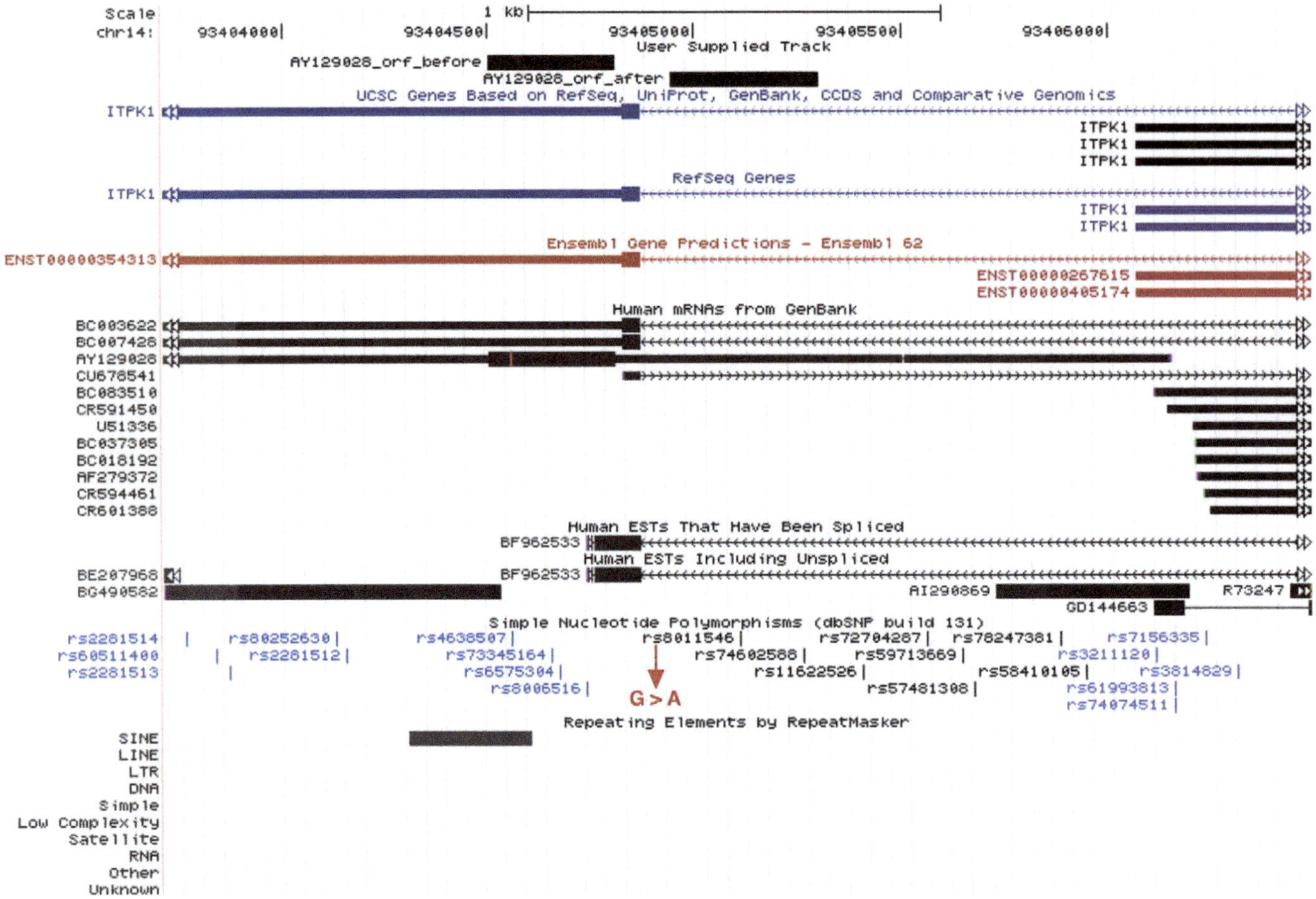

Figure 5. Modification by polymorphism of mRNA AY129028 and its ORFs. The initial longest ORF before modification has length 309 bp, and after has length 357 bp. The polymorphism that effects the modification is a SNP (rs8011546) which replaces the reference allele 'G' with 'A'.

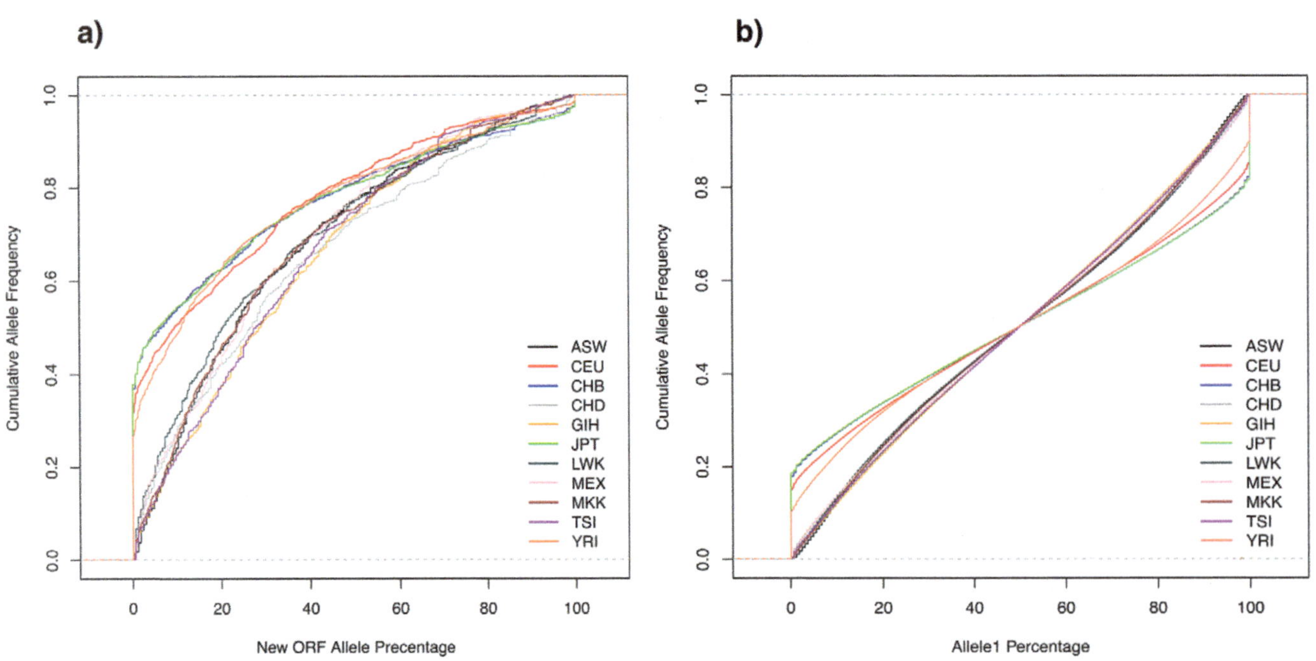

Figure 6. Cumulative allele frequency from 11 populations. In panel (a) we plot the allele percentage of new ORFs and (b) allele percentage of all HapMap data in UCSC Genome Browser. The percentage (x-axis) in panel (b) is based on Allele1, chosen arbitrarily.

Table 2. ORFs and genes that changed after modification.

	ORFs	Genes
Unchanged	202,232	26,491
Changed	18,726	5,737

An ORF is said to be new or undergo changes if it shares no same-frame codons with the initial ORF before modification in the mRNA. The genes are constructed by merging the ORFs (from different mRNAs) that overlap in the same strand of a chromosome.

frequency of all alleles in each population (Figure 6b) directly from the UCSC HapMap data, regardless whether they cause new ORFs or not. The same four populations also stand out as having rare alleles. The reason presumably is an artifact of how the data was collected. These four populations belong to the earlier phase of HapMap project than the other seven populations [19].

For all the ORFs that underwent change, we discover that the majority of them (54%) appear in all 11 populations and around 23% in 4 populations (CEU, CHB, JPT and YRI). In Supplementary Material S1 Section 2 we listed the detailed breakdown of the occurrences of ORFs in every population and additionally in Table S2 we have included lists of ORFs and their occurrence in every population.

Genes and mRNA Counts Affected By Polymorphism

In this work, an ORF is said to be new if it shares no same-frame codons with the initial ORF before modification in mRNA. This implies that these proteins could be completely distinct from one another. In Table 2 we list the numbers of human ORFs and genes affected by polymorphism.

We also looked at the types of polymorphism that cause ORF modification. The table in Supplementary Material S1 Section 3 indicates five types of polymorphisms: deletion (DLT), insertion (INS), insertion/deletion (IND) microsatellite (MIC), multiple nucleotide polymorphism (MNP), and single nucleotide polymorphism (SNP). We found that the primary source of modification that causes ORF change is SNP (79%) followed by insertion (3%), deletion (6%), insertion/deletion (1%), MNP (0.01%) and microsatellite (0.01%).This is consistent with the fact that SNP is the most common in the dbSNP database.

Among all the ORFs and genes, there are 8% a of them are affected by polymorphism. In order to verify whether the new ORF could be significantly affected by other polymorphisms, we perform a further experiment. From the new mRNA after initial modification we re-apply polymorphisms onto it (2nd modification). Although the second modification does change the counts of the disrupted ORFs, the effect is small (Supplementary Material S1 Section 5). One possible future direction following this result is

Table 3. Effect of randomization in ORFs and genes prediction.

Polymorphisms type	ORFs	Genes
Real	18,726	5,737
Random	3,004	1,330

Similar to the previous table, the figures refer to the number of ORFs that changed after modification with real and random polymorphisms. These figures are reported after validating the ORFs through Swiss-Prot homology.

Table 4. Number of new genes after modification by polymorphism with and without overlap with each of 12 other gene sets.

Gene Finder	No Overlap	With Overlap
acembly	365	5,372
ccdsGene	713	5,024
ensGene	473	5,264
geneid	541	5,196
genscan	449	5,288
hinv70Coding	284	5,453
knownGene	452	5,285
nscanGene	521	5,216
refGene	521	5,216
sgpGene	498	5,239
vegaGene	1,762	3,975
xenoRefGene	622	5,115

Table 5. Number of new genes after human polymorphism modification that overlap with other species' genes and not found by Ensembl gene finder (ensGene).

	With Overlap	No Overlap
chicken (galGal3)	65	408
medaka (oryLat2)	68	405
zebrafish (danRer7)	71	402
zebrafinch (taeGut1)	72	401
panda (ailMel1)	76	397
tetraodon (tetNig2)	77	396
frog (xenTro2)	78	395
lizard (anoCar1)	78	395
fugu (fr2)	80	393
stickleback (gasAcu1)	81	392
cat (felCat3)	81	392
pig (susScr2)	83	390
orangutan (ponAbe2)	86	387
chimpanzee (panTro2)	87	386
marmoset (calJac3)	89	384
rhesus (rheMac2)	89	384
rabbit (oryCun2)	90	383
guinea pig (cavPor3)	92	381
horse (equCab2)	93	380
elephant (loxAfr3)	94	379
mouse (mm9)	94	379
cow (bosTau4)	96	377
rat (rn4)	97	376
dog (canFam2)	101	372
opossum (monDom5)	104	369

Table 6. Molecular function gene ontology of new ORFs after modification.

Description	P-Value	FDR q-value	Enrichment (N,B,n,b)
ATP binding	1.49E-33	5.71E-30	1.46 (16846,1442,5262,659)
adenyl ribonucleotide binding	5.77E-33	1.11E-29	1.45 (16846,1466,5262,666)
binding	9.74E-33	1.25E-29	1.09 (16846,11419,5262,3897)
adenyl nucleotide binding	2.80E-32	2.69E-29	1.45 (16846,1475,5262,667)
catalytic activity	1.12E-27	8.63E-25	1.19 (16846,5141,5262,1909)
protein binding	1.40E-25	8.98E-23	1.14 (16846,7073,5262,2519)
purine ribonucleoside triphosphate binding	1.90E-25	1.04E-22	1.35 (16846,1776,5262,751)
purine ribonucleotide binding	4.93E-25	2.36E-22	1.35 (16846,1806,5262,760)
ribonucleotide binding	4.93E-25	2.10E-22	1.35 (16846,1806,5262,760)
purine nucleotide binding	1.67E-24	6.40E-22	1.34 (16846,1818,5262,762)
nucleotide binding	5.94E-21	2.07E-18	1.27 (16846,2318,5262,921)
nucleoside phosphate binding	6.92E-21	2.21E-18	1.27 (16846,2319,5262,921)
small molecule binding	8.24E-21	2.43E-18	1.26 (16846,2485,5262,978)
ion binding	1.70E-19	4.65E-17	1.19 (16846,3833,5262,1426)
cation binding	2.72E-19	6.95E-17	1.19 (16846,3825,5262,1422)
metal ion binding	4.41E-19	1.06E-16	1.19 (16846,3754,5262,1397)
kinase activity	4.11E-18	9.27E-16	1.47 (16846,756,5262,347)
phosphotransferase activity, alcohol group as acceptor	1.79E-17	3.82E-15	1.48 (16846,704,5262,325)
protein kinase activity	7.02E-17	1.42E-14	1.51 (16846,592,5262,280)
transferase act. transferring phosphorus containing grp.	7.95E-17	1.52E-14	1.42 (16846,875,5262,387)

Ranked top 20 terms according to the P-value of overrepresentation against the background set. Last column with 'Enrichment (N, B, n, b)' is defined as follows: N is the total number of genes, B is the total number of genes associated with the corresponding GO term (description), n is the number of genes in the target set, b is the number of genes in the intersection. Enrichment$=(b/n)/(B/N)$. Note that the total number of target genes (n) in the last column could be less or equal to the number of input genes. This is because GOrilla normalised the input gene names with its gene database.

Table 7. Biological process gene ontology of new ORFs after modification.

Description	P-Value	FDR q-value	Enrichment (N,B,n,b)
macromolecule modification	1.23E-13	1.51E-10	1.23 (16846,1977,5262,762)
phosphorylation	5.86E-13	6.48E-10	1.43 (16846,628,5262,280)
Phosphate containing compound metabolic process	9.70E-13	9.75E-10	1.37 (16846,798,5262,342)
phosphorus metabolic process	9.70E-13	8.94E-10	1.37 (16846,798,5262,342)
protein phosphorylation	1.14E-12	9.73E-10	1.44 (16846,575,5262,259)
cellular protein modification process	3.48E-12	2.75E-09	1.23 (16846,1884,5262,721)
protein modification process	3.48E-12	2.57E-09	1.23 (16846,1884,5262,721)
regulation of biological process	1.63E-11	1.12E-08	1.08 (16846,7733,5262,2615)
protein metabolic process	6.10E-11	3.97E-08	1.16 (16846,2921,5262,1061)
regulation of biological quality	3.45E-10	2.12E-07	1.19 (16846,2036,5262,759)
transmem.eceptor protein tyrosine kinase sig. pathway	4.16E-10	2.42E-07	1.42 (16846,489,5262,217)
response to stimulus	6.81E-10	3.77E-07	1.10 (16846,5534,5262,1901)
regulation of cellular process	8.46E-10	4.46E-07	1.08 (16846,7330,5262,2470)
enzyme linked receptor protein signaling pathway	2.26E-09	1.14E-06	1.34 (16846,658,5262,276)
cellular protein metabolic process	3.41E-09	1.64E-06	1.17 (16846,2348,5262,856)
cellular response to stimulus	4.72E-09	2.18E-06	1.12 (16846,4170,5262,1453)
organelle organization	5.27E-09	2.33E-06	1.20 (16846,1654,5262,621)
regulation of response to stimulus	6.76E-09	2.87E-06	1.17 (16846,2110,5262,774)
cellular component organization	7.02E-09	2.88E-06	1.14 (16846,3139,5262,1115)
Peptidyl tyrosine phosphorylation	8.21E-09	3.24E-06	2.17 (16846,59,5262,40)

Ranked top 20 terms according to the P-value of overrepresentation against the background set.

Table 8. Cellular component gene ontology of new ORFs after modification.

Description	P-Value	FDR q-value	Enrichment (N,B,n,b)
cell part	2.57E-32	3.30E-29	1.08 (16846,12147,5262,4108)
intracellular part	6.23E-28	4.00E-25	1.08 (16846,11593,5262,3922)
organelle	2.22E-18	9.52E-16	1.11 (16846,7819,5262,2703)
intracellular organelle	3.61E-18	1.16E-15	1.11 (16846,7802,5262,2696)
cytoplasmic part	2.28E-17	5.85E-15	1.12 (16846,6470,5262,2268)
cytosol	1.01E-16	2.16E-14	1.24 (16846,2261,5262,878)
cytoplasm	1.03E-14	1.88E-12	1.16 (16846,3850,5262,1398)
Membrane bounded organelle	3.36E-14	5.39E-12	1.10 (16846,6897,5262,2377)
intracellular membrane bounded organelle	3.74E-14	5.33E-12	1.10 (16846,6892,5262,2375)
organelle part	8.10E-10	1.04E-07	1.09 (16846,6066,5262,2070)
intracellular organelle part	9.22E-10	1.08E-07	1.09 (16846,5980,5262,2042)
nucleus	1.36E-09	1.46E-07	1.11 (16846,4598,5262,1597)
non-membrane-bounded organelle	1.55E-09	1.53E-07	1.19 (16846,1920,5262,715)
Intracellular non-membrane bounded organelle	1.55E-09	1.42E-07	1.19 (16846,1920,5262,715)
nuclear part	6.90E-08	5.91E-06	1.15 (16846,2412,5262,866)
cytoplasmic vesicle part	7.30E-07	5.86E-05	1.37 (16846,390,5262,167)
cell junction	1.60E-06	1.21E-04	1.26 (16846,694,5262,274)
Golgi apparatus	5.61E-06	4.00E-04	1.26 (16846,629,5262,248)
cell projection	8.08E-06	5.46E-04	1.22 (16846,819,5262,313)
nucleoplasm	1.10E-05	7.05E-04	1.21 (16846,912,5262,344)

Ranked top 20 terms according to the P-value of overrepresentation against the background set.

to investigate the cancelling effect of the above second modification.

Frequency of Random ORFs

We performed a control procedure where the chromosomal positions of polymorphism are randomized. The mRNAs are then modified with these false polymorphisms. Then we applied the same software and parameterization as described in the Method section for determining the longest ORF. Table 3 shows that the number of changes in ORF and gene after modification by random polymorphism is smaller than those caused by real polymorphism. In Supplementary Material S1 Section 1, we detail further the differences of these ORF predictions with respect to their overlap with known genes of other species.

Comparison With Other Gene Catalogs

In general gene prediction approaches can be classified into two categories: sequence similarity based searches and structure or signal based searches. The similarity search approach is based on finding similarity of ESTs (expressed sequence tags) and proteins to the input genome. On the other hand signal based searches rely on information such as motifs, splice sites, start and stop codons to identify the genes. Our method belong to both categories, where we treat the polymorphisms as signal and validation through Swiss-Prot proteins as similarity based approach. We examine the discrepancy of the genes found by our methods with the existing gene finders.

There are fourteen gene finders used by the UCSC Genome Browser to annotate the human genome. They are: Acembly [20], CCDS [21], Ensembl [22], GeneID [23], Genscan [24], KnownGene [25], H-InvDB [26], NSCAN [27], RefGene [28], SGP [29], Vega [30] and Xenoref [28]. We compile the number of our predicted genes that were not found and found by these gene finders in Table 4. The majority of our genes do have overlap, i.e. partially share genomic location, with those genes from other finders. Our finding should complement the results from existing gene finders. On average 10% of our novel coding genes have zero overlap with any given one of these gene finders. It was with H-InvDB genes where we found the greatest number of gene overlaps, around 5% of our genes have no overlap with it.

Gene Comparison With Other Species

Recently there is a growing interest in the origin of human protein coding genes [31]. It was suggested that novel genes regularly appear from messenger RNAs of ancestral genes, and such novel genes significantly affect the evolution of cellular, physiological, morphological, behavioral and reproductive phenotypic traits [32–34].

We compared our novel protein coding human genes subject to polymorphism with their counterpart genes in 25 other genomes. For these genomes we find their genes based on the annotation given by Ensembl gene prediction software [22]. Table 5 shows the number of new protein coding genes discovered by our method that overlap and do not overlap with genes in each of the species. Surprisingly, we discovered that there are some overlaps between predicted genes and the known genes of these species. Moreover, we can distinguish between mammals and non-mammals from the table. The mammals have more overlaps (≥ 81) than non-mammals.

Gene Ontology

Discovery of new genes will be useful if we can elucidate their roles in various biological domains. Gene ontology (GO) provides such a framework. We use Uniprot to estimate gene ontologies of our newly defined ORFs based on their homologs in Swiss-Prot, because Uniprot provides high-quality gene ontology annotation for proteins from multiple species [35].

We tabulate three categories of gene product (molecular function, cellular function and biological process) where the ORFs are over-represented (Table 6, 7, 8). The ontologies are assigned using the web based application GOrilla [36].

In the initial experiment we used a target set of gene names from Uniprot that have homology to our predicted ORFs, and a background set of gene names from Uniprot that have homology with longest ORFs before modification by polymorphism. In this experiment we wish to show what is overrepresented in the new ORFs with respect to old ORFs. We found that in the molecular function category, *ATP binding is the most overrepresented term with significant P-value (1.49E-33)*. For the biological process category we found that *macromolecule modification* is the most overrepresented term with significant P-value (1.23E-13). Finally for the cellular component category we found terms with significant P-value (2.57E-32) related to *cell part*.

We also performed another experiment where we apply gene names from Uniprot that have homology to old ORF and our newly predicted ORFs as target sets separately. All human proteins from Uniprot are used as background set. With this experiment we want to show if the polymorphism does change the GO assignment, and whether we can find more significant terms compared to the old ORFs. The results (Supplementary Material S1 Section 6) showed that polymorphism does indeed change the GO assignment. Moreover, the P-values of GO terms for ORFs after modification are more significant than those before modification.

Conclusion

In this work we describe our findings on the importance of human polymorphism in determining protein-coding genes. Here we indicate that the human population harbors a number of genes that have been missed by previous analyses focusing on one reference genome alone. In our approach we employ strict bioinformatics criteria: ORF length, absence of repeats, E-value of ORF homology to Swiss-Prot, and length of 5′ UTR to verify our findings. We demonstrate that from the new protein-coding genes there were some not found by other gene-finder algorithms, but located in genomic regions where genes are annotated in other species.

The UCSC Genome Browser also continuously updates their data especially with regards to newer human genome assemblies, introduction of new genomes, new gene predictors and annotation of polymorphism. Along with this information, in the short-term when more complete SNP catalogs, frequency data and more distant species will be made available; we hope to explore further the impact of human variations in gene finding. It would be useful for example, to probe for the differences between newly predicted genes and known annotated genes by looking at their sequence similarity in addition to their genomic regions.

And eventually there will be data that include haplotype combinations, more RNAs as well as full-length transcripts. To adapt to these future changes, we believe it would be helpful to make available a gene database based on our approach. A dynamic browsing function for the location of ORFs before and after modification would intuitively convey our results.

Materials and Methods

Source of sequence and polymorphism data

For consistency, all our information described below is obtained from a single resource: the UCSC Genome Browser. First of all, the human mature mRNA sequences from GenBank [37] were used as the source for deriving ORFs. The sequences were obtained from http://hgdownload.cse.ucsc.edu/goldenPath/hg19/bigZips/mrna.fa.gz. In total there are 917,725 mRNA sequences. To construct the genes we utilize the annotation of mRNA location in the human genome. It can be downloaded from http://hgdownload.cse.ucsc.edu/goldenPath/hg19/database/all_mrna.txt.gz.

We applied human polymorphism from the dbSNP build131 database [38] for reconstructing the mRNAs. In total we considered 25,877,929 variations. The annotation of these variants on the human genome (hg19) was obtained from http://hgdownload.cse.ucsc.edu/goldenPath/hg19/database/snp131.txt.gz. The variants consist of: SNP - single nucleotide polymorphism, MNP- multiple nucleotides polymorphism, microsatellite - variation in forms of short tandem repeats, insertion - the polymorphism as insertions relative to the reference assembly, deletion - the polymorphism as deletion relative to the reference assembly and in-del - insertion/deletion. In-del is a special class defined by dbSNP. We compare the length of the reference allele to the length(s) of observed alleles; if the reference allele is shorter than all other observed alleles, then 'in-del' will be considered as 'insertion'. Likewise, if the reference allele is longer than all other observed alleles, it will be considered as 'deletion'. The table in Supplementary Material S1 Section 4 shows the overall frequencies of the above mentioned polymorphisms.

Defining mRNAs and ORFs from polymorphism

The reconstruction of mRNA is done by mutating, inserting or deleting the nucleotides in the mRNA according to the variant alleles in dbSNP, based on the positions of the variants and mRNA in the human genome (hg19). Lacking information on haplotypes (i.e. combinations of polymorphisms), we applied each polymorphism on its own. Thus, it is possible that some haplotypes that we considered do not exist in any individual. For each newly reconstructed mRNA we obtained the longest ORF using getorf from EMBOSS [39], with options -min 300 -norev -find 1. We used 300 bp nucleotide as minimum length for ORFs prediction as suggested by many studies that this is an appropriate threshold [40,41]. Our approach is also based on the standard view of mRNA that it only encodes one distinct protein.

The sequence of triplet codons in ORFs beginning with ATG and ending with a stop codon represents the protein. Here we only look at ORFs with $>=100$ codons, because three out of sixty-four codons encode stops and ORFs greater than 100 codons are unlikely to appear by chance in non-coding sequences of average composition [13]. We define genes by merging the ORFs from different mRNAs that overlap in the same strand of a chromosome. Figure 2 exhibits the procedure for identifying novel coding genes through polymorphism. Notice that in our procedure, there is no re-alignment of the modified mRNA to the human genome involved. For the purpose of determining the genomic loci of exons of new mRNAs, we incorporate the mRNA annotation provided by the UCSC Genome Browser. In this annotation the coordinates of the mRNAs in the genome are specified. Such information enables us to convert the exon locations in new mRNAs into exon locations in the genome.

We also find supporting evidence to show that our newly predicted ORFs are translated into proteins. As a first criterion, we selected ORFs that aligned to the manually curated Swiss-Prot protein database with alignment score ≥ 130. This scoring threshold corresponds to significant homology (E-value ≤ 3.07e-6). At the end of this step we obtained 5,737 protein-coding genes. The majority of the ORFs have low E-values (0) as can be seen from their cumulative distribution in Supplementary Material S1 Section 4. This procedure will allow us to reduce erroneous translation of frameshifted or non-coding nucleotide sequences. The alignment was done using LAST [42]. It contains two sub-programs: lastdb and lastal. We indexed the protein database with lastdb with parameter -p to indicate that the sequences are amino acids and -c to soft-mask lower case letters. The actual alignment was done using lastal with parameter -e 130 which sets the minimum score for gapped alignments to 130. The E-value was computed using the lastex program that comes in the package. Both the translated ORFs and Swiss-Prot proteins were repeat masked using TANTAN [43] before performing the alignment.

As a final criterion we select ORFs based on proximity to the mRNA 5′ end. It has been suggested that the ribosome scanning mechanism favors short 5′ UTRs, whereas long 5′ UTRs are likely to contain encumbrances, e.g. upstream start codons or secondary structure elements [44,45].

Allele Frequency

The percentage of alleles is obtained from the count of individuals who are homozygous and heterozygous for the observed alleles. This information is available from the UCSC Genome Browser HapMap annotation: http://hgdownload.cse.ucsc.edu/goldenPath/hg19/database/hapmapSnps*.txt.gz. Following their nomenclature, the zygote counts are defined as follows: *homoCount*1 and *homoCount*2 are counts of individuals who are homozygous for the first allele and second allele respectively; *heteroCount* is the count of individuals who are *heterozygous*.

Subsequently we define counts of first of and second alleles (*Allele*1*Count* and *Allele*2*Count*):

$$Allele1Count = 2\cdot homoCount1 + heteroCount$$

$$Allele2Count = 2\cdot homoCount2 + heteroCount$$

Finally the allele percentage can be obtained as:

$$Allele1\% = \frac{Allele1Count}{Allele1Count + Allele2Count}\cdot 100\%$$

These percentages are then computed with respect to our predicted ORFs, by looking at the correspondence of modifying variants in our method and the HapMap annotation above.

Comparison with Other Gene Catalogs

There are three sources of information involved in comparing our result to other gene catalogs: 1) list of the new ORFs with their positions in mRNA, 2) alignment of human mRNA with the human genome, and 3) fourteen gene annotations of the human genome. The primary information we employed from these annotations are the predicted starting and ending positions of the exons. We obtained the last two sources from the UCSC Genome Browser.

The key idea in these steps is to compare the overlaps with respect to the genomic location of our predicted ORFs and those predicted by other finders. Initially, we located the positions of the new ORFs in the human genome. We can do that by transferring the positions of new ORFs in mRNA with respect to alignment of human mRNA to the human genome. Since the annotations of all other catalogs already provide the genomic location of the exons, we can easily find their overlap or non-overlap with the modified new ORFs.

Gene Comparison with Other Species

For this task we use the following data sources: 1) list of new ORFs obtained after modification and their corresponding position in the mRNA, 2) alignment of human mRNA with the human genome, 3) gene prediction for human genome (hg19) and all other 25 species provided by Ensembl software [22], 4) human genome alignment with other species. We obtained data sources 2 to 4 from the UCSC genome browser.

Similarly to the previous section, we defined overlap based on the genomic location of new ORFs and gene location of other species aligned to human genome. Hence, the first step we did was to find the positions of the new ORFs in the human genome. This can be done by transferring the positions of data source (1) using data source (2). The next crucial step is to find the location of exons of all other species predicted by Ensembl that align to the human genome. The key data source for this step is (3) and (4). Finally, the overlap and non-overlap of new ORFs in human and all other species can be determined.

Gene Ontology

In finding over-represented gene ontology terms resulting from the newly predicted genes after polymorphism, we applied GOrilla (http://cbl-gorilla.cs.technion.ac.il/), a web-based application. We used *two unranked* lists of genes as the option. We performed two types of experiments for the analysis:

1. For the first experiment, the target list contains 5,630 gene names from Uniprot that have homology to our 18,726 predicted ORFs. The background set consists of 46,961 gene names from Uniprot that have homology with ORFs before modification by polymorphism.
2. In the second experiment, we use all 71,124 human gene names from Uniprot as background set. Then *separately* we use 5,630 gene names from ORFs after and 46,961 gene names from ORFs before modification by polymorphisms as target set.

In both of the above experiments when an ORF has multiple hits to the Uniprot gene names, we only took gene name with highest E-value.

GOrilla uses the minimum hypergeometric score (mHG) statistic for computing the P-value. In total there are 11,109 GO terms used in our computation. To correct the P-value for multiple testing, we further compute the FDR q-value [46]. It is computed as:

$$\text{FDR q-value} = \frac{\text{p-value} * \text{Number of GO Terms}}{\text{Rank of Term}}$$

Supporting Information

Figure S1 GD144643 alignment to hg19.

Figure S2 Effect of polymorphism in the translation of the 3 examples (AK124706, AK127273 and AY129028).

Table S1 List of new genes after modification of mRNA by polymporhisms.

Table S2 List of new ORFs categorized for 11 populations.

Supporting Information S1 Gene Ontology data used for experiment as described in the method section.

Acknowledgments

We thank Dr. Michiaki Hamada, Dr. Toutai Mituyama and Dr. Penmetcha Kumar for constructive comments.

Author Contributions

Conceived and designed the experiments: EW MCF. Performed the experiments: EW. Analyzed the data: EW MCF. Contributed reagents/materials/analysis tools: KA PH. Wrote the paper: EW MCF KA PH.

References

1. The International Human Genome Sequencing Consortium (2004) Finishing the euchromatic sequence of the human genome. Nature 431: 931–45.
2. Mathé C, Sagot M, Schiex T, Rouzé P (2002) Current methods of gene prediction, their strengths and weaknesses. Nucleic Acids Research 30: 4103–17.
3. Clamp M, Fry B, Kamal M, Xie X, Cuff J, et al. (2007) Distinguishing protein-coding and noncoding genes in the human genome. Proc Natl Acad Sci U S A 104: 19428–33.
4. Brent M (2005) Genome annotation past, present, and future: how to define an ORF at each locus. Genome Research 15: 1777–86.
5. Genomes Project Consortium, Durbin R, Abecasis G, Altshuler D, Auton A, et al. (2011) A map of human genome variation from population-scale sequencing. Nature 470: 59–65.
6. Cooper D, Chen J, Ball E, Howells K, Mort M, et al. (2010) Genes, mutations, and human inherited disease at the dawn of the age of personalized genomics. Hum Mutat 31: 631–55.
7. Halvorsen M, Martin J, Broadaway S, Laederach A (2010) Disease-associated mutations that alter the RNA structural ensemble. PLoS Genet 6: e1001074.
8. Ng P, Henikoff S (2002) Accounting for human polymorphisms predicted to affect protein function. Genome Research 12: 436–46.
9. Shimada M, Matsumoto R, Hayakawa Y, Sanbonmatsu R, Gough C, et al. (2008) VarySysDB: a human genetic polymorphism database based on all H-InvDB. Nucleic Acids Research 37: D810–5.
10. Yamaguchi-Kabata Y, Shimada M, Hayakawa Y, Minoshima S, Chakraboty R, et al. (2008) Distribution and effects of nonsense polymorphisms in human genes. PLOS One 3: e3393.
11. Wilson BA, Masel J (2011) Putatively noncoding transcripts show extensive association with ribosomes. Genome Biology and Evolution 3: 1245–1252.
12. Carvunis A, Rolland T, Wapinski I, Calderwood M, Yildirim M, et al. (2012) Proto-genes and denovo gene birth. Nature 487: 370–4.
13. Frith M, Bailey T, Kasukawa T, Mignone F, Kummerfeld S, et al. (2006a) Discrimination of non-protein-coding transcripts from protein-coding mRNA. RNA Biol 3: 40–8.
14. Ota T, Suzuki Y, Otsuki T, Sugiyama T, Irie R, et al. (2004) Complete sequencing and characterization of 21,243 full-length human cDNAs. Nat Genet 36: 40–5.
15. The MGC Project Team (2004) The status, quality, and expansion of the NIH full-length cDNA project: the Mammalian Gene Collection (MGC). Genome Research 14: 2121–2127.
16. Redon R, Ishikawa S, Fitch K, Feuk L, Perry G, et al. (2006) Global variation in copy number in the human genome. Nature 444: 444–454.
17. Spielman R, Bastone L, Burdick J, Morley M, Ewens W, et al. (2007) Common genetic variants account for differences in gene expression among ethnic groups. Nature Genet 39: 226–31.
18. The International HapMap Consortium (2007) A second generation human haplotype map of over 3.1 million SNPs. Nature 449: 851–61.
19. Fujita P, Rhead B, Zweig A, Hinrichs A, Karolchik D, et al. (2011) The UCSC Genome Browser database: update 2011. Nucleic Acids Res 39: D876-D882.
20. Thierry-Mieg D, Thierry-Mieg J (2006) AceView: a comprehensive cDNA-supported gene and transcripts annotation. Genome Biol 7: S12.1–14.
21. Pruitt K, Harrow J, Harte R, Wallin C, Diekhans M, et al. (2009) The consensus coding sequence (CCDS) project: Identifying a common protein-coding gene set for the human and mouse genomes. Genome Research 19: 1316–23.
22. Hubbard T, Barker D, Birney E, Cameron G, Chen Y, et al. (2002) The Ensembl genome database project. Nucl Acids Res 30: 38–41.
23. Blanco E, Parra G, Guigó R (2002) Using geneid to identify genes. In: Baxevanis A, editor, Current Protocols in Bioinformatics, New York: John Wiley & Sons Inc. p. Unit 4.3.
24. Burge C, Karlin S (1997) Prediction of complete gene structures in human genomic DNA. J Mol Biol 268: 78–94.
25. Hsu F, Kent W, Clawson H, Kuhn R, Diekhans M, et al. (2006) The UCSC known genes. Bioinformatics 22: 1036–46.
26. Yamasaki C, Murakami K, Takeda J, Sato Y, Noda A, et al. (2010) H-InvDB in 2009: extended database and data mining resources for human genes and transcripts. Nucleic Acids Research 38: D626–32.
27. van Baren M, Brent M (2006) Iterative gene prediction and pseudogene removal improves genome annotation. Genome Res 16: 678–85.
28. Pruitt K, Tatusova T, Maglott D (2005) NCBI Reference Sequence (RefSeq): a curated non-redundant sequence database of genomes, transcripts and proteins. Nucleic Acids Research 33: D501–4.
29. Wiehe T, Gebauer-Jung S, Mitchell-Olds T, Guigó R (2001) SGP-1: prediction and validation of homologous genes based on sequence alignments. Genome Research 11: 1574–83.
30. Wilming L, Gilbert J, Howe K, Trevanion S, Hubbard T, et al. (2008) The vertebrate genome annotation (Vega) database. Nucleic Acids Research 36: D753–60.
31. Knowles D, McLysaght A (2009) Recent de novo origin of human protein-coding genes. Genome Research 21: 487–93.
32. Kaessmann H (2010) Origins, evolution, and phenotypic impact of new genes. Genome Research 20: 1313–26.
33. Tautz D, Domazet-Lošo T (2011) The evolutionary origin of orphan genes. Nature Reviews Genetics 12: 692–70.
34. Wu D, Irwin D, Zhang Y (2011) De novo origin of human protein-coding genes. Plos Genetics 7: e1002379.
35. The UniProt Consortium (2011) Ongoing and future developments at the Universal Protein Resource. Nucleic Acids Res 39: D214–D219.
36. Eden E, Navon R, Steinfeld I, Lipson D, Yakhini Z (2009) Gorilla: A tool for discovery and visualization of enriched go terms in ranked gene lists. BMC Bioinformatics: 48.
37. Benson D, Karsch-Mizrachi I, Lipman D, Ostell J, Wheeler D (2004) GenBank: update. Nucleic Acids Research 32: D23–6.
38. Sherry S, Ward M, Kholodov M, Baker J, Phan L, et al. (2001) dbSNP: the NCBI database of genetic variation. Nucleic Acids Res 29: 308–11.
39. Rice P, Longden I, Bleasby A (2000) EMBOSS: The European Molecular Biology Open Software Suite. Trends in Genetics 6: 276–277.
40. Sharp P, Cowe E (1991) Synonymous codon usage in saccharomyces cerevisiae. Yeast 7: 657–78.
41. Oliver S, van der Aart Q, Agostoni-Carbone M, Aigle M, Alberghina L, et al. (1992) The complete DNA sequence of yeast chromosome III. Nature 357: 38–46.
42. Kie lbasa S, Wan R, Sato K, Horton P, Frith M (2011) Adaptive seeds tame genomic sequence comparison. Genome Research 21: 487–93.
43. Frith M (2011) A new repeat-masking method enables specific detection of homologous sequences. Nucleic Acids Res 39: e23.
44. Kozak M (2002) Pushing the limits of the scanning mechanism for initiation of translation. Gene 10: 1752–9.
45. Kozak M (2005) Regulation of translation via mRNA structure in prokaryotes and eukaryotes. Gene 361: 13–37.
46. Benjamini Y, Hochberg Y (1995) Controlling the false discovery rate: a practical and powerful approach to multiple testing. Journal of the Royal Statistical Society Series B 57: 289–300.

12

Extensive Variation in Gene Copy Number at the Killer Immunoglobulin-Like Receptor Locus in Humans

Sanne Vendelbosch[1]*, Martin de Boer[1], Remko A. T. W. Gouw[1], Cynthia K. Y. Ho[1], Judy Geissler[1], Wendy T. N. Swelsen[2], Michael J. Moorhouse[1], Neubury M. Lardy[2], Dirk Roos[1], Timo K. van den Berg[1], Taco W. Kuijpers[1,3]

1 Sanquin Research and Landsteiner Laboratory, Department of Blood Cell Research, Amsterdam, The Netherlands, **2** Sanquin Blood Supply Foundation, Department of Immunogenetics, Amsterdam, The Netherlands, **3** Emma Children's Hospital, Department of Pediatric Hematology, Immunology and Infectious Diseases, Academic Medical Center, Amsterdam, The Netherlands

Abstract

Killer immunoglobulin-like receptors (KIRs) are involved in the regulation of natural killer cell cytotoxicity. Within the human genome seventeen KIR genes are present, which all contain a large number of allelic variants. The high level of homology among KIR genes has hampered KIR genotyping in larger cohorts, and determination of gene copy number variation (CNV) has been difficult. We have designed a multiplex ligation-dependent probe amplification (MLPA) technique for genotyping and CNV determination in one single assay and validated the results by next-generation sequencing and with a KIR gene-specific short tandem repeat assay. In this way, we demonstrate in a cohort of 120 individuals a high level of CNV for all KIR genes except for the framework genes *KIR3DL3* and *KIR3DL2*. Application of our MLPA assay in segregation analyses of families from the Centre d'Etude du Polymorphisme Humaine, previously KIR-genotyped by classical techniques, confirmed an earlier reported duplication and resulted in the identification of a novel duplication event in one of these families. In summary, our KIR MLPA assay allows rapid and accurate KIR genotyping and CNV detection, thus rendering improved transplantation programs and oncology treatment feasible, and enables more detailed studies on the role of KIRs in human (auto)immunity and infectious disease.

Editor: Christian Schönbach, Kyushu Institute of Technology, Japan

Funding: This research was supported by the Sanquin Blood Supply Foundation, The Netherlands, Grant number PPOC-09-018. The funders had no role in study design, data collection and analysis, decision to publish, or preparation of the manuscript.

Competing Interests: The authors have declared that no competing interests exist.

* E-mail: s.vendelbosch@sanquin.nl

Introduction

The immune system is a complex network that protects the human body against internal and external threats. It consists of an innate and an adaptive system. Natural killer (NK) cells constitute the first line of innate immunity against viral infections and tumor development. Recent studies have shown that NK cells can also be educated during development and that memory NK cells are able to mount a more effective cytokine response upon reactivation, suggesting adaptive functions [1,2].

The importance of human NK cells has become clear through studies concerning cancer, viral infections and donor/recipient panels in transplantation settings. The cytotoxicity of NK cells is regulated through large families of molecules with a homologous structure, belonging to the C-type lectin-like molecules and immunoglobulin-like receptors [3].

The most prominent family that regulates NK cell cytotoxicity in humans comprises the killer immunoglobulin-like receptors (KIRs). Binding of a KIR to one of its ligands, Human Leukocyte Antigens (HLA) class-I molecules and possibly others, triggers either an activating or an inhibiting signal, thereby controlling the immune reactivity of NK cells.

A widely studied application is the matching of KIRs in donors and recipients in transplantation settings [4–7]. Alloreactive NK cells can lead to rejection of organs after transplantation [7]. In contrast, donor KIR-ligand mismatching with the recipient during hematopoietic stem cell transplantation (HSCT) may be required for the beneficial effects of NK cell alloreactivity against tumor cells [5,8]. Apart from inhibitory KIR-ligand mismatches, the presence of activating KIR genes in the donor grafts has been shown to improve survival after unrelated HLA-matched HSCT [9,10]. These studies and others have shown that a favorable combination of HLA and KIR genotyping of both donor and recipient can strongly enhance the chance of success of transplantation.

The genes for the KIR family are present on human chromosome 19q13.4 [11]. In humans, seventeen highly homologous genes, including two pseudogenes, have been identified to date. For all of these genes a large number of allelic variants exists, which resemble each other up to 98% in sequence homology. Distribution of the genes is thought to occur on two basic haplotypes, designated A and B, both of which contain the same so-called framework genes (*KIR3DL3*, *KIR3DP1*, *KIR2DL4* and *KIR3DL2*). Other KIR genes may or may not be present on one of the haplotypes [12,13]. In addition, individuals may carry more than two copies of a KIR gene, as has been shown so far only in case of *KIR3DL1* and *KIR3DS1* [14–16]. The immunological relevance of copy number variation (CNV) in the *KIR3DL1* and

KIR3DS1 genes was shown by Pelak and co-workers [15], who described that a higher number of genes relates to a better individual resistance against human immunodeficiency virus type 1 (HIV-1). This resistance was related to a higher number of KIR3DS1 protein-expressing NK cells and an inhibition of replication of HIV-1 [15]. The importance of CNV in relation to disease was also supported by the correlation of increased clearance of hepatitis C virus (HCV) in individuals with of two copies of *KIR2DL3* compared to those with one or no copies [17].

Tools to study the extent and functional meaning of this inter-individual KIR locus variation on a larger scale are currently lacking. From other immune receptors we know that extensive CNV exists and that the number of genes can relate to protein levels, leading to a difference in disease outcome. We have observed this for the genes encoding the human Fc-gamma receptors (FcγR) and complement factor 4 [18–20]. To study the level of CNV in the KIR gene cluster, we have used a convenient, sensitive and efficient MLPA-based method to detect all KIR genes in one assay. The introduction of a synthetically derived calibrator has allowed us to accurately quantify KIR gene CNV. Validation of the method by comparing the MLPA method with the standard polymerase chain reaction (PCR) with sequence-specific primers (SSP) in a large cohort of individuals indicated that the MLPA method provided more accurate genotyping. The MLPA method showed an unexpected range of CNV at the KIR locus. Segregation analyses in pedigrees previously genotyped by PCR-SSP confirmed the strength and accuracy of the MLPA method and helped to identify duplication events in the KIR gene cluster in one of these families.

Taken together, CNV at the KIR locus is extensive. The KIR MLPA assay can accurately determine an individual's KIR genotype in a highly efficient manner, allowing routine KIR genotyping in transplantation programs and offering the opportunity to increase the success rates of transplantation or graft-versus-leukemia effects. Also, genotype-phenotype relations may be studied in greater detail to better understand the exact role of KIRs in health and disease.

Results

KIR Genotyping by MLPA Methodology

In designing the set of probes for the KIR MLPA we used the following principles. *First,* making use of publicly available databases (http://www.ebi.ac.uk/ipd/kir/), we selected single nucleotide polymorphisms (SNPs) in each KIR gene that selectively distinguish one gene from all others. In some cases two of those SNPs were combined together in one probe set that consists of three probe parts, to make the probe segregating between the KIR genes. This has previously been applied in the MLPA assay for the complement genes in the HLA class III region [20]. The KIR gene-specific SNPs were chosen such that all allelic variants would be recognized by the corresponding probe. *Secondly,* some specific, additional KIR variant sequences were selected to make a further distinction among the variants of a particular KIR gene. In this way, a probe-based distinction was possible between wild-type *KIR2DS4* and truncated variants of this gene (*KIR2DS4*003-*010, *012, *013*) [21–23]. Next, we designed our probes mainly inside exons of the selected gene sequences (Tables S1 and S2). One allelic variant of *KIR3DL1* (*KIR3DL1*054*) is not detected, because this allele contains a SNP at the specific probe-binding site.

Where possible, two separate independent probes per KIR gene were included in the assay as an internal check on the presence or absence of any of the genes at the KIR gene cluster. The latter is particularly important for excluding putative false-negative results due to SNPs that have not been documented yet. To avoid, on the one hand, possible competition between probes for the same KIR gene and on the other hand to accommodate all probe amplification products in the same assay, we divided the probes over three different probe mixes (Table S3).

Accuracy of the KIR MLPA Assay

The MLPA method was subsequently performed to genotype 120 Caucasian individuals (Table 1). The occurrence frequency of KIR genes in our study population of control individuals was similar to the frequency reported in the "Allele Frequency Net Database" (http://www.allelefrequencies.net/kir6002a.asp; update November 2011 [24]) (Table 1), hence lending credibility to the MLPA-derived genotyping of KIR genes.

PCR with sequence-specific primers (SSP) was performed routinely in parallel to evaluate the accuracy of the MLPA assay by comparing the results in the same cohort of 120 individuals (Table 1). Upon comparison, the datasets of both assays matched for 99.5%. Of the 17 KIR genes per donor typed by both methods, in these 120 individuals 12 individuals showed a mismatch in a single KIR gene.

All KIR genes of 5 individuals were amplified by long-range PCR and sequenced by Next-generation sequencing (NGS) on a 454 FLX Roche Genome Sequencer to resolve these discrepancies (Figure 1). In all cases, the methodological discrepancies appeared to be due to false-positive or false-negative PCR reactions in the PCR-SSP. Upon re-evaluation of the PCR-SSP data, some explanations for the mismatches were found. For one, *KIR2DS1*

Table 1. KIR genotype frequency as determined by MLPA, compared with data from the allele frequency database and determination by PCR-SSP.

KIR Gene	Allele frequency		MLPA *vs* PCR-SSP
	MLPA	Database*	
2DL1	95.8%	87.8–100%	99.2%
2DL2	53.3%	32–67.9%	100%
2DL3	87.5%	84–97%	100%
2DL4	100.0%	99–100%	100%
2DL5	56.3%	41.1–67%	98.3%
2DS1	37.5%	27–54.5%	93.3%
2DS2	53.8%	40–63%	100%
2DS3	31.7%	18.7–41%	100%
2DS4all	96.6%	83.1–98.8%	100%
2DS4wt	39.2%	33–49.6%	100%
2DS4trunc	81.7%	44.2–89%	100%**
2DS5	33.3%	20.8–43%	100%
2DP1	95.8%	94–100%	99.2%
3DL1	96.7%	89–98.5%	100%
3DL2	100.0%	100.0%	100%
3DL3	100.0%	100.0%	100%
3DS1	41.7%	27–63%	100%
3DP1	100.0%	97–100%	100%

n = 120; * Range of Caucasoid populations www.allelefrequencies.net/kir6002a.asp;
**SSP data available for 25 donors.

is represented by a long 1800 base-pair PCR product, leading to some false-negative results for this KIR gene in the PCR-SSP, due to insufficient amplification. Also, the particular PCR product detected for *KIR2DP1* appeared to be too long, leading to an aspecific product and a false-positive result for *KIR2DP1* in the PCR-SSP. From these results we conclude that the KIR MLPA method performed with a higher degree of accuracy than the PCR-SSP method.

Quantification of KIR CNV with a Calibrator

The proposed haplotypes (Figure 2A) contain a variable number of KIR genes within each individual's genome, leading to gene copy number variation (CNV). Within each genome particular genes may be absent or present once, twice and sometimes even more than twice [12,14–16,25]. Our previous experience with the MLPA methodology had already indicated that this technique allows a reliable assessment of the gene copy number in the human genome [18,20]. However, the MLPA also has some disadvantages, one of which is the use of a reference DNA sample that is employed to determine the gene copy number and to reduce inter-assay variation. Since the KIR genotype of this reference sample is unknown, this does not allow an accurate determination of individual KIR gene CNV. To circumvent this problem we created a synthetic control as an internal calibrator for our KIR MLPA method. This internal standard consists of a DNA vector that includes the sequences of all of the probe-binding sites once. The gene-assembly method, as previously described by Stemmer and co-workers, was adapted to create the KIR calibrator [26].

With this calibrator we analyzed the cohort of 120 individuals for CNV (Figure 2B, Tables S4 and S6). According to our previous experience the number of 0, 1, 2, 3 and >3 genes can be discriminated in a highly reproducible manner. Several results should be noted. First, the framework genes *KIR3DL2* and *KIR3DL3* were represented by 2 copies in most individuals, which fits with the current A and B haplotype distribution system [13]. However, the other two framework genes *KIR3DP1* and *KIR2DL4*, separating the two blocks of genes in each haplotype (Figure 2A), have been deleted or duplicated in a small percentage of individuals. *KIR2DL4* occurs 2 times in 88% of the individuals, which suggests that translocation events have taken place in 12% of the individuals. The same seems true for *KIR3DP1*, which occurs 2 times in 84% of the individuals. Furthermore, we observed variation in gene copy numbers for *KIR2DL1*, *KIR2DP1*, *KIR2DL3* and *KIR3DL1*. The *KIR2DL3* and *KIR3DL1* genes, which are considered to be specific for the so-called A haplotype [12], occur in relatively high numbers, with >70% of individuals carrying 2 or 3 copies. Finally, several of the activating genes of the so-called B haplotype [12], such as *KIR2DS1*, *KIR2DS2* and *KIR2DS5*, are rarely seen in 2 or more copies within the genome of the population tested. Within the tested population, the average number of copies per inhibiting receptor was 1.2 (not including the framework genes), while the average number of copies for activating receptors was 0.5.

Quantification of CNV by MLPA Validated by Sequencing

To validate whether our MLPA could determine CNV accurately, we needed an independent assay that could quantify the number of copies of the same gene or the number of the same gene fragments. For this purpose, we used the presence of short tandem repeats (STRs) in the KIR gene locus. STRs are short sequences that have been duplicated multiple times, resulting in stretches of conserved repeated sequences. In KIR genes, several STR regions exist, one of which is a repeat of 4 base pairs, AGAT, located in intron 4 of most KIR genes [25,27,28]. Closer examination of this region showed that each KIR gene contains an invariable number of repeats, with the exception of *KIR2DL4* and *KIR2DL5*, which lack the typical intron 4 of the other KIR genes. Within this STR region single nucleotide polymorphisms (SNPs) are present, which allows for identification of the individual KIR genes even if the number of the repeats is equal.

It is impossible to determine CNV of the KIR genes by these STRs by MLPA methodology. Apart from the many technical issues with respect to the design of such probes, these STRs may not only be present within the KIR gene cluster but might occur

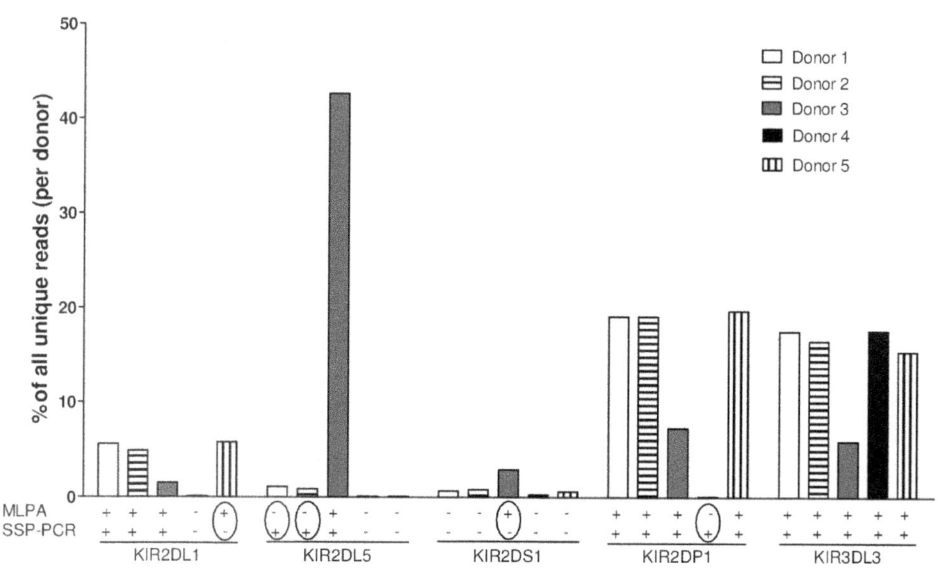

Figure 1. Validation of the accuracy of genotyping with MLPA by next-generation sequencing (NGS). Five of the genomes with mismatches between MLPA and PCR-SSP were analyzed by NGS. The percentage of reads that aligned against a reference for a KIR gene gives an impression about the presence of the KIR gene in that specific donor. Circles indicate the mismatches found between MLPA and PCR-SSP. *KIR3DL3* is a control gene as it is present in all donors.

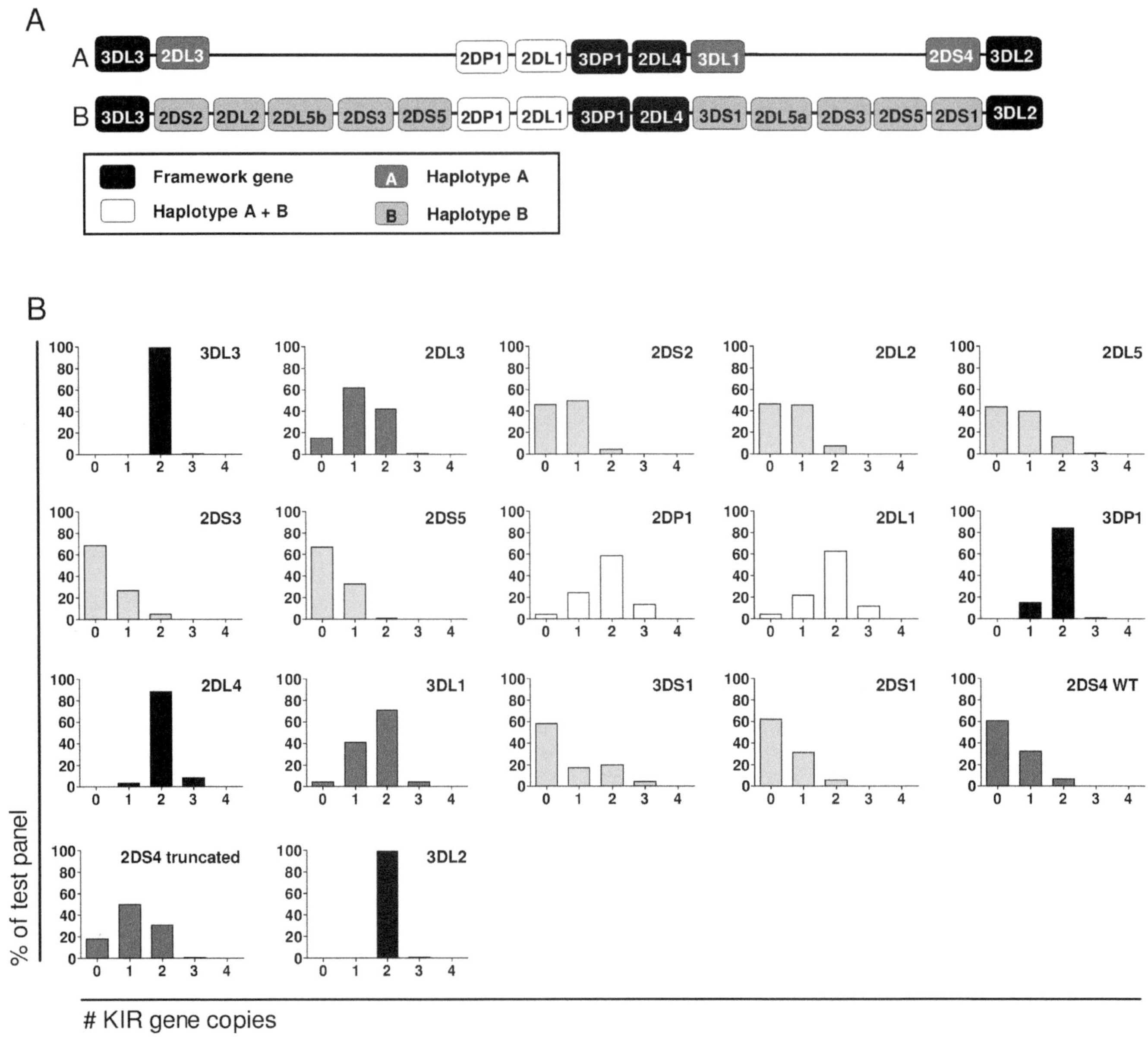

Figure 2. KIR gene copy number determination of a large cohort by MLPA. (A) Overview of the Killer Immunoglobulin-like Receptor gene family. The genes are arranged in two haplotypes (A and B), which both always contain the framework genes as depicted in black. Both haplotypes may or may not contain the genes in white. The genes depicted in grey are supposedly found only on haplotype A or B. (B) KIR gene copy numbers were determined for 120 individuals by MLPA. Shown is a graphical representation of the percentages of donors with 0, 1, 2, 3 or more than 3 copies of each KIR gene in their genome.

more widely throughout the human genome. However, it is possible to use the information about STRs within the KIR gene cluster for an independent validation of KIR CNV quantification obtained by the MLPA method. NGS enabled specific STR detection to determine the presence or absence of KIR genes (except *KIR2DL4* and *KIR2DL5*). CNV can be quantified by determining the coverage of the amplified product, *i.e.* the relative occurrence of a certain STR length and specific sequence, as a signature of the corresponding KIR gene.

Hence, to evaluate the accuracy of CNV determination in the KIR MLPA assay, primers were designed to amplify these repeat stretches in 12 samples. The PCR products were specific for each KIR gene because of the sequence and length (ranging from 390 to 450 bp; Table S5). Subtle sequence differences allow for a distinction between KIR genes with the same number of repeats, except for *KIR2DL1* and *KIR2DS1*, which are inseparable in this region. NGS yielded information about the copy number of the KIR-specific STRs of these 12 samples, which corresponded 100% with the MLPA test results (Figure 3). The STR assay thus confirmed the MLPA data for the KIR genes that can be detected with this assay.

Copy Number Variation Validated within CEPH Pedigrees

The human genome diversity cell line panel (HGDP) and Centre d'Etude du Polymorphisme Humaine (CEPH) bank contain a large number of cultured lymphoblastoid cell lines (LCLs) that are available for family segregation analyses and

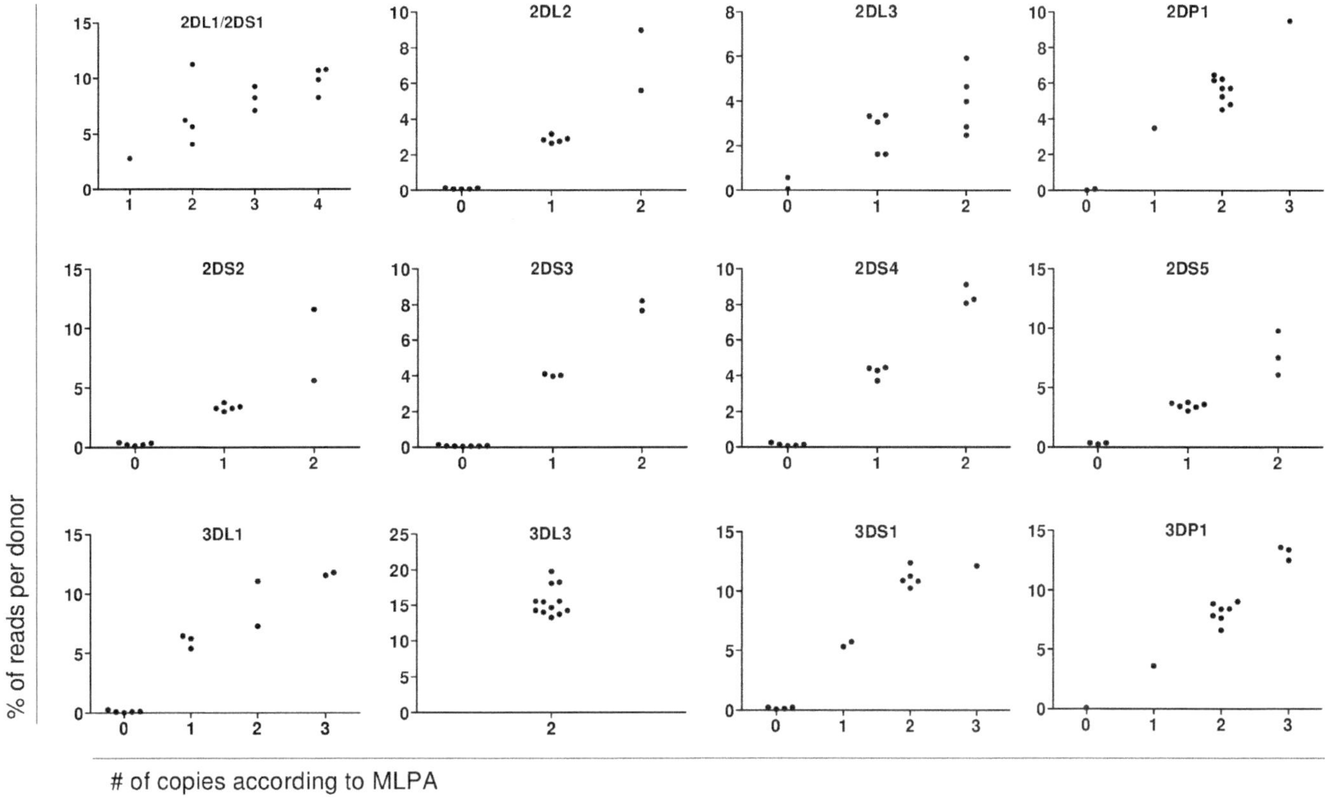

Figure 3. Validation of the KIR MLPA method with an independent short tandem repeat (STR) assay. A primer set was used to amplify a stretch of repeats that occurs in intron 4 of most KIR genes. Next-generation sequencing (NGS) was used to sequence the amplification products based on both specific sequence and number of repeats. The graphs show the percentage of reads in NGS, aligned against the reference sequence of that KIR, in donors with a certain number of copies of that gene as determined by MLPA. The number of reads of framework gene *KIR3DL2* was used as a reference to adjust for the total number of genes per donor. *KIR2DL1* and *KIR2DS1* have the same number of STR and no SNPs to distinguish one from the other and have therefore been combined. Several SNPs within the repeat region of *KIR2DL3* and *KIR3DL3* confirm the copy numbers as found with the MLPA method.

genome research (http://www.cephb.fr/en/hgdp/diversity.php). We typed the KIR genome of two generations of 3 CEPH families and a family from Israel. Family segregation analysis functions as a third form of validation, because the CEPH families have been extensively typed by other groups with PCR-SSP [16,29]. These families were typed in detail by KIR MLPA because of CNV and the indication of duplication events. Using the KIR gene calibrator as our reference sample, we determined the CNV for all the KIR genes of both parents and some or all children in these families and mapped possible inheritance patterns (Figure 4, Figures S1, S2 and S3). For all families tested, the genes of all genotyped children were traced back to the two alleles of both parents.

Interestingly, in one family (Figure 4) we witnessed a cross-over of part of allele B to allele A. The father has 2 copies each of *KIR2DL5* and *KIR2DS3*, while the mother has none. These genes are present twice in the children that inherited the A allele of the father. As the siblings that received the B allele only have one copy of each gene, this excludes the possibility that the genes were already duplicated in the father. This duplication event was confirmed with the STR assay described above for *KIR2DS3* and with a qPCR on DNA for *KIR2DL5* and *KIR2DS3* (Figure S4). The mother of family 1416 (Figure S1) contained a duplication of *KIR3DP1* and *KIR2DL4*, which was subsequently detected in three of the children. These data confirm the results of Martin and co-workers [14] for the same family, who established a crossing-over event of part of a haplotype between *KIR2DL5* and *KIR3DP1*. The translocation event described by Traherne and co-workers [16] in family 1413 (Figure S2) can also be detected by the MLPA method. In addition, when testing another series of untyped control pedigrees we picked up a family with a lack of the framework genes *KIR2DL4* and *KIR3DP1* in all three children (Figure S3), a phenomenon that was previously reported to occur only rarely [30]. These observations lead us to conclude that the KIR gene cluster is subject to more translocation events than is generally assumed.

Materials and Methods

Ethics Statement

The study was approved by the Institutional Medical Ethics Committee of the Academic Medical Center in Amsterdam and was performed in accordance with the Declaration of Helsinki. Participants provided their written informed consent to participate in this study.

DNA Isolation

DNA was purified from 120 healthy donors with a QIAgen Blood Kit (Qiagen, Venlo, The Netherlands) according to the manufacturer's protocol.

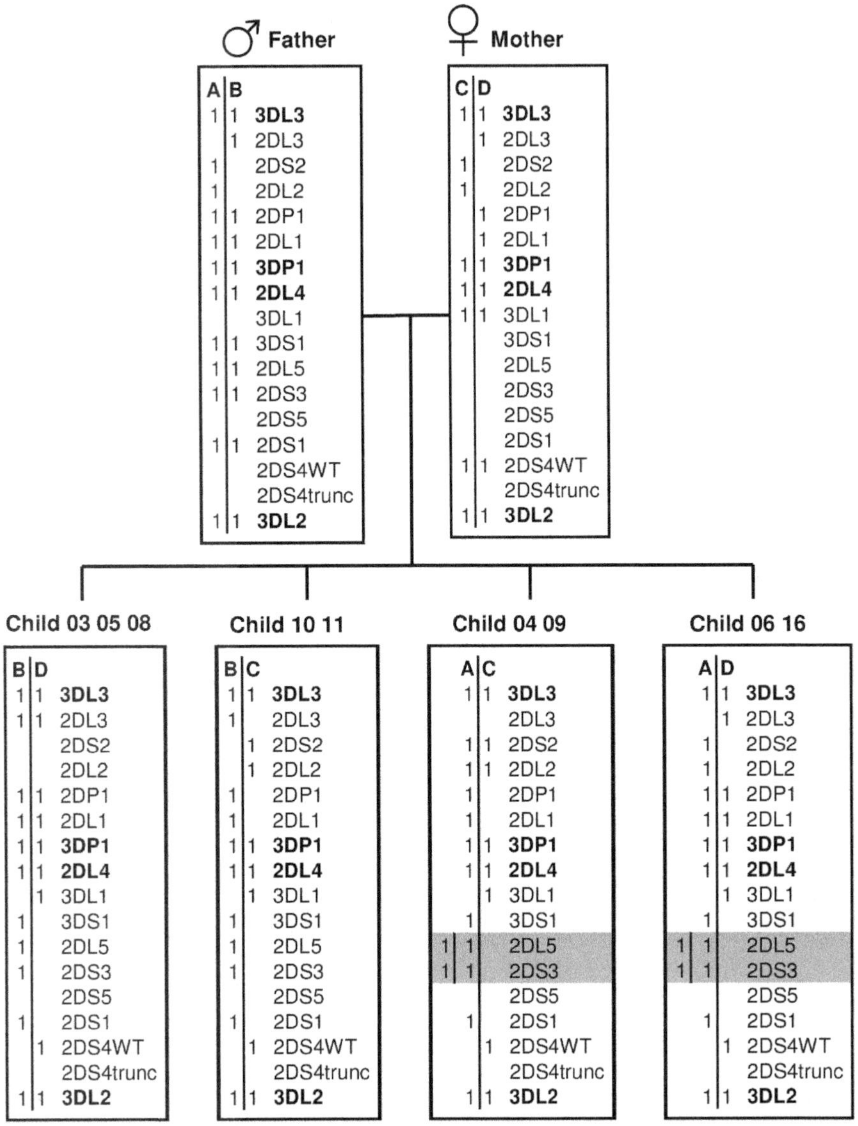

Figure 4. KIR gene pedigree analysis of a Centre d'Etude du Polymorphisme Humaine family by KIR-MLPA. Children 04, 06, 09 and 16 of family 1347 show a translocation event involving *KIR2DL5* and *KIR2DS3*.

Multiplex Ligation-dependent Probe Amplification

The KIR MLPA is based on the method described by Breunis et al. [18]. In short, 100 ng of DNA was denatured at 98°C for 5 min in a thermocycler with heated lid and cooled to 25°C. Three μl of probe mixture, containing KIR-specific probes (Table S1), control probes (Table S2) and competitor probes, was added and denatured at 95°C for 1 min, then hybridized at 60°C for at least 16 hours. The mixture was cooled to 54°C before addition of 32 μl of Ligase 65 mixture (MRC Holland, Amsterdam, The Netherlands). Ligation took place at 54°C for 15 min, deactivation at 98°C for 5 min, and the mixture was cooled to 4°C. Bound probes were amplified by PCR: 10 μl of ligation mixture was added to 5 μl of Accuprime Taq Buffer, 0.4 μl of Accuprime Taq enzymes (AccuPrime Taq DNA Polymerase High Fidelity kit, Life Technologies, Carlsbad, CA, USA) 1.5 μl of 50 mM $MgSO_4$ and 1.5 μl of a single primer pair of which one primer was labeled with a FAM group. PCR conditions included 36 cycles of 30 sec denaturation at 95°C, 30 sec annealing at 60°C and 1 min elongation at 68°C, followed by 20 min elongation at 68°C. Fluorescently labeled products were thus created, allowing for fragment analysis by capillary electrophoresis. 1.5 μl of the PCR reaction was mixed with 9 μl of deionized formamide and 1 μl of Promega Internal Lane Standard 60–600 (Promega, Madison, WI, USA) and incubated at 90°C for 10 min before product separation on an ABI-3130XL (Applied Biosystems by Life Technologies). Fragment analysis was done with the program GeneMarker (version 1.90, SoftGenetics, State College, PA, USA).

Next-Generation Sequencing

For whole KIR genome sequencing, a long-range PCR was performed of each donor. Several primers were designed that amplified the whole KIR genome; FW universal (5′-gccaaataacatcctgtgcgctgctcagct-3′), FW 3DL3 (5′-ctcacaacatcctgtgtgctgctaactga-3′), FW 2DL4 (5′-cacatcgtctgcaccggtcagtcgagccga-3′), REV universal (5′-ttggagaggtgggcaggggtcaagtg-3′) and REV 3DP1 (5′-ctccatctgagggtcccctgaatgtg-3′). The PCR reaction was performed with 500 μM dNTPs, 1x PCR Buffer 2 (Expand Long Template PCR system, Roche Diagnostics, 454 Life Sciences,

Branford, CT, USA), 300 μM of both primers and 3.75 U polymerase enzyme mix (Expand Long Template PCR system, Roche Diagnostics). The PCR program consisted of an initial denaturation of 94°C followed by 10 cycles of 10 sec at 94°C, 30 sec at 60°C and 12 min at 68°C, then another 20 cycles in which the denaturation at 94°C was extended to 15 sec, followed by final elongation of 7 min at 68°C. The resulting PCR products were up to 17 kb in length.

The fragments were sequenced on a 454 FLX Roche Genome Sequencer (Roche Diagnostics, 454 Life Sciences, Branford, CT, USA), with the Shotgun Library method. ReferenceMapper software (Roche Diagnostics) was used to align the reads to a reference sequence.

Synthesis of the KIR-MLPA Calibrator

The KIR-MLPA calibrator was created with an assembly method adjusted from Stemmer *et al.* [26]. The intended synthetic product was divided in six separate sequences. Each sequence was cut up into small oligodeoxyribonucleotides of 60 base pairs that covered either the sense or the anti-sense strand with a 20-base pair overlap of each strand. These oligo's were synthesized by Life Technologies (Carlsbad, CA, USA). Multiple 60-mers were fused together in a PCR reaction: 2 μM oligo's, 0.2 μl of Accuprime Taq polymerase, 0.4 μl of Accuprime Pfx polymerase, 2 μl of Buffer I (Life Technologies) and de-ionized water were combined to a volume of 20 μl. This reaction contained a denaturation step at 94°C for 15 sec, then 55 cycles of 30 sec at 94°C, 30 sec at 52°C, 60 sec at 68°C. Subsequent amplification of 1 μl of this product with 5 μl of the two outer oligo's of each synthesized sequence, 0.5 μl of Accuprime Taq polymerase, 1 μl of Accuprime Pfx polymerase, 5 μl of Buffer I and de-ionized water up to 50 μl. This second PCR program included 24 cycles of 30 sec at 94°C, 30 sec at 52°C, 90 sec at 68°C.

With this method, products of around 680 base pairs were synthesized. For larger sequences, 1 μl of two 680-mers were pooled together (with an overlap of 180 bp) and subjected to another round of PCR. Genes larger than 1180 bp were synthesized in multiple steps of these assembly rounds, until the right size was achieved.

These products were sequenced by Sanger sequencing, to check for mutations. The mutations were mutated back to their original sequence by means of the Quickchange protocol (Stratagene, La Jolla, CA, USA) as described by the manufacturer. Transformation in *E. coli* and subsequent sequencing of the plasmids was used to check the final product.

Copy Number Determination with the KIR-MLPA Calibrator

The calibrator was added in a comparable number of molecules as present in an average human genome of 100 ng. Because the calibrator has a length of approximately 11 kb, there is relatively little DNA present in each sample. To prevent the DNA from sticking to the plastic storage tube, we used 400 ng/ml herring sperm DNA as a decoy. Also, LoBind tubes (Eppendorf, Nijmegen, The Netherlands) were used to prevent the DNA from sticking to the plastic. The calibrator served as a reference and was run in parallel to the test samples in every MLPA run as an extra sample. To minimize variation within one experiment, one calibrator was present for every 7 samples.

Analysis was performed with GeneMarker software (Softgenetics). The amounts of amplification product that were measured for all calibrator samples within one run were averaged and used as one reference sample. The software calculated a ratio for the peak height or area of each test sample against the corresponding (averaged) calibrator peak. The peak pattern of the calibrator represented 2 copies of each KIR gene, after normalization with the software. Therefore, a ratio of 1 means that 2 copies of this gene were present in the test sample. The control probes provided a quality control for each test, as each gene targeted by these probes is present twice in each individual.

Not all probes in our MLPA gave a good approximation of the copy number. The probes that were used for copy number determination are listed in Table S4. When multiple probe sets were used for copy number determination, an average was calculated of all ratios. We apply as a rule of thumb that an (average) ratio of 0.25–0.75 represents 1 copy, 0.75–1.25 represents 2 copies, 1.25–1.75 represents 3 copies and >1.75 represents 4 or more copies. For border cases multiple runs were taken into account.

PCR-SSP

The polymerase chain reaction with sequence-specific primers (PCR-SSP) was performed as previously described [4,12].

Sanger Sequencing

Sequence analysis was performed according to the manufacturer's protocol of the BigDyeTerminator cycle sequencing kit on an ABI-3130XL (Applied Biosystems by Life Technologies).

Short Tandem Repeat Assay

Primers were designed around the short tandem repeat region in intron 4, present in all intron 4 containing KIR genes (sense 5′-ccaaagagaactagagagaccgagaggc-3′, 5′ccaaatagaactagagagactga-gaggc-3′, 5′-ccaaagagagctagagagaccgagaggc-3′, 5′-ccaaaagggaac-tagagagactgagaggc-3′ and anti-sense 5′-tgtgtccttgtgtcctgttca-taactttctgc-3′, 5′-tgtgtccttgtgtcctgtccataactttctgc-3′, 5′-tgtgtccttctgtcttgctcataactttctgc-3′, 5′-tgtgtccttctgtcttgttca-taactttctgc-3′, 5′-tgtgtccttgtgtcccgttcataactttctgc-3′). One PCR reaction amplified all short tandem repeat regions (Table S5): 5 μl of 20 ng/μl DNA was mixed with 12.6 μl of de-ionized water, 3 μl of SalsaBuffer (MRC Holland), 1.2 μl of 10 mM dNTP, 1 μl of each primer mix with a 10 pmol/primer/μl concentration. 0.4 μl of Taq Start Antibody (Clontech, Saint-Germain-en-Laye, France) and 0.8 μl of Salsa DNA polymerase (MRC Holland) were combined on ice and added to the DNA mix. The PCR program consisted of 2 min denaturation at 94°C, followed by 36 cycles of 15 sec denaturation at 95°C, 1 min annealing at 58°C and 5 min elongation at 68 °C. The final step was an elongation at 68°C of 20 min.

The amplified fragments were processed on a 454 FLX Roche Genome Sequencer, by the Shotgun Library method, followed by analysis with ReferenceMapper. The percentage of reads per donor per KIR gene was normalized against the framework gene *KIR3DL2* to compensate for the difference in total gene copy numbers between donors. The percentage of reads that were found per KIR gene represented a relative number of copies present of that gene. A reference sequence was used to sort the reads, based upon the sequences within the STR region of the KIRs, available from the IPD KIR database: http://www.ebi.ac.uk/ipd/kir/(Release 2.4.0, 15 April 2011).

Quantitative DNA PCR

A quantitative PCR on DNA was designed to validate gene copy number determination by MLPA. Primers were designed for *KIR2DL5* (sense 5′-gctggctccacatcctcgtt-3′ and anti-sense 3′-cccaagacgagagcgactca-3′) and for *KIR2DS3* (sense 5′-ccaagatcag-caagtgtgggttt-3′ and anti-sense 5′-cttggcaggaggtatgaactcaa-3′).

The qPCR was performed as described earlier [31]. The relative ratio is a result of the absolute numbers from the KIR genes compared to the absolute numbers of a PCR on exon 8 of the Cytochrome b_{558} beta chain, gene (*CYBB*) (Sense: 5′-atgtcaaatatt-taagcaagcctac-3′ and anti-sense 5′-acttgtccatgatatagttagacac-3′). The *CYBB* numbers are corrected for sex, as this gene is present on the X-chromosome.

Discussion

In this study we have demonstrated the presence of extensive CNV at the KIR gene cluster. To determine KIR genotype and copy number, we designed a novel MLPA assay, using a synthetic calibrator as an internal reference. We found more gene copies for inhibiting than for activating KIRs. Similar gene copy variation was unexpectedly found in a number of the framework genes. This suggests either a more complex role for these supposed framework genes in the haplotype system than assumed until now, and/or crossing-over events, deletions and duplications in the KIR gene cluster occur at a far higher extent than previously thought. Because more variation exists in the middle framework gene block containing *KIR2DL4* and *KIR3DP1* than in the outer framework genes, *KIR3DL2* and *KIR3DL3*, frequent duplications within the KIR gene cluster seem likely. Although it has been shown before that duplications, crossing-over events and deletions in the KIR gene cluster can occur [14,16], the frequency of these events has never been evaluated in detail.

To confirm our findings, next-generation sequencing was applied to validate our assay for KIR genotyping. No false-positive or false-negative genotyping was apparent when checked by 454 sequencing. An STR assay was used as an independent technique to measure CNV in the KIR gene cluster. This assay confirmed the copy numbers as found by the MLPA method for those KIRs that could be identified with the STR assay.

The analysis of hereditary patterns of KIRs in several families supports the importance of copy number determination. Using the MLPA assay, we detected several independent duplication/deletion events in previously typed CEPH families and in one additional family. One of these duplications represents a previously unreported event. To our knowledge, there are no analytical techniques available to easily detect these events at the KIR gene cluster, explaining the fact that one of these events went undetected in these 'standard' pedigrees until now.

The MLPA assay is designed to detect the presence of entire genes, based on the current knowledge. Novel genes and hybrid genes are not specifically targeted when using this technique. From our 454-sequencing data and the probe analysis of our MLPA method, we know that the hybrid gene *KIR2DL5A/3DP1* [14] (*3DP1*004*) is detected in our MLPA method as *KIR3DP1*. The hybrid gene *KIR2DL3/2DP1* [16] is not detected at all and the hybrid gene *KIR2DL1/2DS1* [16] is detected by both the *KIR2DL1* and *KIR2DS1* probes. As the first two genes are most likely pseudogenes, only the latter causes concern for proper detection with the current MLPA assay. The hybrid gene *KIR2DL1/2DS1* likely resembles *KIR2DS1* the most, as fusion of the two sequences occurs before the first transcribed domain, which would result in a false-positive *KIR2DL1* scoring by the KIR MLPA technique. Nevertheless, the frequency of this hybrid gene is expected to be very low, as was reported by Traherne et al. [16]. In addition to the 2 families in which this gene was originally found, the hybrid gene was detected in only 3 (non-Caucasian) individuals of 1214 samples from various ethnic origins. To the best of our knowledge, no protein expression has been confirmed to date.

Also, specific allelic variants were not separately targeted. There are several KIR genes with allelic variants that are not expressed. Specific targeting of these allelic variants could give an even more accurate KIR genotype. Although the MLPA assay allows further expansion of probe sets when required, we have to consider the additional relative value of such probes to detect rare allelic variants because of the low frequency in cohort studies.

From an evolutionary perspective, one would expect an immunological effect of CNV at the KIR gene cluster, as shown for other immune receptors [18]. In fact, studies have been published that indicate a relationship between CNV and transcription levels for several KIRs [32], and also the relevance of CNV in the KIR gene cluster in relation to certain diseases has been suggested for *KIR3DL1* and *KIR3DS1* [15,17,33,34].

An NK cell resides in a delicate balance between activated and inhibited state. Expression seems to be regulated in a clonal fashion, in which each cell expresses its own selection of receptors. From this notion, we can hypothesize that it is also a selection of several receptors that decides cell fate during disease. Indeed, the presence of some activating KIR-HLA pairs in combination with a lack of inhibitory receptor pairs has been associated with an increased risk of developing certain autoimmune diseases like Diabetes type 1 and Ankylosing Spondylitis [35,36]. Also, the total number of activating KIR genes might influence NK cell effector function, as has been suggested for CMV reactivation during HSCT [37], even though we still do not know the nature of the ligands for most of these activating KIRs. The use of our KIR MLPA creates interesting extra dimensions to the analysis of such data panels, adding information on the presence or absence of CNV.

KIRs are mostly expressed by NK cells, which have a role in cancer surveillance and host defense against viral infections. A pivotal role for NK cells in human disease was shown for the first time in a rare severe immunodeficiency disease (SCID) patient who was able to clear CMV infection in the absence of T cells [38]. A similar situation can be encountered early after myeloablative HSCT, since donor-derived NK cells reconstitute earlier than T cells [5,8,39]. In experimental models and in clinical trials, NK cells have received much attention in transplantation settings during cancer treatment. NK cells have been demonstrated not to cause Graft-versus-host-disease (GVHD), even when alloreactive across major histocompatibility barriers [5,40]. Unlike T cells that carry the risk of causing lethal GVHD, NK cells are safe as a form of cellular immunotherapy in the allogeneic HSCT setting. Indeed, alloreactive NK cells are exploited as mediators of graft-versus-leukemia (GVL) effects in T-cell-depleted haploidentical HSCT for acute myeloid leukemia following HSCT, or as potential cellular treatment in high-risk leukemia patients instead of HLA-identical HSCT [5,8].

In summary, there is a clear need for simple but highly accurate tools to study KIR genotype-phenotype relations. We have therefore developed our current KIR MLPA assay for this purpose. Irrespective of the relative contribution of inhibitory or activating KIRs or lectins such as NKG2C or NKG2D in viral clearance and/or transplantation outcome [9,39], realizing the potential role of CNV on KIR expression and functional activity as indicated in the present study, our current KIR MLPA assay will also enable us to relate disease severity to CNV in the KIR gene cluster in more detail on a larger scale. In addition, our MLPA assay for KIRs may be helpful in providing a rapid and efficient method for complete genotyping and at the same time provide data on CNV as a selection basis for donors in transplantation programs to improve the prognosis of graft survival and outcome of patients.

Supporting Information

Figure S1 KIR gene pedigree analysis of a Centre d'Etude du Polymorphisme Humaine family by KIR MLPA. The mother of family 1416 has a duplication of two KIR genes on allele D, which was transferred to three of her genotyped children.

Figure S2 KIR gene pedigree analysis of a Centre d'Etude du Polymorphisme Humaine family by KIR MLPA. The father of family 1413 has only one copy of both *KIR2DL4* and *KIR3DP1*, which can be traced back to some of his children.

Figure S3 KIR gene pedigree analysis of an Israeli family by KIR MLPA. Both parents carry only one copy of the framework genes *KIR3DP1* and *KIR2DL4*, which results in a complete absence of those genes in all three children.

Figure S4 Graphical representation of the quantitative PCR on DNA from Centre d'Etude du Polymorphisme Humaine family 1347 and some control donors. (A) The relative product of KIR2DL5 compared to the occurrence of *CYBB* (corrected for sex). (B) The relative product of KIR2DS3 compared to the occurrence of *CYBB* (corrected for sex).

Table S1 Overview probe sequences of the KIR MLPA technique.

Table S2 Overview control probe sequences of the KIR MLPA technique.

Table S3 KIR MLPA technique probe distribution over 3 mixes.

Table S4 Copy number determination by KIR MLPA technique.

Table S5 A sequence alignment of the short tandem repeat region in intron 4 of most KIRs.

Table S6 KIR gene copy number distribution in a cohort of healthy individuals.

Acknowledgments

We are grateful to Karin van Leeuwen for her ideas and assistance during the development of the KIR MLPA and for technical assistance with the 454 analysis. We thank Paul Verkuijlen for his assistance in the design of the KIR MLPA calibrator. Both contributors work at the Department of Blood Cell Research, Sanquin Research and Landsteiner Laboratory, Amsterdam, The Netherlands.

We thank Menno Hofman (Department of Immunogenetics, Sanquin Blood Supply Foundation, Amsterdam, The Netherlands), for technical support of the PCR-SSP experiments.

Finally, we thank Ronit Gavrieli from the Meir Medical Center in Irsael, for sending us DNA from the Israeli family.

Author Contributions

Conceived and designed the experiments: SV MDB RATWG. Performed the experiments: SV RATWG CKYH JG. Analyzed the data: SV MJM WTNS NML. Wrote the paper: SV TWK. Provided intellectual input: DR TKVDB TWK.

References

1. Sun JC, Lanier LL (2011) NK cell development, homeostasis and function: parallels with CD8+ T cells. Nat Rev Immunol 11: 645–657. 10.1038/nri3044.
2. Vivier E, Raulet DH, Moretta A, Caligiuri MA, Zitvogel L, et al. (2011) Innate or adaptive immunity? The example of Natural Killer cells. Science 331: 44–49.
3. Parham P, Abi-Rached L, Matevosyan L, Moesta AK, Norman PJ, et al. (2010) Primate-specific regulation of natural killer cells. J Med Primatol 39: 194–212. 10.1111/j.1600-0684.2010.00432.x.
4. Kunert K, Seiler M, Mashreghi MF, Klippert K, Schönemann C, et al. (2007) KIR/HLA Ligand incompatibility in kidney transplantation. Transplantation 84: 1527–1533.
5. Ruggeri L, Capanni M, Urbani E, Perruccio K, Shlomchik WD, et al. (2002) Effectiveness of donor Natural Killer cell alloreactivity in mismatched hematopoietic transplants. Science 295: 2097–2100.
6. Stern M, de Angelis C, Urbani E, Mancusi A, Aversa F, et al. (2010) Natural killer-cell KIR repertoire reconstitution after haploidentical SCT. Bone Marrow Transplant 45: 1607–1610.
7. van Bergen J, Thompson A, Haasnoot GW, Roodnat JI, de Fijter JW, et al. (2011) KIR-Ligand mismatches are associated with reduced long-term graft survival in HLA-compatible kidney transplantation. Am J Transplant 11: 1959–1964. 10.1111/j.1600-6143.2011.03621.x.
8. Leung W (2011) Use of NK cell activity in cure by transplant. Br J Haematol 155: 14–29. 10.1111/j.1365-2141.2011.08823.x.
9. Gallez-Hawkins GM, Franck AE, Li X, Thao L, Oki A, et al. (2011) Expression of Activating KIR2DS2 and KIR2DS4 Genes after Hematopoietic Cell Transplantation: Relevance to Cytomegalovirus Infection. Biol Blood Marrow Transplant 17: 1662–1672. doi: 10.1016/j.bbmt.2011.04.008.
10. Venstrom JM, Gooley TA, Spellman S, Pring J, Malkki M, et al. (2010) Donor activating KIR3DS1 is associated with decreased acute GVHD in unrelated allogeneic hematopoietic stem cell transplantation. Blood 115: 3162–3165.
11. Trowsdale J (2001) Genetic and functional relationships between MHC and NK receptor genes. Immunity 15: 363–374. doi: 10.1016/S1074-7613(01)00197-2.
12. Uhrberg M, Valiante NM, Shum BP, Shilling HG, Lienert-Weidenbach K, et al. (1997) Human diversity in killer cell inhibitory receptor genes. Immunity 7: 753–763. doi: 10.1016/S1074-7613(00)80394-5.
13. Parham P (2005) MHC class I molecules and KIRs in human history, health and survival. Nat Rev Immunol 5: 201–214. 10.1038/nri1570.
14. Martin MP, Bashirova A, Traherne J, Trowsdale J, Carrington M (2003) Cutting edge: Expansion of the KIR locus by unequal crossing over. J Immunol 171: 2192–2195.
15. Pelak K, Need AC, Fellay J, Shianna KV, Feng S, et al. (2011) Copy number variation of KIR genes influences HIV-1 control. PLoS Biol 9: e1001208. doi:10.1371/journal.pbio.1001208.
16. Traherne JA, Martin M, Ward R, Ohashi M, Pellett F, et al. (2010) Mechanisms of copy number variation and hybrid gene formation in the KIR immune gene complex. Hum Mol Genet 19: 737–751.
17. Khakoo SI, Thio CL, Martin MP, Brooks CR, Gao X, et al. (2004) HLA and NK cell inhibitory receptor genes in resolving hepatitis C virus infection. Science 305: 872–874.
18. Breunis WB, van Mirre E, Bruin M, Geissler J, de Boer M, et al. (2008) Copy number variation of the activating FCGR2C gene predisposes to idiopathic thrombocytopenic purpura. Blood 111: 1029–1038.
19. Breunis WB, van Mirre E, Geissler J, Laddach N, Wolbink G, et al. (2009) Copy number variation at the FCGR locus includes FCGR3A, FCGR2C and FCGR3B but not FCGR2A and FCGR2B. Hum Mutat 30: E640–E650. 10.1002/humu.20997.
20. Wouters D, van Schouwenburg P, van der Horst A, de Boer M, Schooneman D, et al. (2009) High-throughput analysis of the C4 polymorphism by a combination of MLPA and isotype-specific ELISA's. Mol Immunol 46: 592–600. doi: 10.1016/j.molimm.2008.07.028.
21. Crum KA, Logue SE, Curran MD, Middleton D (2000) Development of a PCR-SSOP approach capable of defining the natural killer cell inhibitory receptor (KIR) gene sequence repertoires. Tissue Antigens 56: 313–326. 10.1034/j.1399-0039.2000.560403.x.
22. Hsu KC, Liu XR, Selvakumar A, Mickelson E, O'Reilly RJ, et al. (2002) Killer Ig-like Receptor haplotype analysis by gene content: Evidence for genomic diversity with a minimum of six basic framework haplotypes, each with multiple subsets. J Immunol 169: 5118–5129.
23. Middleton D, Gonzalez A, Gilmore PM (2007) Studies on the expression of the deleted KIR2DS4*003 gene product and distribution of KIR2DS4 deleted and nondeleted versions in different populations. Hum Immunol 68: 128–134. doi: 10.1016/j.humimm.2006.12.007.

24. Gonzalez-Galarza FF, Christmas S, Middleton D, Jones AR (2011) Allele frequency net: a database and online repository for immune gene frequencies in worldwide populations. Nucleic Acids Res 39: D913–D919.
25. Trowsdale J, Barten R, Haude A, Stewart CA, Beck S, et al. (2001) The genomic context of natural killer receptor extended gene families. Immunol Rev 181: 20–38. 10.1034/j.1600-065X.2001.1810102.x.
26. Stemmer WPC, Crameri A, Ha KD, Brennan TM, Heyneker HL (1995) Single-step assembly of a gene and entire plasmid from large numbers of oligodeoxyribonucleotides. Gene 164: 49–53. doi: 10.1016/0378-1119(95)00511-4.
27. Martin AM, Freitas EM, Witt CS, Christiansen FT (2000) The genomic organization and evolution of the natural killer immunoglobulin-like receptor (KIR) gene cluster. Immunogenet 51: 268–280.
28. Sambrook J, Bashirova A, Palmer S, Sims S, Trowsdale J, et al. (2005) Single haplotype analysis demonstrates rapid evolution of the killer immunoglobulin-like receptor (KIR) loci in primates. Gen Res 15: 25–35.
29. Martin M, Single R, Wilson M, Trowsdale J, Carrington M (2008) KIR haplotypes defined by segregation analysis in 59 Centre d'Etude Polymorphisme Humain (CEPH) families. Immunogenet 60: 767–774.
30. Nowak I, Majorczyk E, Płoski R, Senitzer D, Sun JY, et al. (2011) Lack of KIR2DL4 gene in a fertile Caucasian woman. Tissue Antigens 78: 115–119. 10.1111/j.1399-0039.2011.01711.x.
31. van Mirre E, Breunis WB, Geissler J, Hack CE, de Boer M, et al. (2006) Neutrophil responsiveness to IgG, as determined by fixed ratios of mRNA levels for activating and inhibitory FcγRII (CD32), is stable over time and unaffected by cytokines. Blood 108: 584–590.
32. McErlean C, Gonzalez A, Cunningham R, Meenagh A, Shovlin T, et al. (2010) Differential RNA expression of KIR alleles. Immunogenet 62: 431–440.
33. Espeli M, Niederer HA, Traherne JA, Trowsdale J, Smith KG (2010) Genetic variation, Fcγ receptors, KIRs and infection: the evolution of autoimmunity. Curr Opin Immunol 22: 715–722. doi: 10.1016/j.coi.2010.10.003.
34. Alter G, Heckerman D, Schneidewind A, Fadda L, Kadie CM, et al. (2011) HIV-1 adaptation to NK-cell-mediated immune pressure. Nature 476: 96–100. 10.1038/nature10237.
35. Tajik N, Shahsavar F, Poormoghim H, Radjabzadeh MF, Mousavi T, et al. (2011) KIR3DL1+HLA-B Bw4Ile80 and KIR2DS1+HLA-C2 combinations are both associated with ankylosing spondylitis in the Iranian population. Int J Immunogenet 38: 403–409.
36. van der Slik AR, Koeleman BPC, Verduijn W, Bruining GJ, Roep BO, et al. (2003) KIR in Type 1 Diabetes: Disparate Distribution of Activating and Inhibitory Natural Killer Cell Receptors in Patients Versus HLA-Matched Control Subjects. Diabetes 52: 2639–2642.
37. Zaia JA, Sun JY, Gallez-Hawkins GM, Thao L, Oki A, et al. (2009) The Effect of Single and Combined Activating Killer Immunoglobulin-like Receptor Genotypes on Cytomegalovirus Infection and Immunity after Hematopoietic Cell Transplantation. Biol Blood Marrow Transplant 15: 315–325.
38. Kuijpers TW, Baars PA, Dantin C, van den Burg M, van Lier RAW, et al. (2008) Human NK cells can control CMV infection in the absence of T cells. Blood 112: 914–915.
39. Foley B, Cooley S, Verneris MR, Pitt M, Curtsinger J, et al. (2012) Cytomegalovirus reactivation after allogeneic transplantation promotes a lasting increase in educated NKG2C+ natural killer cells with potent function. Blood 119: 2665–2674.
40. Ruggeri L, Mancusi A, Capanni M, Urbani E, Carotti A, et al. (2007) Donor natural killer cell allorecognition of missing self in haploidentical hematopoietic transplantation for acute myeloid leukemia: challenging its predictive value. Blood 110: 433–440.

13

Imputation of Variants from the 1000 Genomes Project Modestly Improves Known Associations and Can Identify Low-frequency Variant - Phenotype Associations Undetected by HapMap Based Imputation

Andrew R. Wood[1], John R. B. Perry[1,2,3], Toshiko Tanaka[4], Dena G. Hernandez[5,6], Hou-Feng Zheng[7], David Melzer[8], J. Raphael Gibbs[5,6], Michael A. Nalls[5], Michael N. Weedon[1], Tim D. Spector[3], J. Brent Richards[3,7], Stefania Bandinelli[9,10], Luigi Ferrucci[4], Andrew B. Singleton[5], Timothy M. Frayling[1]*

1 Genetics of Complex traits, Institute of Biomedical and Clinical Sciences, Peninsula College of Medicine and Dentistry, University of Exeter, Exeter, United Kingdom, **2** Wellcome Trust Centre for Human Genetics, University of Oxford, Oxford, United Kingdom, **3** Department of Twin Research and Genetic Epidemiology, King's College London, London, United Kingdom, **4** Longitudinal Studies Section, Clinical Research Branch, Gerontology Research Center, National Institute on Aging, Baltimore, Maryland, United States of America, **5** Laboratory of Neurogenetics, National Institute of Aging, Bethesda, Maryland, United States of America, **6** Department of Molecular Neuroscience and Reta Lila Laboratories, Institute of Neurology, UCL, London, United Kingdom, **7** Departments of Medicine, Human Genetics, Epidemiology and Biostatistics, Jewish General Hospital, McGill University, Montreal, Quebec, Canada, **8** Epidemiology and Public Health, Institute of Biomedical and Clinical Sciences, Peninsula College of Medicine and Dentistry, University of Exeter, Exeter, United Kingdom, **9** Tuscany Regional Health Agency, Florence, Italy, I.O.T. and Department of Medical and Surgical Critical Care, University of Florence, Florence, Italy, **10** Geriatric Unit, Azienda Sanitaria di Firenze, Florence, Italy

Abstract

Genome-wide association (GWA) studies have been limited by the reliance on common variants present on microarrays or imputable from the HapMap Project data. More recently, the completion of the 1000 Genomes Project has provided variant and haplotype information for several million variants derived from sequencing over 1,000 individuals. To help understand the extent to which more variants (including low frequency ($1\% \leq$ MAF $<5\%$) and rare variants ($<1\%$)) can enhance previously identified associations and identify novel loci, we selected 93 quantitative circulating factors where data was available from the InCHIANTI population study. These phenotypes included cytokines, binding proteins, hormones, vitamins and ions. We selected these phenotypes because many have known strong genetic associations and are potentially important to help understand disease processes. We performed a genome-wide scan for these 93 phenotypes in InCHIANTI. We identified 21 signals and 33 signals that reached $P<5\times10^{-8}$ based on HapMap and 1000 Genomes imputation, respectively, and 9 and 11 that reached a stricter, likely conservative, threshold of $P<5\times10^{-11}$ respectively. Imputation of 1000 Genomes genotype data modestly improved the strength of known associations. Of 20 associations detected at $P<5\times10^{-8}$ in both analyses (17 of which represent well replicated signals in the NHGRI catalogue), six were captured by the same index SNP, five were nominally more strongly associated in 1000 Genomes imputed data and one was nominally more strongly associated in HapMap imputed data. We also detected an association between a low frequency variant and phenotype that was previously missed by HapMap based imputation approaches. An association between rs112635299 and alpha-1 globulin near the *SERPINA* gene represented the known association between rs28929474 (MAF = 0.007) and alpha1-antitrypsin that predisposes to emphysema ($P=2.5\times10^{-12}$). Our data provide important proof of principle that 1000 Genomes imputation will detect novel, low frequency-large effect associations.

Editor: Dan E. Arking, Johns Hopkins University, United States of America

Funding: This work was supported by the Wellcome Trust 083270/Z/07/Z. The funders had no role in study design, data collection and analysis, decision to publish, or preparation of the manuscript.

Competing Interests: The authors have declared that no competing interests exist.

* E-mail: tim.frayling@pms.ac.uk

Introduction

Genome-wide association (GWA) studies have identified many novel associations between common genetic variants and human traits. These studies capture a large proportion of SNP-based common variation in the human genome but are less efficient at capturing low frequency and rare variants.

Imputation is the process of inferring missing data based on known data [1]. Imputation is useful for performing GWA meta-analyses across studies with the same traits but different genotyping arrays. Imputation can also potentially identify signals of association not detected by direct genotypes. In the simplest scenario, a causal variant, or the most strongly associated genetic variant, may not be directly genotyped and may exist on a haplotype that is optimally captured by two rather than one directly genotyped SNP (**Figure 1**).

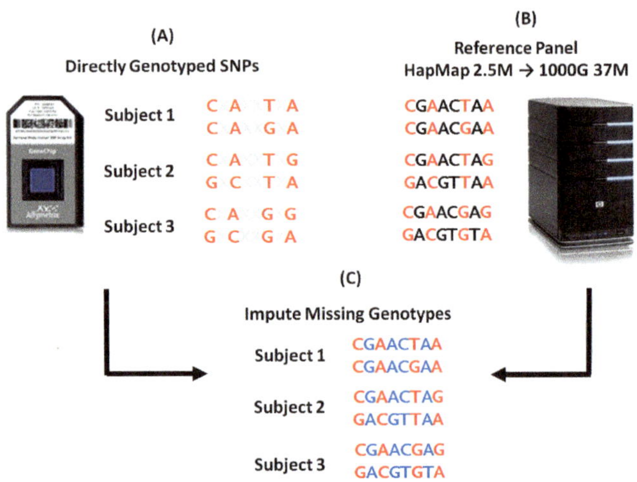

Figure 1. Schematic overview of imputation. (**A**) Directly genotyped SNPs are phased to estimate haplotypes that alleles reside on. (**B**) Publically available reference haplotypes from projects including the International HapMap Project and the 1000 Genomes Project are downloaded that contain variants to be imputed. (**C**) Phased haplotypes from the study are assessed and likened to publically available haplotypes. The most likely genotypes are imputed into the study based on the alleles found in reference haplotype panel.

There are now two main sources of reference genotypes for imputation of missing variants. The International HapMap Project Phases 1 to 3 [2–5] consists of 4.1 million directly genotyped SNPs in 1,486 individuals and has facilitated the imputation of approximately 2.5 million SNPs in standard GWA studies. More recently, the 1000 Genomes Project has completed the low-pass sequencing of the whole genome from 1,094 individuals from a range of ethnic groups [6]. The project has annotated >37 million variants. The sequencing based approach and the large number of individuals means these variants include many low frequency (1% ≤ MAF <5%) and rare (MAF<1%) variants.

In this study we tested the hypothesis that imputation of variants identified by the 1000 Genomes Project (including low frequency and rare variants) would enhance previously identified associations and identify novel associations. We selected 93 traits measured in ~1,200 individuals in the InCHIANTI study (**Table 1** and **Table S1**). All traits were continuous measures of circulating factors [7], [8]. These traits included circulating lipids, proteins, ions and vitamins and were selected for two reasons. First, for many traits, strong genetic associations have previously been observed in relatively small sample sizes, and so we potentially had good power to detect novel effects and stronger signals at known loci. These signals include those far exceeding genome-wide significance in our dataset alone and therefore offer a good opportunity to assess whether or not 1000 Genomes imputation can improve and refine known associations. Second, identifying genetic associations with circulating factors is potentially important to help understand disease processes. Many circulating factors are altered in disease cases compared to controls, but it is rarely understood if these factors are causal, confounded or secondary to disease processes. Genetic variants can be used to help identify causal pathways using Mendelian randomisation [9–12].

Results

Variation captured by HapMap and 1000 Genomes

After applying quality control metrics we captured 2,493,682 and 10,879,115 SNPs from HapMap (release 22) and 1000 Genomes imputation (November 2010, Phase1-α), respectively (**Table S2**, **Figure S1**).

The 1000 Genomes based imputation captured many more variants at high quality (defined as $r^2_{imp}>0.3$ in HapMap imputation and $r^2_{imp}>0.5$ in 1000 Genomes imputation) compared to the HapMap based imputation (**Figure 2**).

We captured 2,172,804 and 5,737,929 common SNPs (MAF≥5%) based on HapMap and 1000 Genomes based imputation, respectively. For low frequency variants (1% ≤ MAF <5%) we captured 274,995 and 1,953,185 based on HapMap and 1000 Genomes imputation respectively. For rare variants (MAF<1%) we captured 45,883 and 3,188,001 rare SNPs based on the HapMap and 1000 Genomes based imputation, respectively.

Identification of association signals using HapMap and 1000 Genomes Imputation

Associations identified by HapMap imputation. Using HapMap based imputation we identified 21 signals of association at $P<5\times10^{-8}$ including 19 circulating factors (**Table 2**). Nine of these signals reached the stricter statistical threshold of $P<5\times10^{-11}$. Of these 21 signals, 14 were known, based on the same or an $r^2>0.8$ proxy to our index SNP reaching $P<5\times10^{-8}$

Table 1. Basic characteristics of the 1210 InCHIANTI subjects at baseline.

Characteristic	Mean (Range) or%
Age (years)	68.2 (21–102)
Sex (% male)	44.6%
BMI	27.2 (18–46.6)
Current smokers (% case)	18.8%
History of hypertension (% case)	38.4%
Treatment for hypertension - former and current (% case)	30.1%
History of diabetes (% case)	9.5%
Treatment for diabetes - former and current (% case)	6.58%
History of myocardial infarction (% case)	3.6%
Treatment for myocardial infarction - former and current (% case)	2.0%

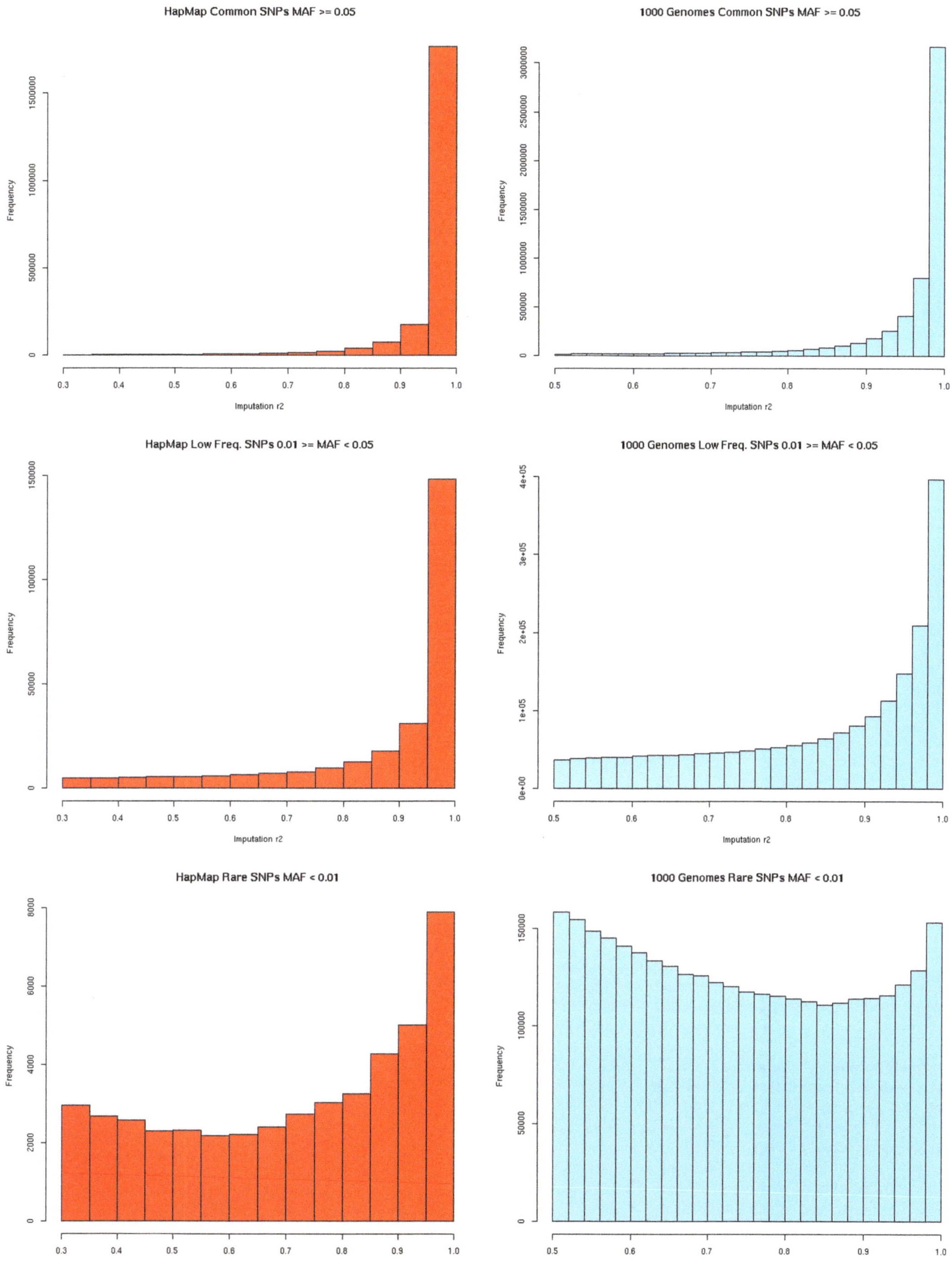

Figure 2. Distributions of imputation quality (r^2) across different minor allele frequency (MAF) bins. Note differences in Y axes scales.

Table 2. Associations of circulating factors based on HapMap imputation ordered by minor allele frequency group (<5% or ≥5), chromosome and base-pair position.

Trait	GC	Chr	SNP	BP (b37)	MAF	P*	Nearest Gene	Annotation
Alkaline phosphatase	1.055	12	rs1880889	41721235	0.03	1.96E-08	*PDZRN4*	Known
Homocysteine	1.043	1	rs1801133	11856378	0.495	1.75E-11	*MTHFR*	Known
Vitamin B6	1.017	1	rs4654748	21786068	0.499	8.84E-09	*NBPF3*	Known
Interleukin-6R	1.134	1	rs8192284	154426970	0.385	3.60E-53	*IL6R*	Known
C-reactive protein	1.035	1	rs12741825	159670145	0.301	1.08E-08	*CRP*	Known
Oxidized LDL	1.029	2	rs676210	21231524	0.239	4.05E-30	*APOB*	Known (other trait)
Macrophage inflammatory protein-1b	1.016	3	rs1500004	46150937	0.104	2.06E-08	*XCR1/CCR1*	Novel
Lipoprotein a	1.057	6	rs10455872	161010118	0.056	1.18E-20	*LPA*	Known
Insulin-like growth factor bp-3	1.038	7	rs10260816	46010100	0.456	4.42E-14	*IGFBP3*	Known
90K Protein	1.016	8	rs406652	16042507	0.159	1.56E-09	*MSR1*	Novel
Interleukin-18	1.047	11	rs2250417	112085316	0.44	2.94E-14	*BCO2*	Known
Vitamin E alpha tocopherol	1.066	11	rs964184	116648917	0.152	1.33E-08	*ZNF259*	Known
Triglycerides	1.052	11	rs964184	116648917	0.152	9.03E-09	*ZNF259*	Known
Soluble transferrin receptor	1.04	11	rs1242229	117062370	0.129	6.55E-09	*SIDT2*	Known
HDL cholesterol	1.033	16	rs1800775	56995236	0.486	8.01E-13	*CETP*	Known
Alpha-2 globulin (%)	1.058	16	rs2000999	72108093	0.228	2.13E-08	*HPR*	Known (other trait)
Lutein	1.026	16	rs11645428	81258896	0.468	1.17E-13	*PKD1L2*	Known
Beta carotene	1.051	16	rs6564851	81264597	0.364	1.70E-10	*PKD1L2/BCMO1*	Known
Free thyroxine (FT4)	1.03	17	rs7212734	6340460	0.444	1.10E-08	*AIPL1*	Novel
Macrophage inflammatory protein-1b	1.016	17	rs1015673	34477386	0.443	1.25E-22	*TBC1D3B*	Known
Vitamin E alpha tocopherol	1.066	19	rs10401969	19407718	0.057	9.32E-09	*SUGP1*	Known (other trait)

GC = genomic control factor used to adjust *P*-value, Chr = chromosome, BP (b37) = human genome build 37 base pair position, SNP = index SNP representing signal, MAF = minor allele frequency, *P** = *P*-value derived after applying genomic control.
Annotations refer to whether associations represent known (present in NHGRI catalogue at $P<5\times10^{-8}$ or GWAS literature) signals for same trait, different traits or represent putative novel signals.

in the NHGRI catalogue or in other published GWA studies with the same or very similar trait.

A further 3 signals were classified as known signals but with different traits. A signal near the *HPR* gene was associated with alpha-2 globulin and was previously reported as associated with haptoglobulin. A signal near the *APOB* gene was associated with oxidized LDL and was previously associated with HDL and triglycerides, and a signal near the *SUGP1* gene was associated with vitamin E and was previously associated with LDL cholesterol, total cholesterol and triglycerides.

Of the remaining 4 signals, 3 were putative novel associations (**Table 2**), and one was an association with hsCRP and represented a previously known locus – our index SNP was correlated with an r^2 of 0.32 with a published hsCRP associated variant (rs876537 [13]) ~7 kb upstream of the *CRP* gene.

Associations identified by 1000 Genomes imputation. Using 1000 Genomes based imputation we identified 33 signals of association at $P<5\times10^{-8}$ across 30 circulating factors (**Table 3**). Eleven of these signals reached a stricter statistical threshold of $P<5\times10^{-11}$. These 33 signals included 14 of the 21 identified by HapMap based imputation (r^2 between Index HapMap and index 1000 Genomes SNP >0.8 of which 6 were the same lead SNP in both analyses) and 6 signals that represented the same loci as detected by HapMap based imputation ($0.2<r^2<0.8$ between the 1000 Genomes and HapMap index SNPs) (**Table 4**). There was one HapMap based signal of association not detected by 1000 Genomes based imputation – rs12741825 representing CRP which did not reach genome-wide significance having passed 1000 Genomes imputation quality control ($P=1\times10^{-7}$, r^2_{imp} >0.99).

Of the 13 signals of association detected by 1000 Genomes imputation that were not detected by HapMap imputation (at $P<5\times10^{-8}$) in our study (**Table 5**), eight represented putative novel signals. The remaining five signals represented known loci ($0.2<r^2<0.8$ with index SNPs in NHGRI GWAS catalogue). Plots of six of the 22 associations detected at $P<5\times10^{-8}$ in the 1000 genomes imputation analysis and that represent previously known signals or loci ($r^2>0.2$ with SNPs in the NHGRI GWAS catalogue) are shown in **Figure 3** and **Figure 4**.

Known and putative novel signals identified by 1000 Genomes that were missed by imputation with Hapmap

For the 13 signals detected by the 1000 Genomes imputation but not by the HapMap imputation analysis, we calculated linkage disequilibrium between the index 1000 Genomes SNP and the most strongly associated HapMap SNPs within a 2 Mb window centred on the 1000 Genomes SNP (**Table 5**). One of the 13 signals of association was present in the HapMap data with a proxy SNP at $r^2>0.8$, four with a proxy SNP at $0.5<r^2<0.8$ and one with a proxy SNP at $r^2=0.26$. As expected, HapMap associations for these six signals were close to genome-wide significance ($P=1.2\times10^{-5}$ to 7.9×10^{-8}).

The remaining 7 signals detected in the 1000 Genomes imputation analyses but not by the HapMap based analysis

Table 3. Genome-wide significant associations of circulating factors based on 1000 Genomes imputation ordered by minor allele frequency group (<5% or ≥5), chromosome and base-pair position.

Trait	GC	Chr	SNP	BP (b37)	MAF	P**	Nearest Gene	Annotation
Total cholesterol	1.049	1	rs116435220	46890686	0.011	2.18E-08	*FAAH*	Novel
Tumor necrosis factor-a	1.024	2	rs188141385	14975369	0.004	4.16E-10	*FAM84A*	Novel
Beta globulins (%)	1.034	4	rs28645201	91528927	0.02	1.22E-09	*FAM190A*	Novel
Vitamin B12	1.055	5	rs146226203	150185826	0.012	3.01E-08	*C5ORF62/IRGM*	Novel
Lipoprotein a	1.057	6	rs55730499	161005610	0.048	3.43E-25	*LPA*	Known
Cortisol:DHEAS ratio	1.006	7	rs34670419	99130834	0.028	2.35E-08	*ZKSCAN5*	Known
Dehydroepiandrosterone	1.023	7	rs34670419	99130834	0.028	2.07E-09	*ZKSCAN5*	Known
Cl-	1.075	7	rs2371549	144172841	0.005	5.38E-09	*TPK1*	Novel
Alkaline phosphatase	1.055	12	rs1880889	41721235	0.029	3.08E-08	*PDZRN4*	Known
Alpha-1 globulin (%)	1.054	14	rs112635299	94838142	0.007	2.51E-12	*SERPINA1*	Novel
Alpha-2 globulin (%)	1.058	17	rs181929163	63740545	0.01	1.57E-08	*CEP112*	Novel
Homocysteine	1.043	1	rs1801133	11856378	0.497	2.01E-11	*MTHFR*	Known
Vitamin B6	1.017	1	rs4654932	21809436	0.417	7.39E-09	*NBPF3*	Known
Interleukin-6R	1.134	1	rs12730935	154419892	0.377	2.28E-53	*IL6R*	Known
Oxidized LDL	1.029	2	rs676210	21231524	0.239	4.12E-30	*APOB*	Known (other trait)
Macrophage inflammatory protein-1b*	1.016	3	rs113341849	46384204	0.09	2.14E-11	*CCR2*	Novel
Insulin-like growth factor bp-3	1.038	7	rs71550311	45987430	0.459	1.39E-14	*IGFBP3*	Known
90K Protein*	1.016	8	rs28491433	16050871	0.148	6.03E-10	*MSR1*	Novel
Beta Cryptoxanthin	1.021	8	rs75226183	77295977	0.087	1.69E-09	*LOC100192378*	Novel
Creatine phosphokinase	1.029	10	rs61871700	101828261	0.28	4.18E-08	*CPN1*	Known
Interleukin-18	1.047	11	rs2250417	112085316	0.44	2.97E-14	*BC02*	Known
Vitamin E alpha tocopherol	1.066	11	rs964184	116648917	0.151	6.66E-09	*ZNF259*	Known
Triglycerides	1.052	11	rs964184	116648917	0.151	1.01E-08	*ZNF259*	Known
Soluble transferrin receptor	1.04	11	rs7940310	117024481	0.13	7.17E-09	*PAFAH1B2*	Known
HDL cholesterol	1.033	16	rs711752	56996211	0.436	8.32E-13	*CETP*	Known
Alpha-2 globulin (%)	1.058	16	rs2287997	72140553	0.227	1.45E-10	*DHX38*	Known (other trait)
Beta carotene	1.051	16	rs12926540	81258987	0.362	6.98E-11	*PKD1L2/BCMO1*	Known
Lutein	1.026	16	rs9708919	81260786	0.391	7.84E-14	*PKD1L2/BCMO1*	Known
Free thyroxine (FT4)*	1.03	17	rs58926603	6341870	0.393	5.62E-09	*AIPL1/FAM64A*	Novel
Macrophage inflammatory protein-1b	1.016	17	rs4796217	34819191	0.345	6.62E-21	*TBC1D3G*	Known
Vitamin E alpha tocopherol	1.066	19	rs58542926	19379549	0.064	1.18E-08	*TM6SF2*	Known (other trait)
LDL cholesterol	1.041	19	rs1065853	45413233	0.075	1.76E-09	*APOE*	Known
Gamma-glutamyltransferase	1.043	22	rs2006227	24995756	0.386	1.70E-08	*GGT1*	Known

* = associations that represent novel signals in both HapMap and 1000 Genomes-based analyses, P** = P-value derived after applying genomic control.

appeared to represent signals specifically detected by 1000 Genomes imputation. These seven signals had very low correlations ($r^2<0.05$) with the most strongly associated HapMap SNP (**Table 5**). Six of these seven signals represented putative low frequency – large effect variants, with minor allele frequencies ≤ 0.02 and per allele effect sizes of 1.1 – 2.1 standard deviations (SDs). We therefore examined these loci in more detail. We checked imputed dosages and trait values for the small number of heterozygous individuals that were driving these associations, but found no obvious sources of error (**Figure S2**). All but one of these 7 signals had a proxy SNP that also showed strong evidence of association, even if the association did not reach $P<5\times10^{-8}$.

Does 1000 Genomes imputation identify stronger associations at loci identified by HapMap based imputation?

We identified 14 loci associated with 13 traits where both the HapMap and 1000 Genomes index SNPs were $P<5\times10^{-8}$ and differed from each other (**Table 4 and Table 6**). In the multivariable SNP analyses that included the two SNPs at each locus against the relevant phenotype, the 1000 Genomes SNP was more strongly associated in 13 of 14 loci (one-tail Binomial $P=0.0009$). The P-value improvements were marginal for 8 of these associations and the 1000 genomes signal was not statistically stronger, but 5 of the 14 associations, showed evidence of stronger signals at $P<0.05$ and three at $P<0.01$, **Table 6**).

Table 4. The 20 signals and loci found in both HapMap and 1000 Genomes ordered by minor allele frequency group (<5% or ≥5) and chromosome.

Trait	Chr	HapMap-based Analysis			1000 Genomes-based Analysis			BP Dist.	LD (r^2)
		SNP	MAF	P*	SNP	MAF	P*		
Alkaline Phosphatase	12	rs1880889	0.03	1.96E-08	rs1880889	0.029	3.08E-08	–	–
Interleukin-6R	1	rs8192284	0.385	3.60E-53	rs12730935	0.377	2.28E-53	7078	0.957
Homocysteine	1	rs1801133	0.495	1.75E-11	rs1801133	0.497	2.01E-11	–	–
Vitamin B6	1	rs4654748	0.499	8.84E-09	rs4654932	0.417	7.39E-09	23368	0.744
Oxidized LDL	2	rs676210	0.239	4.05E-30	rs676210	0.239	4.12E-30	–	–
Macrophage inflammatory protein-1b	3	rs1500004	0.104	2.06E-08	rs113341849	0.09	2.14E-11	233267	0.667
Lipoprotein a	6	rs10455872	0.056	1.18E-20	rs55730499	0.048	3.43E-25	4508	0.888
Insulin-like growth factor bp-3	7	rs10260816	0.456	4.42E-14	rs71550311	0.459	1.39E-14	22670	0.871
90K Protein	8	rs406652	0.159	1.56E-09	rs28491433	0.148	6.03E-10	8364	0.932
Soluble transferrin receptor	11	rs1242229	0.129	6.55E-09	rs7940310	0.13	7.17E-09	37889	0.956
Triglycerides	11	rs964184	0.152	9.03E-09	rs964184	0.151	1.01E-08	–	–
Vitamin E alpha tocopherol	11	rs964184	0.152	1.33E-08	rs964184	0.151	6.66E-09	–	–
Interleukin-18	11	rs2250417	0.44	2.94E-14	rs2250417	0.44	2.97E-14	–	–
Alpha-2 globulin (%)	16	rs2000999	0.228	2.13E-08	rs2287997	0.227	1.45E-10	32460	0.937
Beta Carotene	16	rs6564851	0.364	1.70E-10	rs12926540	0.362	6.98E-11	5610	0.954
Lutein	16	rs11645428	0.468	1.17E-13	rs9708919	0.391	7.84E-14	1890	0.53
HDL cholesterol	16	rs1800775	0.486	8.01E-13	rs711752	0.436	8.32E-13	975	0.774
Macrophage inflammatory protein-1b	17	rs1015673	0.443	1.25E-22	rs4796217	0.345	6.62E-21	341805	0.577
Free thyroxine (FT4)	17	rs7212734	0.444	1.10E-08	rs58926603	0.393	5.62E-09	1410	0.547
Vitamin E alpha tocopherol	19	rs10401969	0.057	9.32E-09	rs58542926	0.064	1.18E-08	28169	1

SNP = index SNP representing signal, MAF = minor allele frequency, P^* = P-value derived after applying genomic control correction, BP Dist. = base pair distance between the HapMap and 1000 Genomes index SNPs, LD (r^2) = linkage disequilibrium measure based on r^2 correlation.
Fourteen associations represent the same signals (r^2>0.8) of which 6 are represented by the same index SNP. The remaining 6 represent the same locus (0.2<r^2<0.8).

Replication and validation

For three of the putative novel associations we attempted to replicate associations in additional data. We used 476 and 529 individuals from the BLSA study to test associations with TNF-alpha and vitamin B12, respectively, and 232, 1554 and 406 individuals from the TwinsUK study to test associations with TNF-alpha, vitamin B12 and Cl^-, respectively. We did not find any evidence of association. However, one of the putative novel variants does represent a real association. On closer inspection of the literature and comparison of linkage disequilibrium statistics we showed that the SNP rs112635299 represents the known signal between an amino acid changing variant in the *SERPINA* gene and alpha-1-globulin [14]. This signal is present in our data at 0.007 MAF, is a strong risk factor for the lung disease emphysema and was missed by imputation of HapMap genotypes (the SNP identified is not present in the HapMap reference panel, best *P*-value in the region 2×10^{-4}) (**Table 7** and **Figure 5**).

Discussion

Our results show that imputation of genotype data from the sequenced based 1000 Genomes reference panel successfully captures many more variants than imputation of genotype data from the microarray based HapMap reference panel. As expected, this increase in variants captured was much greater for low frequency and rare SNPs compared to common SNPs.

The main purpose of our study was to assess how imputation of missing genotypes from the sequenced based 1000 Genomes reference panel improves our ability to detect genotype-phenotype associations compared to the microarray-genotype based HapMap reference panel. Our results provide two main conclusions. First, that imputation of 1000 Genomes genotype data marginally improves the strength of association signals at known loci. Of 20 associations detected at $P<5\times10^{-8}$ in both analyses (17 of which represent well replicated signals in the NHGRI catalogue), six were captured by the same index SNP, five were more strongly associated in 1000 Genomes imputed data and one was nominally more strongly associated in HapMap imputed data (based on multivariable statistics **Table 6**). This result was not surprising given that most known associations involve common variants that are likely to be well imputed by both reference datasets. Nevertheless, it is possible that associations undetected by HapMap can be appreciably stronger when using 1000 Genomes imputation, as previously reported in *cis*-eQTL data [15]. We also detected previously unreported associations between known signals and traits. The signals between variants near *SUGP1/TM6SF2* and vitamin E and *APOB* and oxidised LDL were both previously reported with triglycerides and LDL cholesterol. Vitamin E is a lipid-soluble vitamin and associations between lipid variants and vitamin levels have been described before [16].

The second main conclusion is that imputation from 1000 Genomes data can detect associations between low frequency variants and phenotypes that were previously missed by HapMap based imputation approaches. The association between rs112635299 and alpha-1 globulin (%) near the *SERPINA* gene represents the same signal as that detected between rs28929474

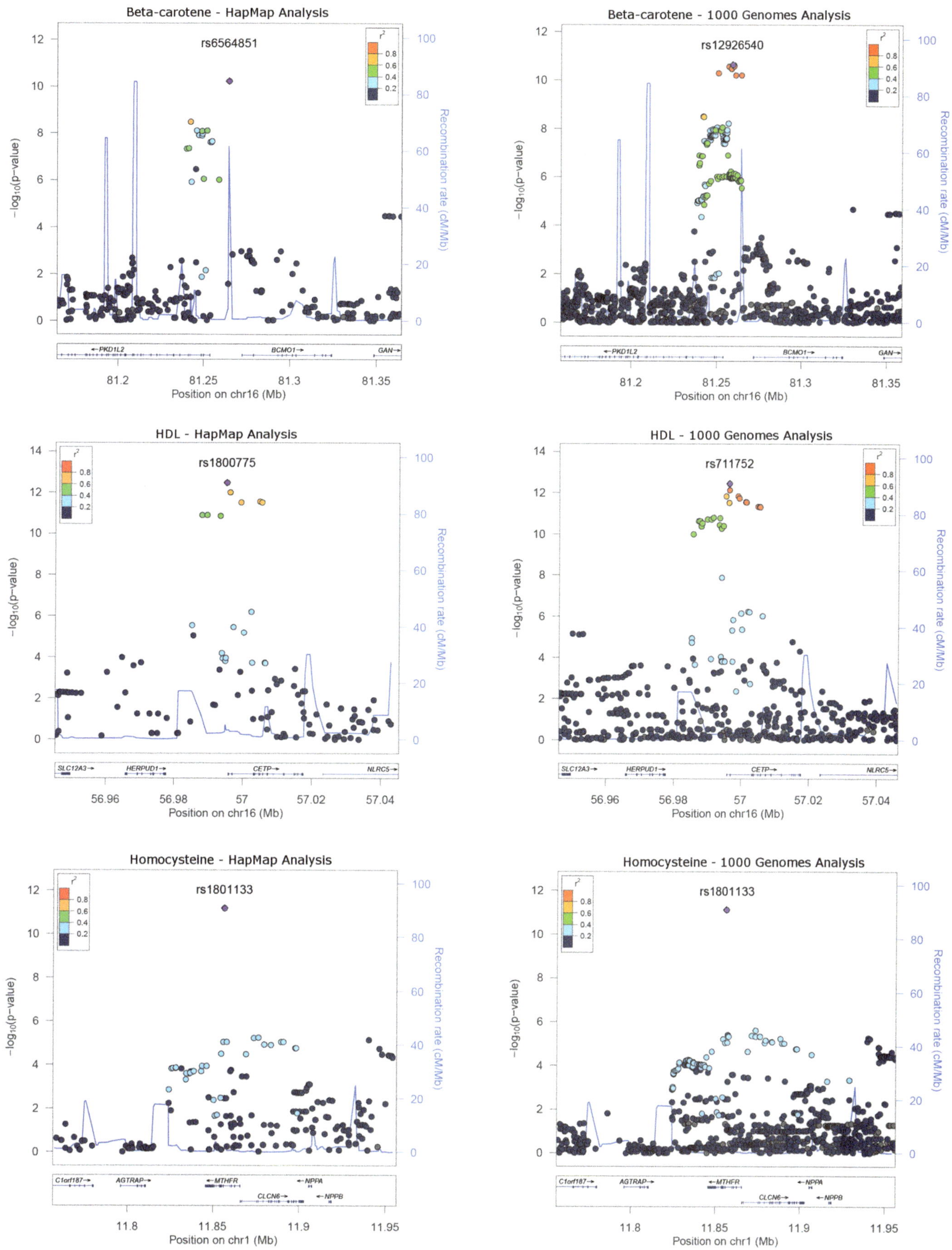

Figure 3. Locus zoom plots of 6 loci reaching $p<5\times10^{-8}$ in both the HapMap- and the 1000 Genomes imputed data and that represent known associations (r^2>0.2 in the NHGRI GWAS catalogue).

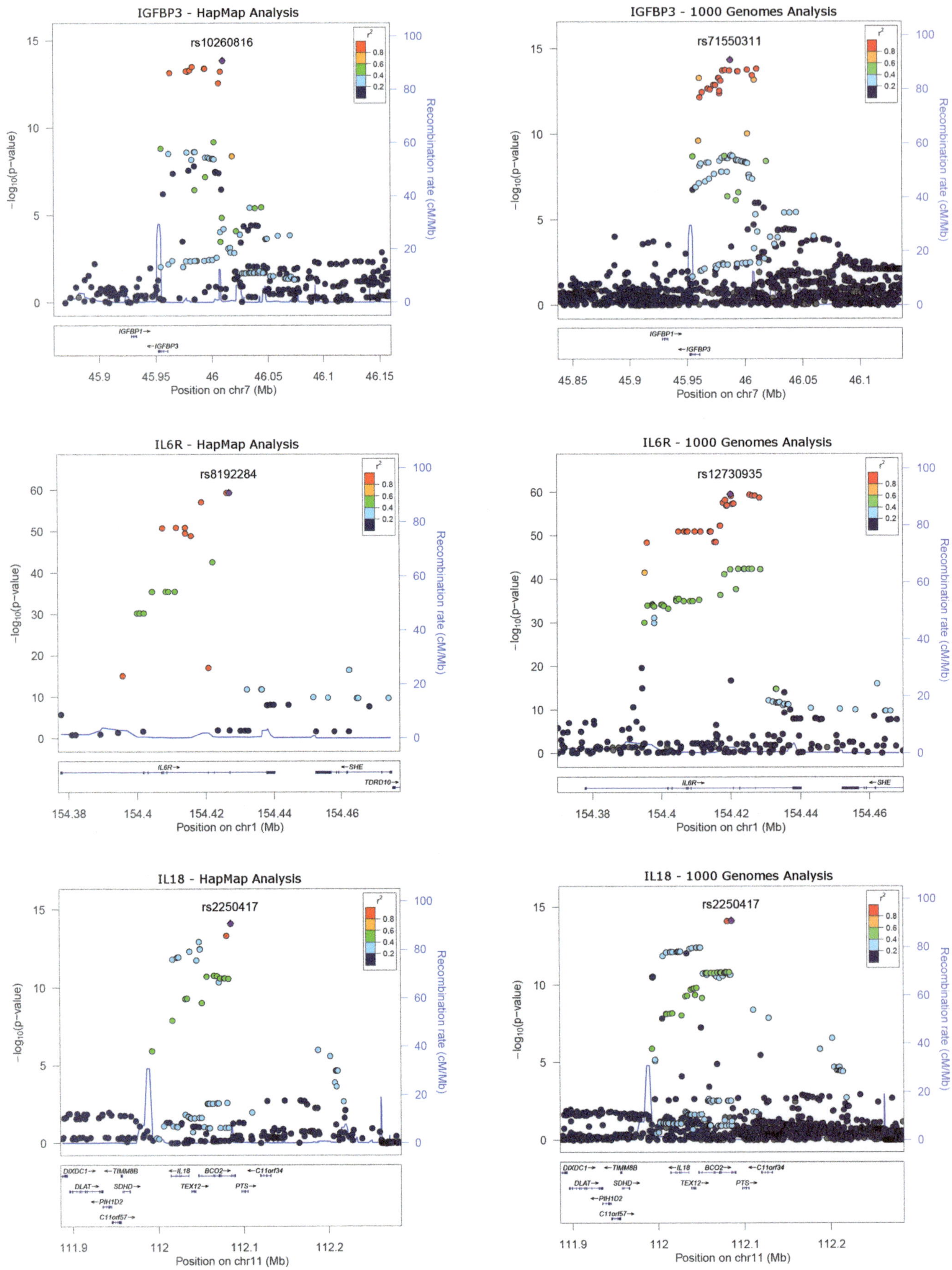
IGFBP3 - HapMap Analysis
rs10260816
IGFBP1→
←IGFBP3
Position on chr7 (Mb)
IGFBP3 - 1000 Genomes Analysis
rs71550311
Position on chr7 (Mb)
IL6R - HapMap Analysis
rs8192284
IL6R→
←SHE
TDRD10→
Position on chr1 (Mb)
IL6R - 1000 Genomes Analysis
rs12730935
Position on chr1 (Mb)
IL18 - HapMap Analysis
rs2250417
DIXDC1→ ←TIMM8B ←IL18 BCO2→ ←C11orf34
DLAT→ SDHD→ TEX12→ PTS→
←PIH1D2
C11orf57→
Position on chr11 (Mb)
IL18 - 1000 Genomes Analysis
rs2250417
Position on chr11 (Mb)
−log10(p−value)
Recombination rate (cM/Mb)
r^2
0.8
0.6
0.4
0.2

Figure 4. Locus zoom plots of 6 loci reaching $p<5\times10^{-8}$ in both the HapMap- and the 1000 Genomes imputed data and that represent known associations ($r^2>0.2$ in the NHGRI GWAS catalogue).

(MAF = 0.007; $r^2=0.88$) and alpha1-antitrypsin that predisposes to emphysema [14]. Although previously known, this association provides important proof of principle that 1000 Genomes imputation can detect novel, low frequency-large effect associations, and potentially save costs of sequencing within studies. One additional association reached our stricter cut-off of $P<5\times10^{-11}$ in 1000 Genomes based imputation that was not previously reported. Although we have no replication data, the signal between rs113341849 near the *CCR1* and *CCR2* genes and Macrophage inflammatory protein-1b is likely to be real given that these genes encode the receptors to which the MIP1b proteins bind.

A third point emerging from our data is that the selection of appropriate statistical thresholds is important when testing more than one trait and many more variants. Of three putative low-frequency-large effect signals we were able to follow up in additional samples none showed evidence of replication. We note that 7 of the 8 putative low frequency-large effect novel associations identified in the 1000 Genomes based data did not reach our stricter cut-off of $P<5\times10^{-11}$, and suspect that most represent false positives due to multiple testing.

The main strength of our study was that we tested a large number of circulating biomarkers. Of these phenotypes, 20 had known signals for the same trait, detectable at genome-wide significance in our single study of 1210 individuals. This large number of phenotypes allowed us to test the extent to which 1000 Genomes imputation enhances known association signals. One limitation to our study is the relatively small sample size. Imputation from the 1000 Genomes reference panel may be more beneficial for strengthening associations at known loci in larger sample sizes. Nevertheless, although we only used 1210 individuals, 7 of the known signals were associated at $P<5\times10^{-12}$ and 3 at $P<\times10^{-15}$, and therefore provided similarly strong association signals to those detected in larger sample sizes for disease traits. A second limitation to our study is that we did not use the very latest version of 1000 genomes and we only tested one method of imputation (MACH). Our study used the version I panel that includes 1094 individuals and 27,713,623 non-singleton SNPs whereas the latest version includes 1092 individuals and 28,681,763 non-singleton SNPs plus indels. Our data therefore provide a lower bound on the gains to be made from 1000 genomes imputation, although a comparison of association statistics for one biomarker and one chromosome with a known strong signal (chromosome 14 and alpha 1 globulin; **Figure S3**) suggests that any improvement from version I to version III will not be large.

In conclusion, imputation of missing genotypes from sequenced based 1000 Genomes reference data can detect novel genotype-phenotype associations but may not appreciably enhance known signals.

Methods

Samples

We used DNA from 1,210 individuals from the population based InCHIANTI study; a study of aging from the Chianti region in Tuscany, Italy [7], [8] (**Table 1**). All summary statistics based association data is available on request from the authors or from the InCHIANTI study door website at http://www.inchiantistudy.net/.

Phenotypes

We selected 93 circulating factors measured in the InCHIANTI study (**Table S1**). These circulating factors included lipids, proteins, (including binding proteins and hormones), vitamins and ions.

Genotyping and Imputation

Genome-wide genotyping was performed using the Illumina Infinium HumanHap550 genotyping chip. Standard quality control procedures were used to exclude individuals with discordant sex and call rates less than 98% and filter out SNPs with MAF <1%, Hardy Weinberg $P<1\times10^{-4}$ and a call rate <99% [8]. There were no ethnic outliers based on a principle components analysis [8]. This resulted in data from 495,343 directly genotyped SNPs in 1,210 individuals. These 495,343 SNPs were used as the scaffold for HapMap and 1000 Genomes based imputation.

For HapMap based imputation we used MACH 1.0.16 to impute missing genotypes not captured by the Illumina chip. We used the HapMap r22 build-36 reference panel (CEU) and imputed 2,543,887 SNPs. The reference panel comprised of 120 haplotypes from 60 parents of 30 Caucasian trios. Of these 2,543,887 SNPs, we excluded 50,205 that were imputed with an $r^2_{imp}<0.3$ (a measure of imputation quality). This left 2,493,682 SNPs for association analyses.

To perform 1000 Genomes based imputation we estimated phase of contiguous variants in the InCHIANTI subjects using the haplotypes calculated using data from the 1000 Genomes Project consisting of 1,094 individuals and 2,188 haplotypes and the program MACH 1.0.16. We then imputed the variants in the build-37 November 2010 release of 1000 Genomes (Phase 1-α interim) into the phased haplotypes using MINIMAC. This resulted in 37,426,733 imputed SNPs. Of these 37,426,733 SNPs, we excluded 26,547,618 that were imputed with an $r^2_{imp}<0.5$. We used a more stringent cut-off for 1000 Genomes imputation as recommended at time of analysis. This left 10,879,115 SNPs for association analyses. We used a multi-ethnic reference panel that included 381 Europeans (including 98 Tuscans), 181 Americans, 246 Africans and 286 Asians in an attempt to capture variants that may be rare in Europeans but more common on haplotypes from different ethnic backgrounds.

Association analysis

Each phenotype measured at baseline was inverse normalised before age and sex adjusted residuals were generated. A second inverse-normalisation was subsequently applied to the residuals. We ran genome-wide association analysis using MACH2QTL for each trait using both HapMap and 1000 Genomes imputed datasets. We performed a univariable test with dosages representing the index HapMap or index 1000 Genomes SNP as independent variables in an additive genetic model and the normalized residuals of the relevant phenotypes as dependent variables in a linear regression analysis.

Linkage disequilibrium estimates

We estimated linkage disequilibrium within the InCHIANTI data. We used 1000 Genomes imputed dosages (except for two SNPs, where we used HapMap-based dosages because either the SNP was not available in 1000 Genomes imputation reference

Table 5. The 13 signals identified as having an associated 1000 Genomes-based imputed SNP reaching genome-wide significance where the best HapMap SNP did not, ordered by minor allele frequency group (<5% or ≥5) and the minor allele frequency of the 1000 Genomes SNP.

Trait	HapMap-based Analysis			1000 Genomes-based Analysis				BP Dist.	LD (r^2)	Annotation
	SNP	MAF	P*	SNP	MAF	P*	Beta			
Tumor necrosis factor-a	rs10206283	0.289	5.40E-03	rs188141385	0.004	4.16E-10	−2.101	218450	0.002	Novel
Cl-	rs10262256	0.091	2.21E-02	rs2371549	0.005	5.38E-09	−2.059	508371	0.001	Novel
Alpha-1 globulin (%)	rs10149388	0.054	2.25E-04	rs112635299	0.007	2.51E-12	−1.891	371308	0.034	Novel
Alpha-2 globulin (%)	rs16959655	0.079	1.58E-03	rs181929163	0.01	1.57E-08	1.472	656607	0.004	Novel
Total cholesterol	rs17102098	0.021	1.16E-05	rs116435220	0.011	2.18E-08	−1.243	149475	0.526	Novel
Vitamin B12	rs3797617	0.146	1.58E-03	rs146226203	0.012	3.01E-08	−1.38	566327	0.009	Novel
Beta globulins (%)	rs10516879	0.163	1.51E-03	rs28645201	0.02	1.22E-09	1.116	211623	0.032	Novel
Dehydroepiandrosterone	rs11761528	0.092	9.67E-07	rs34670419	0.028	2.07E-09	−0.78	12033	0.263	Known
Cortisol:DHEAS ratio	rs10278040	0.045	1.17E-05	rs34670419	0.028	2.35E-08	0.721	10539	0.588	Known
LDL cholesterol	rs445925	0.114	3.60E-05	rs1065853	0.075	1.76E-09	−0.526	2407	0.714	Known
Beta Cryptoxanthin	rs1464093	0.178	5.05E-06	rs75226183	0.087	1.69E-09	−0.514	821919	0.049	Novel
Creatine phosphokinase	rs17112705	0.295	6.92E-07	rs61871700	0.28	4.18E-08	0.257	108933	0.78	Known
Gamma-glutamyltransferase	rs2017869	0.394	7.94E-08	rs2006227	0.386	1.70E-08	0.287	1553	0.973	Known

P* - *P*-value derived after applying genomic control correction.
Details of the best HapMap SNP within a 2 MB window centred on the 1000 Genomes SNP are provided. Annotation "Novel" refers to signals not previously reported. Annotation "known" refers to signals not identified by HapMap but identified in the NHGRI GWAS catalogue.

Table 6. Results from the univariable and multivariable analysis using the Index HapMap and 1000 Genomes SNPs for 13 traits across 14 loci where both index SNPs were genome-wide significant and differed from each other.

Trait	Index HM SNP	HM Uni *P**	HM Multi *P**	Index 1 KG SNP	1 KG Uni *P**	1 KG Multi *P**
90 K Protein	rs406652	2.33E-09	0.90	rs28491433	4.29E-10	0.07
Alpha-2 globulin (%)	rs2000999	2.79E-10	0.78	rs2287997	1.01E-10	0.14
Vitamin E alpha tocopherol	rs10401969	1.07E-08	0.80	rs58542926	9.38E-09	0.58
Beta carotene	rs6564851	1.26E-10	0.86	rs12926540	4.73E-11	0.1625
HDL cholesterol	rs1800775	1.98E-12	0.17	rs711752	4.77E-13	0.0305
Free thyroxine (FT4)	rs7212734	8.49E-09	0.04	rs58926603	4.39E-09	0.0342
Insulin-like growth factor bp-3	rs10260816	2.07E-14	0.39	rs71550311	6.16E-15	0.0774
Interleukin-6R	rs8192284	8.84E-60	0.19	rs12730935	4.87E-60	0.0882
Lipoprotein a	rs10455872	5.06E-25	0.44	rs55730499	2.34E-26	0.0097
Lutein	rs11645428	5.96E-14	0.01	rs9708919	4.03E-14	0.0072
Macrophage inflammatory protein-1b	rs1015673	1.37E-23	0.002	rs4796217	1.09E-21	0.2842
Macrophage inflammatory protein-1b	rs1500004	9.33E-09	0.68	rs113341849	1.37E-11	0.0003
Soluble transferrin receptor	rs1242229	7.51E-09	0.71	rs7940310	5.66E-09	0.4052
Vitamin B6	rs4654748	6.73E-09	0.20	rs4654932	4.33E-09	0.1922

HM = HapMap, 1 KG = 1000 Genomes, Uni *P** = *P*-value derived from univariable analysis after applying genomic control correction. Multi *P** = *P*-value derived from multivariable analysis after applying genomic control correction.

panel or it failed 1000 Genomes quality control based on the imputation-r^2) and converted to best-guess genotypes. PLINK was then used to obtain linkage disequilibrium estimates of the SNPs of interest within a given locus.

Statistical thresholds

We report associations at two statistical thresholds. First, we used a *P*-value of 5×10^{-11} to correct the usual threshold of 5×10^{-8} for the ~100 traits analysed and the ~10 fold increase in signals tested (based on 10 million tests arising from 1000 Genomes imputation). This threshold is likely to be conservative given that the 10 million variants are not independent. There is also a Bayesian argument that reasonably presumes that more causal variants exist for more traits and that capturing more variants should result in more true positive signals of association. This argument, combined with the fact that many associations were known means we also report results at the traditional 5×10^{-8}.

Known vs. novel Signals

We downloaded the catalogue of all association signals reaching genome-wide significance (available from the National Human Genome Research Institute (NHGRI), part of National Institutes of Health). We obtained lists of all known SNPs correlated with

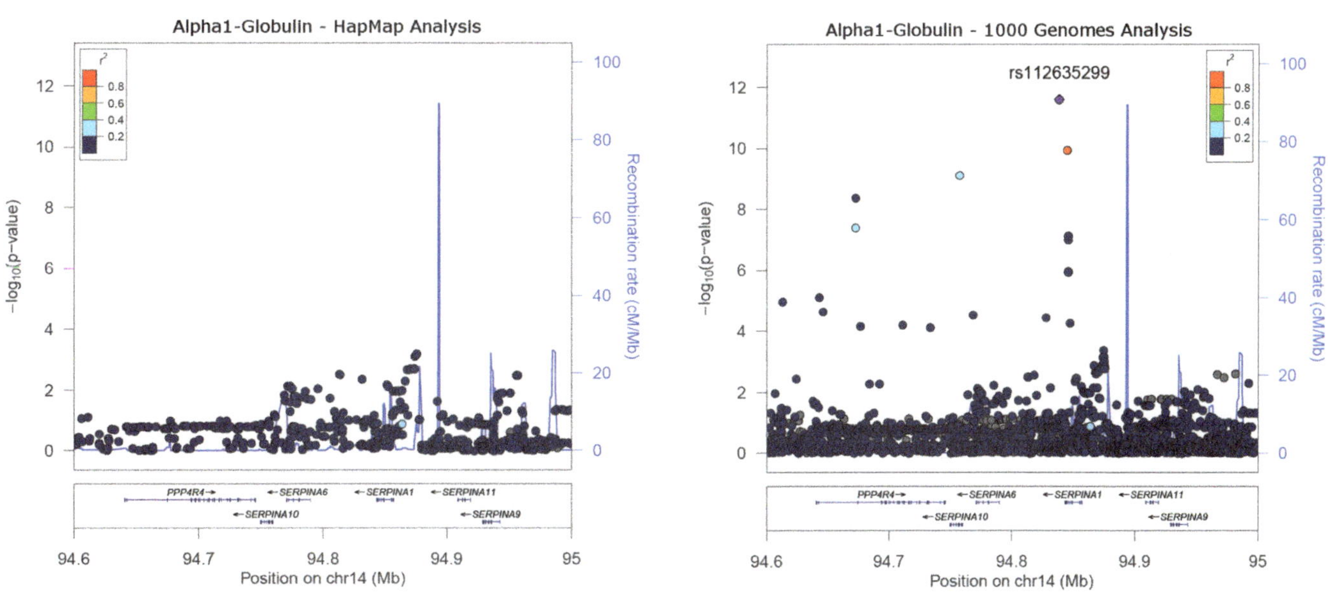

Figure 5. Locus zoom plots of the associations between HapMap and 1000 genomes imputed variants and alpha-1 globulin.

Table 7. Replication data from three biomarker phenotypes available in the BLSA and or TwinsUK studies.

Trait	1000 G Index SNP	MAF	*P*-value	Beta	BLSA Replication *P*	TwinsUK Replication *P*
TNF-a	rs188141385	0.004	4.20E-10	2.1	0.38	0.56
Cl-	rs2371549	0.005	5.40E-09	2.06	–	0.9
Vitamin B12	rs146226203	0.01	3.00E-08	1.38	0.35	0.41

Beta = standard deviations from the double inverse-normalised phenotype values.

$r^2>0.2$ in Europeans with our index genome wide significant SNPs using the program SNAP (available from the BROAD Institute). We linked this list to phenotype-associated index SNPs using both the HapMap and 1000 Genomes based imputation results.

We classified a *signal* as known if our index SNP or an $r^2>0.8$ proxy was associated with the same phenotype or a closely related phenotype (e.g. fatty acids and other lipids) in the NHGRI catalogue (at $P<5\times10^{-8}$). We classified a *locus* as known if our index SNP or an $r^2>0.2$ and $r^2<0.8$ proxy was associated with the same phenotype in the NHGRI catalogue ($P<5\times10^{-8}$).

As the NHGRI catalogue is not 100% complete, we also conducted literature searches using the name of the circulating factor and other key terms including "GWAS" or "Genome wide".

Does 1000 Genomes imputation identify stronger associations at loci identified by HapMap based imputation?

For all loci reaching genome-wide significance in the HapMap and 1000 Genomes based analysis we performed additional statistical tests to assess whether or not 1000 Genomes based imputation identifies stronger associations. We performed a multivariable test with dosages representing both the index HapMap and index 1000 Genomes SNP as independent variables (where the two index SNPs were different) in an additive genetic model and the normalized residuals of the relevant phenotypes as dependent variables in a linear regression analysis. We used 1000 Genomes imputed dosages to represent SNPs identified by both HapMap and 1000 Genomes analyses in STATA. Where 1000 Genomes based dosages were not available we extracted the dosage from the HapMap imputation dataset. We compared the strength of association explained between univariable and multivariable SNP analyses.

Replication and validation

For three of the putative novel associations we attempted to replicate associations in additional data. We used 476 and 529 individuals from the BLSA study [17] to test associations with TNF-alpha and vitamin B12, respectively. In addition, 232, 1554 and 4068 individuals from the TwinsUK study [18] were used to test associations with TNF-alpha, vitamin B12 and chlorine ion levels, respectively.

Supporting Information

Figure S1 Distribution of SNP minor allele frequencies (MAFs) within imputation r^2 categories.

Figure S2 Distributions of the inverse normalised residual values of the 6 traits with evidence of low-frequency/large effect variants as captured by 1000 Genomes imputation.

Figure S3 A comparison of P-values when testing Alpha 1 globulin using overlapping SNPs present in both 1000 Genomes Phase 1 versions 1 and 3 on chromosome 14. Correlation coefficient = 0.931.

Table S1 The 93 circulating factors analysed.

Table S2 HapMap and 1000 Genomes based SNP counts per minor allele frequency (MAF) and imputation quality metric (r^2) bin.

Author Contributions

Conceived and designed the experiments: ARW TMF MNW. Performed the experiments: ARW JRBP TT HFZ TDS JBR. Analyzed the data: ARW JRBP TT HFZ TDS JBR. Contributed reagents/materials/analysis tools: DM. Wrote the paper: ARW TMF. Genotyped the InCHIANTI study: DGH JRG MAN ABS. Lead the InCHIANTI study: SB LF.

References

1. Marchini J, Howie B (2010) Genotype imputation for genome-wide association studies. Nat Rev Genet 11: 499–511.
2. The International HapMap Consortium (2003) The International HapMap Project. Nature 426: 789–796.
3. The International HapMap Consortium (2005) A haplotype map of the human genome. Nature 437: 1299–1320.
4. Altshuler DM, Gibbs RA, Peltonen L, Altshuler DM, Gibbs RA, et al. (2010) Integrating common and rare genetic variation in diverse human populations. Nature 467: 52–58.
5. Frazer KA, Ballinger DG, Cox DR, Hinds DA, Stuve LL, et al. (2007) A second generation human haplotype map of over 3.1 million SNPs. Nature 449: 851–861.
6. The 1000 Genomes Project Consortium (2010) A map of human genome variation from population-scale sequencing. Nature 467: 1061–1073.
7. Ferrucci L, Bandinelli S, Benvenuti E, Di Iorio A, Macchi C, et al. (2000) Subsystems contributing to the decline in ability to walk: bridging the gap between epidemiology and geriatric practice in the InCHIANTI study. J Am Geriatr Soc 48: 1618–1625.
8. Melzer D, Perry JR, Hernandez D, Corsi AM, Stevens K, et al. (2008) A genome-wide association study identifies protein quantitative trait loci (pQTLs). PLoS Genet 4: e1000072.
9. Freathy RM, Timpson NJ, Lawlor DA, Pouta A, Ben-Shlomo Y, et al. (2008) Common variation in the FTO gene alters diabetes-related metabolic traits to the extent expected given its effect on BMI. Diabetes 57: 1419–1426.

10. Perry JR, Ferrucci L, Bandinelli S, Guralnik J, Semba RD, et al. (2009) Circulating beta-carotene levels and type 2 diabetes-cause or effect? Diabetologia 52: 2117–2121.
11. Perry JR, Weedon MN, Langenberg C, Jackson AU, Lyssenko V, et al. (2010) Genetic evidence that raised sex hormone binding globulin (SHBG) levels reduce the risk of type 2 diabetes. Hum Mol Genet 19: 535–544.
12. Rafiq S, Melzer D, Weedon MN, Lango H, Saxena R, et al. (2008) Gene variants influencing measures of inflammation or predisposing to autoimmune and inflammatory diseases are not associated with the risk of type 2 diabetes. Diabetologia 51: 2205–2213.
13. Wu Y, McDade TW, Kuzawa CW, Borja J, Li Y, et al. (2012) Genome-wide association with C-reactive protein levels in CLHNS: evidence for the CRP and HNF1A loci and their interaction with exposure to a pathogenic environment. Inflammation 35: 574–583.
14. Ferrarotti I, Thun GA, Zorzetto M, Ottaviani S, Imboden M, et al. (2012) Serum levels and genotype distribution of alpha1-antitrypsin in the general population. Thorax 67: 669–674.
15. Wood AR, Hernandez DG, Nalls MA, Yaghootkar H, Gibbs JR, et al. (2011) Allelic heterogeneity and more detailed analyses of known loci explain additional phenotypic variation and reveal complex patterns of association. Human molecular genetics 20: 4082–4092.
16. Major JM, Yu K, Wheeler W, Zhang H, Cornelis MC, et al. (2011) Genome-wide association study identifies common variants associated with circulating vitamin E levels. Hum Mol Genet 20: 3876–3883.
17. Terracciano A, Balaci L, Thayer J, Scally M, Kokinos S, et al. (2009) Variants of the serotonin transporter gene and NEO-PI-R Neuroticism: No association in the BLSA and SardiNIA samples. Am J Med Genet B Neuropsychiatr Genet 150B: 1070–1077.
18. Spector TD, Williams FM (2006) The UK Adult Twin Registry (TwinsUK). Twin Res Hum Genet 9: 899–906.

14

Deletion of the LTR Enhancer/Promoter Has No Impact on the Integration Profile of MLV Vectors in Human Hematopoietic Progenitors

Arianna Moiani[1], Annarita Miccio[2], Ermanno Rizzi[3], Marco Severgnini[3], Danilo Pellin[4], Julia Debora Suerth[5], Christopher Baum[5], Gianluca De Bellis[3], Fulvio Mavilio[2,6]*

1 Division of Genetics and Cell Biology, Istituto Scientifico H. San Raffaele, Milan, Italy, **2** Department of Life Sciences, University of Modena and Reggio Emilia, Modena, Italy, **3** Institute for Biomedical Technologies, Consiglio Nazionale delle Ricerche, Milan, Italy, **4** CUSSB, Vita-Salute San Raffaele University, Milan, Italy, **5** Institute of Experimental Hematology, Hannover Medical School, Hannover, Germany, **6** Genethon, Evry, France

Abstract

Moloney murine leukemia virus (MLV)-derived gamma-retroviral vectors integrate preferentially near transcriptional regulatory regions in the human genome, and are associated with a significant risk of insertional gene deregulation. Self-inactivating (SIN) vectors carry a deletion of the U3 enhancer and promoter in the long terminal repeat (LTR), and show reduced genotoxicity in pre-clinical assays. We report a high-definition analysis of the integration preferences of a SIN MLV vector compared to a wild-type-LTR MLV vector in the genome of CD34$^+$ human hematopoietic stem/progenitor cells (HSPCs). We sequenced 13,011 unique SIN-MLV integration sites and compared them to 32,574 previously generated MLV sites in human HSPCs. The SIN-MLV vector recapitulates the integration pattern observed for MLV, with the characteristic clustering of integrations around enhancer and promoter regions associated to H3K4me3 and H3K4me1 histone modifications, specialized chromatin configurations (presence of the H2A.Z histone variant) and binding of RNA Pol II. SIN-MLV and MLV integration clusters and hot spots overlap in most cases and are generated at a comparable frequency, indicating that the reduced genotoxicity of SIN-MLV vectors in hematopoietic cells is not due to a modified integration profile.

Editor: Yasuhiro Ikeda, Mayo Clinic, United States of America

Funding: This work was supported by grants from the European Commission (PERSIST), the European Research Council (GT-SKIN), and the Italian National Research Council (CNR – EPIGEN). The funders had no role in study design, data collection and analysis, decision to publish, or preparation of the manuscript.

Competing Interests: The authors have declared that no competing interests exist.

* E-mail: fmavilio@genethon.fr

Introduction

Retroviral integration is a non-random process, whereby the viral RNA genome, reverse transcribed into double-stranded DNA and assembled in pre-integration complexes (PICs), associates with the host cell chromatin and integrates through the activity of the viral integrase [1]. Large-scale surveys of retroviral integration in murine and human cells uncovered some genomic features systematically and specifically associated with retroviral insertions, and revealed that each retrovirus type has a unique, characteristic pattern of integration within mammalian genomes [2]. Target site selection depends on both viral and cellular determinants, poorly defined for most retroviruses. The Moloney murine leukemia virus (MLV) and its derived vectors integrate preferentially in transcriptionally active promoters and regulatory regions [2,3,4], while the simian (SIV) and human immunodeficiency virus (HIV) and their derived lentiviral vectors target gene-dense regions and the transcribed portion of expressed genes, away from regulatory elements [2,4,5].

Recent clinical studies have shown that transplantation of stem cells genetically modified by retroviral vectors may cure severe genetic diseases such as immunodeficiencies [6,7,8], skin adhesion defects [9] and lysosomal storage disorders [10]. Some of these studies showed also the genotoxic consequences of retroviral gene transfer technology. Insertional activation of proto-oncogenes by MLV-derived vectors caused T-cell lymphoproliferative disorders in patients undergoing gene therapy for X-linked severe combined immunodeficiency [11,12] and Wiskott-Aldrich syndrome [13], and pre-malignant expansion of myeloid progenitors in patients treated for chronic granulomatous disease (CGD) [14,15]. The strong transcriptional enhancers present in the MLV LTR played a major role in deregulating gene expression. Pre-clinical studies showed that enhancer-less, self-inactivating (SIN) MLV-derived vectors are less prone to insertional oncogenesis and cell immortalization than their full-LTR counterparts, with a genotoxic profile closer to that of SIN-HIV vectors [16,17,18,19]. The MLV U3 enhancer contains repeated binding sites for cellular transcription factors (TF), which may play a role in tethering retroviral pre-integration complexes to transcriptionally active regulatory regions and contribute to the MLV genotoxic characteristics [20].

In this study, we have analyzed the integration profiles of a MLV and SIN MLV vectors in the genome of a clinically relevant target cell population, cord blood-derived CD34$^+$ hematopoietic stem/progenitor cells (HSPCs), by ligation-mediated PCR (LM-PCR) and high-throughput sequencing. We show that SIN-MLV and MLV vectors have very similar integration preferences, with the typical clustering around enhancer and promoter regions associated to specific histone modifications, specialized chromatin configurations and binding of RNA Pol II. Strikingly, SIN-MLV and MLV integration clusters and hot spots overlap in most cases and are generated at a similar frequency, indicating that the U3 enhancer has no role in targeting MLV PICs to the genome, at least in hematopoietic cells.

Results

MLV and SIN-MLV vectors share the same integration profile in the genome of human HSPCs

To generate a high-definition integration profile of SIN MLV integrations in human HSPCs, we transduced umbilical cord blood-derived $CD34^+$ cells with a previously described SIN-MLV vector carrying a GFP expression cassette under the control of the human elongation factor 1α (EFS) promoter [16,21]. Cells were transduced at 40 to 60% efficiency, and were selected for GFP expression by cell sorting 10 days after infection, to dilute unintegrated vectors. Vector-genome junctions were amplified from genomic DNA by ligation-mediated (LM)-PCR and pyrosequenced as previously described [4]. Raw sequences (available at the NCBI Sequence Read Archive with the accession number SRA061405) were processed by a previously described bioinformatic pipeline [4] and mapped on the University of California at Santa Cruz (UCSC) hg19 release of the human genome (http://genome.ucsc.edu), to obtain 13,011 unique insertion sites. Two previously generated datasets of full-LTR MLV vector integrations (32,574) and random sites (40,000) normalized for a number of parameters [4] were re-annotated on the hg19 release of the human genome and used for comparison.

To identify differences in the integration preferences of MLV and SIN-MLV in HSPCs, we first analyzed the relationship between integration sites and Known Genes (UCSC definition) in the human genome: integration were annotated as TSS-proximal when occurring in an interval of ±2.5 kb from the TSS of any Known Gene, intragenic when occurring inside a Known Gene >2.5 kb from the TSS, and intergenic in all other cases. Intergenic and intragenic integrations were <40% for both MLV and SIN-MLV, while TSS-proximal integrations were 22.9% and 23.8% respectively ($P>0.1$ for all comparisons) (**Table 1**). Plotting the relative distance of all integration sites in an interval of ±50 kb from any TSS showed two virtually overlapping distributions with the characteristic enrichment in the ±2.5 kb interval and, at 50-bp resolution, the characteristic drop in frequency in close proximity (±0.2 kb) of a TSS (**Figure 1**). Similarly, integrations were enriched around annotated CpG islands (UCSC track) for both MLV and SIN-MLV vectors (12.8 and 12.4% respectively, $P>0.1$), and showed again overlapping, bimodal distributions in the ±2.5 kb interval around the island midpoint (**Figure S1A**). Finally, the moderate enrichment in the frequency of integration around mammalian, evolutionarily conserved non-coding sequences (CNC) [22] was the same for both vectors (8.0 and 7.7% respectively, $P>0.1$), and showed a virtually identical distribution at ±2.5 kb around the feature midpoint (**Figure S1B**). In all cases, there were significant differences between the distributions of the two MLV vectors and that of the random controls ($P<10^{-15}$).

We previously reported that MLV integrations are strongly associated with histone modifications marking transcriptionally active promoters and enhancers, with the specialized H2A.Z histone variant, and with binding sites for RNA Pol II and transcription factors in both HSCs and T cells [4,20,23]. Taking advantage of published ChIP-Seq datasets on epigenetic features in the chromatin of human $CD34^+$ HSPCs [24], we analyzed the association of MLV and SIN-MLV integrations with histone modifications marking active enhancers and promoters (H3K4me1, H3K4me3) and heterochromatin (H3K27me3), H2A.Z and binding of Pol II. In all cases, there was no obvious difference between the two vectors: we observed a strong association with H3K4me1, H3K4me3, Pol II and H2A.Z, and no correlation with H3K27me3 (**Figure 1C** and **Figure S1C**).

All together, these data indicate that the absence of the U3 region of the LTR of the SIN-MLV vector causes no significant change in the integration preferences of MLV in the genome of human HSPCs.

MLV and SIN-MLV vectors integrate in the same hot spots in the HSPC genome

MLV and SIN-MLV integrations showed the same non-random, highly clustered distribution in the human genome, with integration hot and cold spots. Integration clusters were statistically defined as described [4], obtaining a numerosity-adjusted threshold for cluster definition of 3 integrations within 31,525 bp for SIN-MLV and 12,587 bp for MLV at a P-value<0.01. We identified 1,415 clusters containing 56.0% (7,318) of the total SIN-MLV integration sites, an overall clustering highly comparable ($P>0.1$) to that showed by the MLV vector (3,497 clusters containing 65.3% of the integrations). Most of the SIN-MLV clusters (75.9%) overlapped for at least 1 bp with MLV clusters, and up to 69% overlapped for at least 1,000 bp. The non-overlapping clusters contained only 3 integrations and mapped less than 200-bp apart in the MLV and SIN-MLV datasets.

We then looked at the integration clusters of both vectors in a number of individual loci characteristically hit at high frequency by MLV integration. Most of the loci highly targeted by MLV were targeted also by SIN-MLV. Whenever the numerosity was sufficient, we observed a striking overlap between the integration hot spots of the two vectors within the same loci. As examples, SIN-MLV integrations faithfully reproduced the MLV integration patterns in the LMO2, EVI2A/B, RUNX1, RUNX2, ZNF217-BCAS1, CD34, ELF1 and NFE2 loci, which were targeted at the same overall frequency and at the same hot spots within each locus (**Figure 2** and **Figure S2**). We also found SIN-MLV integrations mapping very closely to the MLV integrations in two hot spots (MECOM and PRDM16) observed in the CGD and WAS clinical trials [13,14,15] (**Figure S2**).

Discussion

Retroviruses select their target integration sites by tethering their PICs to the host cell chromatin through protein-protein interactions that appear to be specific for each retrovirus type [2]. The chromatin component LEDGF/p75 has a major role in tethering HIV PICs to the body of active transcription units, associated with H4K12ac, H2BK5me1, H3K27me1, H3K36me3, and H4K20me1 histone modifications [5,25]. LEDGF/p75 is directly bound by the HIV integrase [26], a major viral determinant of target site selection [27]. Much less is known for the MLV PICs: regions preferred by MLV are marked by acetylations of histones H2A, H2B, H3 and H4, methylations of H3 (H3K4me1, H3K4me2, H3K4me3), and binding of Pol II, CTCF, histone acetyl transferases (p300 and CBP) and H2A.Z [4,23], a histone variant enriched at targets of the Polycomb complex and marking elements involved in the regulation of cell commitment and differentiation [28]. Most of these associations are statistically significant for all MLV integrations, independently

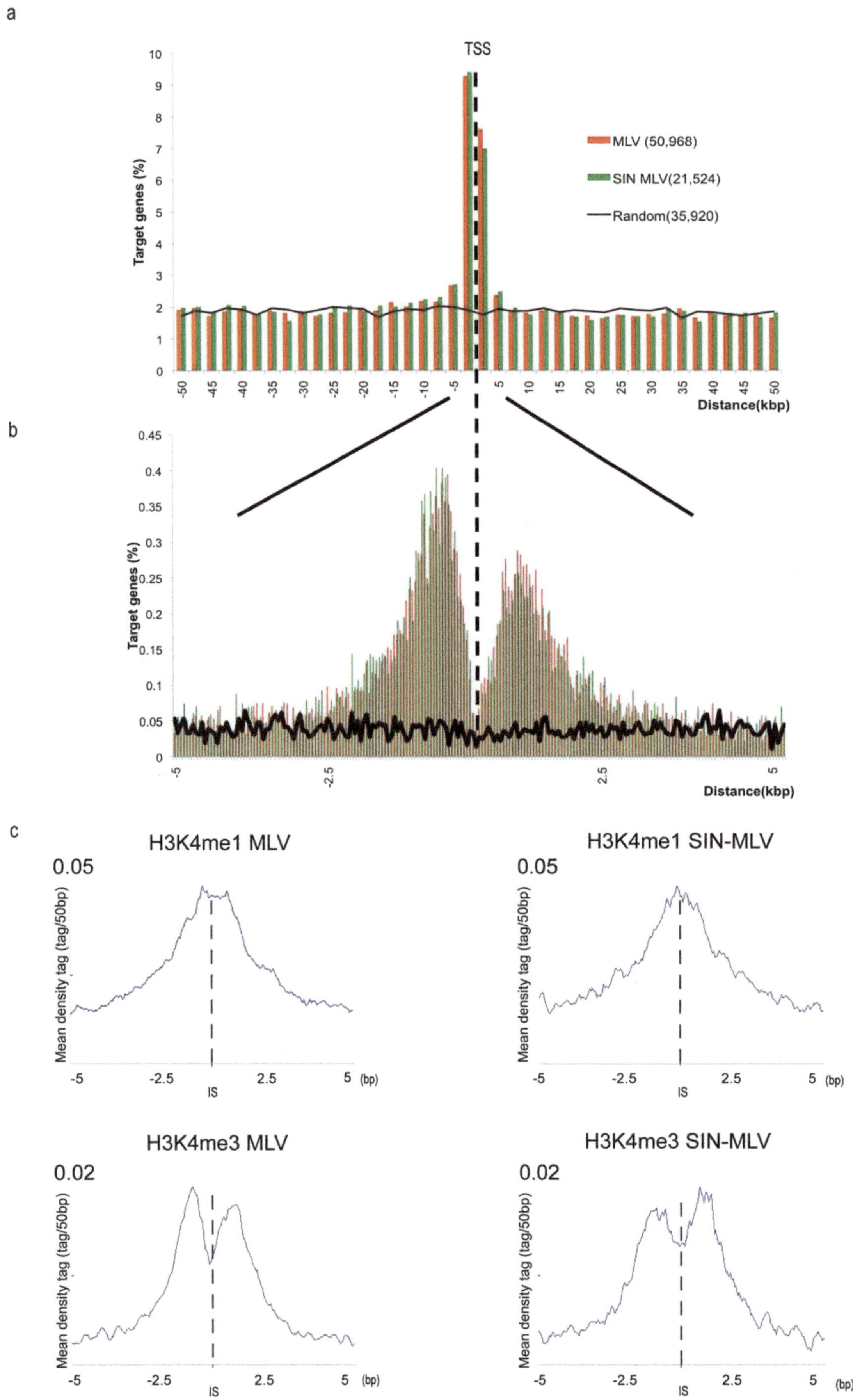
a
TSS
Target genes (%)
MLV (50,968)
SIN MLV(21,524)
Random(35,920)
Distance(kbp)
b
Target genes (%)
Distance(kbp)
c
H3K4me1 MLV
H3K4me1 SIN-MLV
H3K4me3 MLV
H3K4me3 SIN-MLV
0.05
0.02
Mean density tag (tag/50bp)
IS
(bp)

Figure 1. Genomic distribution and association with histone modifications of MLV and SIN-MLV integrations in human HSPCs. Distribution of the distance of SIN-MLV (green bars) and MLV (red bars) integration sites from the TSS of targeted genes at 2,500 bp (**a**) or 50 bp (**b**) resolution. The percentage of genes targeted at each position is plotted on the y axis. The number of plotted sites is higher than the actual number of mapped integrations sites, since each site may relate to more than one gene. The black line indicates the distribution of random control sites. (**c**) The distribution of H3K4me1 (top panels) and H3K4me3 (lower panels) epigenetic marks in a 10-kb window around vector integration sites (IS) is shown for MLV integrations (left panels) or SIN-MLV integrations (right panels). The mean density of tags (tag/50 bp) of the reference dataset (i.e. each chromatin feature) is plotted on the y axis. The scale of the graph is shown at the top left of each panel.

from their location with respect to promoters and in all analyzed cell types, possibly reflecting the engagements of PICs by basal components of the enhancer-binding and/or RNA Pol II transcriptional machinery. A yeast two-hybrid analysis of proteins potentially interacting with the MLV integrase provided biochemical evidence in this direction [29].

The peculiar characteristics of MLV integration, coupled with the strong transcriptional enhancer activity of the LTR U3 region, explain the high risk of insertional gene activation and genotoxicity observed in pre-clinical [19,30,31,32] as well as clinical [11,12,13,14,15] studies. SIN vectors have been developed to reduce the genotoxic potential of MLV vectors, and have indeed shown a reduced cell immortalization and tumorigenic activity by sensitive pre-clinical assays [18,19,33]. In this study, we compared the integration preferences of a SIN and a traditional MLV vector in human CD34$^+$ HSPCs, the most used target cell in clinical applications of retroviral transgenesis. By an LM-PCR coupled to pyrosequencing approach we show that the lack of the U3 enhancer/promoter in the LTR has no impact on the integration pattern of MLV in human HSPCs. Indeed, the SIN-MLV integration map reproduced with remarkable precision that of the unmodified MLV, including the association with TSSs, CpG islands, CNCs and representative epigenetic marks of active and highly regulated enhancers and promoters. SIN-MLV and MLV integrations cluster into hot spots at approximately the same frequency, and generate almost overlapping integration maps at the level of highly targeted loci, including the gene involved in most of the severe adverse events observed in clinical trials, i.e., LMO2 [11,12]. On the basis of the integration pattern, a SIN-MLV vector therefore maintains the same genotoxic potential of a traditional MLV vector. Insertion of either provirus has a high chance of altering gene regulation by disrupting the physical continuity of enhancers and promoters, and by altering the chromatin configuration induced by the binding of transcription factors and the basal transcriptional machinery (the enhanceosome). This type of effect is not expected to differ between SIN-MLV and MLV vectors. On the contrary, the lack of the two copies of the U3 enhancer may significantly reduce the dominant activity in overcoming gene regulation typical of oncogenic retroviruses *in vivo* [19], although studies based on *in vitro* immortalization of bone marrow-derived cells provided conflicting evidence on this point [18,34]. The choice of a cellular, possibly restricted enhancer to drive the internal transgene cassette may therefore overcome the LTR-specific component of the MLV genotoxicity. Indeed, most of the gene deregulation and genotoxic activity of SIN vectors appears to be due to the characteristics of the enhancer driving transgene expression more than by the SIN design *per se* [16,17,18,19].

Genotoxicity of retroviral vectors has many components, including the vector design, the nature of the target cell and the genetic background of the patient, all ultimately affecting the risk of a specific gene therapy approach [35]. Target site selection is just one of these components. Based on current knowledge, SIN lentiviral vectors appear to combine an integration profile that does not target regulatory elements with the lack of strong viral enhancers. SIN-MLV vectors share with SIN lentiviral vectors only the latter component. On the other hand, the lower propensity to integrate within transcribed regions may reduce the recently emerged post-transcriptional component of insertional gene deregulation [23,36,37,38,39,40]. For SIN-MLV vectors, the designs of the transgene expression cassette, and particularly the choice of its transcriptional regulatory elements, appear to be the most relevant determinants of their biosafety characteristics.

Materials and Methods

Vectors and cells

Human CD34$^+$ HSPCs were purified form umbilical cord blood, pre-stimulated for 48 hours in serum-free Iscove modified Dulbecco medium supplemented with 20% FCS, 20 ng/ml human thrombopoietin, 100 ng/ml Flt-3 ligand, 20 ng/ml interleukin-6, and 100 ng/ml stem cell factor, as previously described [3]. HSPCs were transduced with the SIN-MLV vector pSRS11.EFS.GFP.pre, expressing GFP under the control of the elongation factor 1α promoter, pseudotyped in an amphotropic envelope by three-plasmid transfection in 293 cells, as previously described [16,21]. Cells were infected by 3 rounds of spinoculation (1,500 rpm for 45 min) in the presence of 4 μg/ml polybrene. Transduction efficiency was evaluated by cytofluorimetric analysis of GFP expression 48 hrs after infection. All human studies were approved by the San Raffaele Scientific Institute Ethical Committee. Written informed consent was received from participants prior to inclusion in the study.

Table 1. Distribution of SIN-MLV and MLV integrations in the genome of human HSPCs.

	Intergenic (%)	TSS-proximal (%)	Intragenic (%)	CpG islands (%)	CNCs (%)	Total hits
SIN-MLV	38.0	23.8	38.1	12.4	7.7	13,011
MLV	38.2	22.9	38.8	12.8	8.0	32,574
Random	59.1	3.0	37.8	1.2	5.4	40,000

Percentage of intergenic, TSS-proximal or intragenic integration sites in the SIN-MLV, MLV and random control site datasets, together with the percentage of sites at a distance of ±1,000 bp from at least one CpG island or conserved non coding region (CNC).

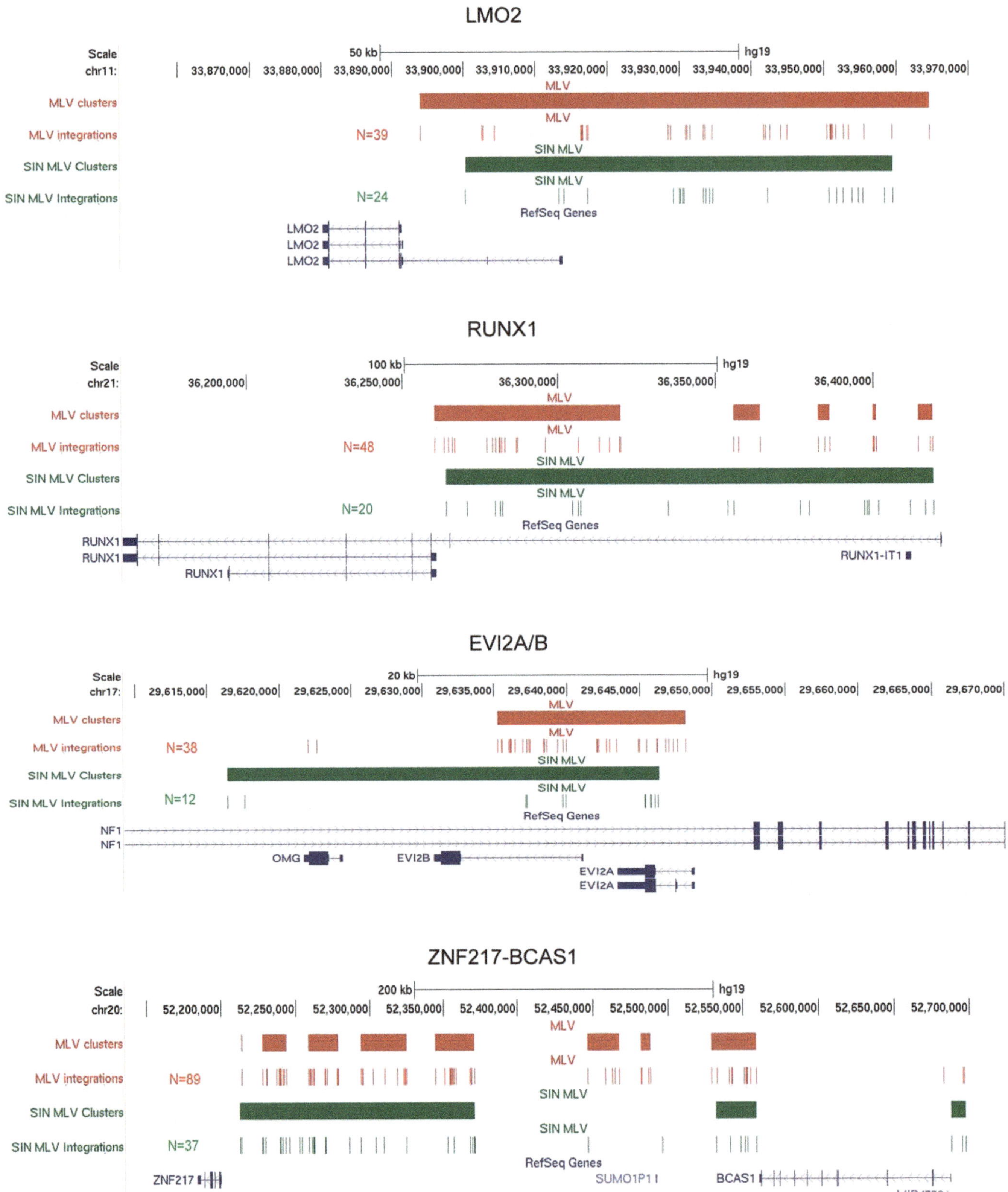

Figure 2. MLV and SIN-MLV integration sites and clusters in CD34^{+} HSPC-specific loci. Distribution of MLV (red) and SIN-MLV (green) integration clusters (horizontal solid bars) and integrations (vertical marks) in the LMO2, RUNX1, EVI2A/B, and ZNF217-BCAS1 loci as displayed by the UCSC Genome Browser. The base position feature at the top (scale bar and chromosome number) identifies the genomic coordinates of the displayed region.

Amplification, sequencing, and analysis of retroviral integration sites

Genomic DNA was extracted from a pool of 2×10^6 $CD34^+$/GFP^+ cells enriched by fluorescence-activated cell sorting, after a brief period in culture to dilute unintegrated vectors. 3′-LTR vector-genome junctions were amplified by LM-PCR adapted to the GS-FLX Genome Sequencer (Roche/454 Life Sciences) pyrosequencing platform, as previously described [3,4]. Raw sequence reads were processed by an automated bioinformatic pipeline that eliminated small and redundant sequences [4] and mapped on the UCSC hg19 release of the human genome. All UCSC Known Genes having their TSS at ±50 kb from an integration site were annotated as targets. Genomic features were annotated when their genomic coordinates overlapped for ≥1 nucleotide with a ±50-kb interval around each integration site. We used UCSC tracks for both cytosine-phosphate-guanosine (CpG) islands and conserved TFBSs. The genomic coordinates of 82,335 mammalian conserved non-coding sequences (CNCs) were previously described [22]. For the association of the integrations with epigenetic marks, we re-annotated published ChIP-Seq data [24] in the UCSC hg19 release of the human genome, and analyzed the distribution of histone modifications (H3K4me1, H3K4me3, H3K27me3) H2A.Z and Pol II binding sites around the integrations, using the seqMINER platform [41]. Previously generated MLV integrations and random control sequences datasets [4] were also re-annotated on the UCSC hg19 genome. For all pairwise comparisons we applied a 2-sided Fisher's exact test. The threshold for statistical significance was set at a P value<0.05.

Supporting Information

Figure S1 Association between MLV and SIN-MLV integration sites and CpG islands, CNCs, PolII and histone modifications. Distribution of the distance of SIN-MLV (green bars) and MLV wt (red bars) integrations from the midpoint of CpG islands (A) or CNCs (B) in a 20 kb window. In the y axis is plotted the percentage of the total number of CpG islands or CNCs located at ±50 kb distance from the integrations. The black line indicates the distribution of random control sites. (C) The distribution of epigenetic marks in a 10 kb window around vector integration sites (IS) shown for H3K27me3 (top panels), H2A.Z (middle panels), PolII (lower panels) with respect to MLV integrations (left panels) or SIN-MLV integrations (right panels). See legend of Figure 1C for explanation of the graphs.

Figure S2 MLV and SIN-MLV integration sites and clusters in $CD34^+$ HSPC-specific loci. Distribution of MLV (red) and SIN-MLV (green) integration clusters (horizontal solid bars) and integrations (vertical marks) in the CD34, ELF1, NFE2, and RUNX2, MECOM and PRDM16 loci as displayed by the UCSC Genome Browser. The base position feature at the top (scale bar and chromosome number) identifies the genomic coordinates of the displayed region.

Acknowledgments

A.M. is a student of the Ph.D. Program in Cellular and Molecular Biology of the Vita-Salute San Raffaele University, Milan, Italy.

Author Contributions

Conceived and designed the experiments: A. Moiani FM. Performed the experiments: A. Moiani ER JDS. Analyzed the data: A. Moiani A. Miccio MS DP GDB. Contributed reagents/materials/analysis tools: JDS CB. Wrote the paper: A. Moiani FM.

References

1. Coffin JM, Huges SH, Varmus HE (1997) Retroviruses. Cold Spring Harbor, NY: Cold Spring Harbor Laboratory Press.
2. Bushman F, Lewinski M, Ciuffi A, Barr S, Leipzig J, et al. (2005) Genome-wide analysis of retroviral DNA integration. Nat Rev Microbiol 3: 848–858.
3. Cattoglio C, Facchini G, Sartori D, Antonelli A, Miccio A, et al. (2007) Hot spots of retroviral integration in human CD34+ hematopoietic cells. Blood 110: 1770–1778.
4. Cattoglio C, Pellin D, Rizzi E, Maruggi G, Corti G, et al. (2010) High-definition mapping of retroviral integration sites identifies active regulatory elements in human multipotent hematopoietic progenitors. Blood 116: 5507–5517.
5. Wang GP, Ciuffi A, Leipzig J, Berry CC, Bushman FD (2007) HIV integration site selection: analysis by massively parallel pyrosequencing reveals association with epigenetic modifications. Genome Res 17: 1186–1194.
6. Aiuti A, Cattaneo F, Galimberti S, Benninghoff U, Cassani B, et al. (2009) Gene therapy for immunodeficiency due to adenosine deaminase deficiency. N Engl J Med 360: 447–458.
7. Hacein-Bey-Abina S, Le Deist F, Carlier F, Bouneaud C, Hue C, et al. (2002) Sustained correction of X-linked severe combined immunodeficiency by ex vivo gene therapy. N Engl J Med 346: 1185–1193.
8. Boztug K, Schmidt M, Schwarzer A, Banerjee PP, Diez IA, et al. (2010) Stem-cell gene therapy for the Wiskott-Aldrich syndrome. N Engl J Med 363: 1918–1927.
9. Mavilio F, Pellegrini G, Ferrari S, Di Nunzio F, Di Iorio E, et al. (2006) Correction of junctional epidermolysis bullosa by transplantation of genetically modified epidermal stem cells. Nat Med 12: 1397–1402.
10. Cartier N, Hacein-Bey-Abina S, Bartholomae CC, Veres G, Schmidt M, et al. (2009) Hematopoietic stem cell gene therapy with a lentiviral vector in X-linked adrenoleukodystrophy. Science 326: 818–823.
11. Hacein-Bey-Abina S, Garrigue A, Wang GP, Soulier J, Lim A, et al. (2008) Insertional oncogenesis in 4 patients after retrovirus-mediated gene therapy of SCID-X1. J Clin Invest 118: 3132–3142.
12. Howe SJ, Mansour MR, Schwarzwaelder K, Bartholomae C, Hubank M, et al. (2008) Insertional mutagenesis combined with acquired somatic mutations causes leukemogenesis following gene therapy of SCID-X1 patients. J Clin Invest 118: 3143–3150.
13. Krause D (2011) Gene Therapy for Wiskott-Aldrich Syndrome: Benefits and Risks. The Hematologist 8: 10.
14. Ott MG, Schmidt M, Schwarzwaelder K, Stein S, Siler U, et al. (2006) Correction of X-linked chronic granulomatous disease by gene therapy, augmented by insertional activation of MDS1-EVI1, PRDM16 or SETBP1. Nat Med 12: 401–409.
15. Stein S, Ott MG, Schultze-Strasser S, Jauch A, Burwinkel B, et al. (2010) Genomic instability and myelodysplasia with monosomy 7 consequent to EVI1 activation after gene therapy for chronic granulomatous disease. Nat Med 16: 198–204.
16. Zychlinski D, Schambach A, Modlich U, Maetzig T, Meyer J, et al. (2008) Physiological Promoters Reduce the Genotoxic Risk of Integrating Gene Vectors. Mol Ther.
17. Maruggi G, Porcellini S, Facchini G, Perna SK, Cattoglio C, et al. (2009) Transcriptional Enhancers Induce Insertional Gene Deregulation Independently From the Vector Type and Design. Mol Ther 17: 851–856.
18. Modlich U, Navarro S, Zychlinski D, Maetzig T, Knoess S, et al. (2009) Insertional transformation of hematopoietic cells by self-inactivating lentiviral and gammaretroviral vectors. Mol Ther 17: 1919–1928.
19. Montini E, Cesana D, Schmidt M, Sanvito F, Bartholomae CC, et al. (2009) The genotoxic potential of retroviral vectors is strongly modulated by vector design and integration site selection in a mouse model of HSC gene therapy. J Clin Invest 119: 964–975.
20. Felice B, Cattoglio C, Cittaro D, Testa A, Miccio A, et al. (2009) Transcription factor binding sites are genetic determinants of retroviral integration in the human genome. PLoS ONE 4: e4571.
21. Thornhill SI, Schambach A, Howe SJ, Ulaganathan M, Grassman E, et al. (2008) Self-inactivating gammaretroviral vectors for gene therapy of X-linked severe combined immunodeficiency. Mol Ther 16: 590–598.
22. Kim SY, Pritchard JK (2007) Adaptive evolution of conserved noncoding elements in mammals. PLoS Genet 3: 1572–1586.
23. Cattoglio C, Maruggi G, Bartholomae C, Malani N, Pellin D, et al. (2010) High-definition mapping of retroviral integration sites defines the fate of allogeneic T cells after donor lymphocyte infusion. PLoS One 5: e15688.

24. Cui K, Zang C, Roh TY, Schones DE, Childs RW, et al. (2009) Chromatin signatures in multipotent human hematopoietic stem cells indicate the fate of bivalent genes during differentiation. Cell Stem Cell 4: 80–93.
25. Wang GP, Levine BL, Binder GK, Berry CC, Malani N, et al. (2009) Analysis of lentiviral vector integration in HIV+ study subjects receiving autologous infusions of gene modified CD4+ T cells. Mol Ther 17: 844–850.
26. Engelman A, Cherepanov P (2008) The lentiviral integrase binding protein LEDGF/p75 and HIV-1 replication. PLoS Pathog 4: e1000046.
27. Lewinski MK, Yamashita M, Emerman M, Ciuffi A, Marshall H, et al. (2006) Retroviral DNA integration: viral and cellular determinants of target-site selection. PLoS Pathog 2: e60.
28. Creyghton MP, Markoulaki S, Levine SS, Hanna J, Lodato MA, et al. (2008) H2AZ is enriched at polycomb complex target genes in ES cells and is necessary for lineage commitment. Cell 135: 649–661.
29. Studamire B, Goff SP (2008) Host proteins interacting with the Moloney murine leukemia virus integrase: multiple transcriptional regulators and chromatin binding factors. Retrovirology 5: 48.
30. Kustikova O, Fehse B, Modlich U, Yang M, Dullmann J, et al. (2005) Clonal dominance of hematopoietic stem cells triggered by retroviral gene marking. Science 308: 1171–1174.
31. Modlich U, Kustikova OS, Schmidt M, Rudolph C, Meyer J, et al. (2005) Leukemias following retroviral transfer of multidrug resistance 1 (MDR1) are driven by combinatorial insertional mutagenesis. Blood 105: 4235–4246.
32. Montini E, Cesana D, Schmidt M, Sanvito F, Ponzoni M, et al. (2006) Hematopoietic stem cell gene transfer in a tumor-prone mouse model uncovers low genotoxicity of lentiviral vector integration. Nat Biotechnol 24: 687–696.
33. Modlich U, Bohne J, Schmidt M, von Kalle C, Knoss S, et al. (2006) Cell-culture assays reveal the importance of retroviral vector design for insertional genotoxicity. Blood 108: 2545–2553.
34. Bosticardo M, Ghosh A, Du Y, Jenkins NA, Copeland NG, et al. (2009) Self-inactivating retroviral vector-mediated gene transfer induces oncogene activation and immortalization of primary murine bone marrow cells. Mol Ther 17: 1910–1918.
35. Kustikova O, Brugman M, Baum C (2010) The genomic risk of somatic gene therapy. Semin Cancer Biol 20: 269–278.
36. Cavazzana-Calvo M, Payen E, Negre O, Wang G, Hehir K, et al. (2010) Transfusion independence and HMGA2 activation after gene therapy of human beta-thalassaemia. Nature 467: 318–322.
37. Almarza D, Bussadori G, Navarro M, Mavilio F, Larcher F, et al. (2011) Risk assessment in skin gene therapy: viral-cellular fusion transcripts generated by proviral transcriptional read-through in keratinocytes transduced with self-inactivating lentiviral vectors. Gene Ther 18: 674–681.
38. Cesana D, Sgualdino J, Rudilosso L, Merella S, Naldini L, et al. (2012) Whole transcriptome characterization of aberrant splicing events induced by lentiviral vector integrations. J Clin Invest 122: 1667–1676.
39. Heckl D, Schwarzer A, Haemmerle R, Steinemann D, Rudolph C, et al. (2012) Lentiviral vector induced insertional haploinsufficiency of Ebf1 causes murine leukemia. Mol Ther 20: 1187–1195.
40. Moiani A, Paleari Y, Sartori D, Mezzadra R, Miccio A, et al. (2012) Lentiviral vector integration in the human genome induces alternative splicing and generates aberrant transcripts. J Clin Invest 122: 1653–1666.
41. Ye T, Krebs AR, Choukrallah MA, Keime C, Plewniak F, et al. (2010) seqMINER: an integrated ChIP-seq data interpretation platform. Nucleic Acids Res 39: e35.

15

SOAP3-dp: Fast, Accurate and Sensitive GPU-Based Short Read Aligner

Ruibang Luo[1]⁹, Thomas Wong[1]⁹, Jianqiao Zhu[1,5]⁹, Chi-Man Liu[1]⁹, Xiaoqian Zhu[2], Edward Wu[1], Lap-Kei Lee[1], Haoxiang Lin[3], Wenjuan Zhu[3], David W. Cheung[1], Hing-Fung Ting[1], Siu-Ming Yiu[1], Shaoliang Peng[2], Chang Yu[3], Yingrui Li[3]*, Ruiqiang Li[4]*, Tak-Wah Lam[1]*

1 HKU-BGI Bioinformatics Algorithms and Core Technology Research Laboratory & Department of Computer Science, University of Hong Kong, Hong Kong, China, **2** School of Computer Science, National University of Defense Technology, Changsha, Hunan, China, **3** BGI-Shenzhen, Shenzhen, Guangdong, China, **4** Peking-Tsinghua Center for Life Sciences, Biodynamic Optical Imaging Center and School of Life Sciences, Peking University, Beijing, China, **5** Department of Computer Sciences, University of Wisconsin-Madison, Madison, Wisconsin, United States of America

Abstract

To tackle the exponentially increasing throughput of Next-Generation Sequencing (NGS), most of the existing short-read aligners can be configured to favor speed in trade of accuracy and sensitivity. SOAP3-dp, through leveraging the computational power of both CPU and GPU with optimized algorithms, delivers high speed and sensitivity simultaneously. Compared with widely adopted aligners including BWA, Bowtie2, SeqAlto, CUSHAW2, GEM and GPU-based aligners BarraCUDA and CUSHAW, SOAP3-dp was found to be two to tens of times faster, while maintaining the highest sensitivity and lowest false discovery rate (FDR) on Illumina reads with different lengths. Transcending its predecessor SOAP3, which does not allow gapped alignment, SOAP3-dp by default tolerates alignment similarity as low as 60%. Real data evaluation using human genome demonstrates SOAP3-dp's power to enable more authentic variants and longer Indels to be discovered. Fosmid sequencing shows a 9.1% FDR on newly discovered deletions. SOAP3-dp natively supports BAM file format and provides the same scoring scheme as BWA, which enables it to be integrated into existing analysis pipelines. SOAP3-dp has been deployed on Amazon-EC2, NIH-Biowulf and Tianhe-1A.

Editor: Frederick C. C. Leung, University of Hong Kong, China

Funding: TL was partially supported by RGC General Research Fund 10612042. R. Luo was partially supported by Hong Kong ITF Grant GHP/011/12. Both R. Luo and TL are partially supported by the GRF Grant HKU-713512E. The funders had no role in study design, data collection and analysis, decision to publish, or preparation of the manuscript.

Competing Interests: The authors have declared that no competing interests exist.

* E-mail: liyr@genomics.cn (YL); lirq@pku.edu.cn (RL); twlam@cs.hku.hk (TL)

⁹ These authors contributed equally to this work.

Introduction

With the rapid advancement of Next-Generation Sequencing technologies, modern sequencers like Illumina HiSeq 2500 can sequence a human genome into 600 million pairs of reads of 100 bp in length (total 120 Gigabases) in merely 27 hours. The cost is also decreasing fast. By 2013 year's end, sequencing a human genome is projected to cost less than $1,000. Bioinformatics research using sequencing data often starts with aligning the data onto a reference genome, followed by various downstream analyses. Alignment is computationally intensive; the 1000 genomes pilot paper [1] published in 2010 reported that a 1192-processor cluster was used to align the reads using MAQ [2]. This kind of computing resources is not available to most laboratories, let alone clinical settings. Although considerable advances have been made on new aligners, alignment still remains a bottleneck in bioinformatics analyses. Thus, ultra-fast alignment tools without relying on extensive computing resources are needed.

There are quite a few software tools for aligning short reads onto a reference genome. The more popular ones include MAQ, Bowtie [3], BWA [4] and SOAP2 [5]. The faster ones [3–5] index the reference genome using the Burrows-Wheeler Transform (BWT), which is efficient for aligning short reads with limited mismatches, but inefficient for alignment with gaps. These tools (running on a quad-core processor) take tens of hours to align 120 Gigabases with limited (or even none for Bowtie and SOAP2) gapped alignment found. Alignment gaps can result from insertions and deletions (Indels), which are thought to comprise over 20% of genetic variations [6] and contribute to human traits [7]. Hence, a successor is expected to be faster and more sensitive to gaps.

SeqAlto [8], CUSHAW2 [9], and GEM [10] were published recently. SeqAlto is a hash-based aligner that improves an earlier hash-based aligner SNAP [11] (reported to have relatively poor sensitivity for real reads and provide no mapping quality [8]) using additional global and local alignments. SeqAlto is slower than SNAP, yet SeqAlto is still faster than the BWT-based aligners except Bowtie2. CUSHAW2 [9] uses the seed-and-extend approach and maximal-exact-match seeds to enable gapped alignment of long reads. GEM mapper leverages string matching with filtration to search the alignment space more efficiently [10]. GEM is faster than comparable state-of-the-art aligners. Yet it does not provide PHRED [12] compliant mapping quality score;

this saves some tedious computation, but prohibiting it from integrating into existing analysis pipelines.

Nowadays, general-purpose computing on graphics processing units (GPUs) is becoming popular. A GPU is a piece of low-cost hardware providing massive parallelism but with limited memory and restricted usage. A number of GPU-based bioinformatics tools have emerged last year [13]. CUSHAW [14] is the first to introduce a complete alignment pipeline utilizing GPU power for paired-end short reads (note that CUSHAW2, mentioned above, is CPU-based). BarraCUDA [15] implements BWA to align reads in parallel on a GPU; limited by the branch and bound trie algorithm that requires extensive decisions making, BarraCUDA works sub-optimally on GPU and gains a 4-time boost than a single-thread BWA. SOAP3 [16] successfully exploits the massive parallelism of a GPU with tailor-made GPU-BWT and read-characteristics sensitive load balancing to effectively align short reads. Albeit not supporting gapped alignment, which makes it unsuitable for production, it is to date the most competitive aligner for ungapped alignment.

Results

Here we present a GPU-based alignment software SOAP3-dp that allows multiple mismatches and gaps, which is well suited for production environments (real data alignments) than the predecessor SOAP3. A simple approach to extend mismatch alignment to gapped alignment is to first identify candidate regions by exact or mismatch alignment of short substrings (seeds) in the reads, then use dynamic programming to perform a detailed alignment of the read to the regions. Such an approach has been widely used (e.g., Bowtie2). The bottleneck occurs as substring alignment often results in a large number of candidates, especially when mismatches are allowed. As a result, reads with too many candidates are often ignored due to time constraint. On the other hand, the parallelism of GPU apparently would allow many candidates to be verified in parallel; yet dynamic programming is memory consuming, and the limited-memory of GPU becomes a prohibiting factor to fully utilize the parallelism. SOAP3-dp gives a pragmatic realization of this approach (Figure 1, Methods). By exploiting compressed indexing and memory-optimizing dynamic programming on a GPU, SOAP3-dp can efficiently tackle a large number of candidates in parallel, and thus can examine gapped alignments extensively and achieve a drastic improvement in both speed and sensitivity over other tools.

See the Methods about the design of the dynamic programming which attempts to minimize the memory usage for each candidate so as to let a GPU to handle hundreds of candidates in parallel while using limited shared memory. We also show the details of SOAP3-dp's intricate engineering solution to finding the optimal way to align different reads using either the CPU or the GPU.

Experiments & Performance

We compared SOAP3-dp to other short-read alignment software in terms of the speed, sensitivity and accuracy. We used both real and simulated Illumina data. Furthermore, we tested out SOAP3-dp's alignment quality for variant calling using real data. In particular, 41-fold of 100 bp (PE100) and 77-fold of 150 bp (PE150) Illumina paired-end reads of YH [17] samples have been generated (Table S1 in File S1) for the testing for variant calling.

Alignment Performance

We first used real data to test SOAP3-dp with BWA (BWA-SW [18] for SE reads), Bowtie2, SeqAlto, CUSHAW2, GEM, BarraCUDA, and CUSHAW. The aim was to compare the time and alignment rate when each runs in the default setting. Next, to assess the accuracy and sensitivity, we used simulated reads whose correct alignments were known. We then considered more detailed comparison with the software running in different settings. We also attempted to compare SOAP3-dp against its predecessor SOAP3.

In our experiments, we assume that input reads are plain text instead of in gzip-compressed format. This is because GEM (to the date of paper submission) does not accept gzip-compressed FASTQ file. All other software can handle compressed input, which is getting common nowadays. Regarding output format, we require all software to use SAM format, which is mandatory for downstream analysis software including GATK [19] and SAM-TOOLS [20]. All software except GEM can output directly in SAM format; GEM first outputs in a simple format and then takes an extra step to convert to SAM format. To test GEM's efficiency when using a simple format, we include a comparison to SOAP3-dp also using a simple format (see the Remark section).

Real data

We used three real datasets for benchmarking of alignment performance: (1) a lane (122.43M reads) from PE100, and (2) a lane (374.87M reads) from PE150, and (3) SRR211279 (25.23M 100 bp paired-end reads generated by Illumina GAIIx) from the Washington University Genome Sequencing Center. We tested SOAP3-dp and seven other aligners (CPU-based: BWA, Bowite2, SeqAlto, CUSHAW2, GEM; GPU-based: Barracuda, and CU-SHAW; see Supplementary Note for receipts), all using 4 CPU threads and one GPU device (for GPU aligner). As shown in Table 1, SOAP3-dp is much faster than all others (Tables S2–S10 in File S1 for more details). It is at least 3.5 times faster than GEM, and 7 to 15 times faster than the other six. SOAP3-dp also gave better alignment rate consistently. SeqAlto comes closest, aligning 0.48% to 3.6% less reads than SOAP3-dp, and the others are in the range of 2% to 8% less than SOAP3-dp. Notice that except SOAP3-dp, aligners usually have an obvious drop in alignment rate for longer reads (dataset 2). The two GPU-based aligners, Barracuda and CUSHAW, are relatively primitive in optimizing GPU's utilization and overheads, and their performance was dominated by new CPU-based aligners like GEM and Bowtie2. For SOAP3, its alignment rate, as expected, is much poorer than SOAP3-dp and the others (due to lack of gapped alignment); furthermore, SOAP3 is slower than SOAP3-dp for 100 bp reads. We did not include Barracuda, CUSHAW and SOAP3 for further experiments.

Simulated data

To assess the accuracy and sensitivity of SOAP3-dp, we used the short read simulator Mason [21] to obtain 5 sets of 6M Illumina-style paired-end (PE) reads with 500 bp insert size from GRCh37 major build, with length ranging from 50–250 bp.

Notably, Bowtie2, SeqAlto and GEM were designed with switches to favor speed at the expense of accuracy and sensitivity. We applied "very-fast", "sensitive", and "very-sensitive" switches to Bowtie2, "fast (-f)" to SeqAlto and "fast adaptive (–fast-mapping)", "fastest (–fast-mapping = 0)" to GEM. For SOAP3-dp, we tested three versions whose indices are based on 1/4 sampled, 1/2 sampled and full suffix array (SA), respectively. Different sized SAs still deliver identical alignment results, but a smaller one consumes less memory at the expense of slightly longer alignment time. All parameters of SOAP3-dp and SOAP3 remained as default (for a 100 bp read, one gap up to 68 bp, to 23 one-bp gaps) while parameters for other aligners were set to favor different read types and lengths as suggested by previous studies. Detailed command lines and descriptions of critical parameters were

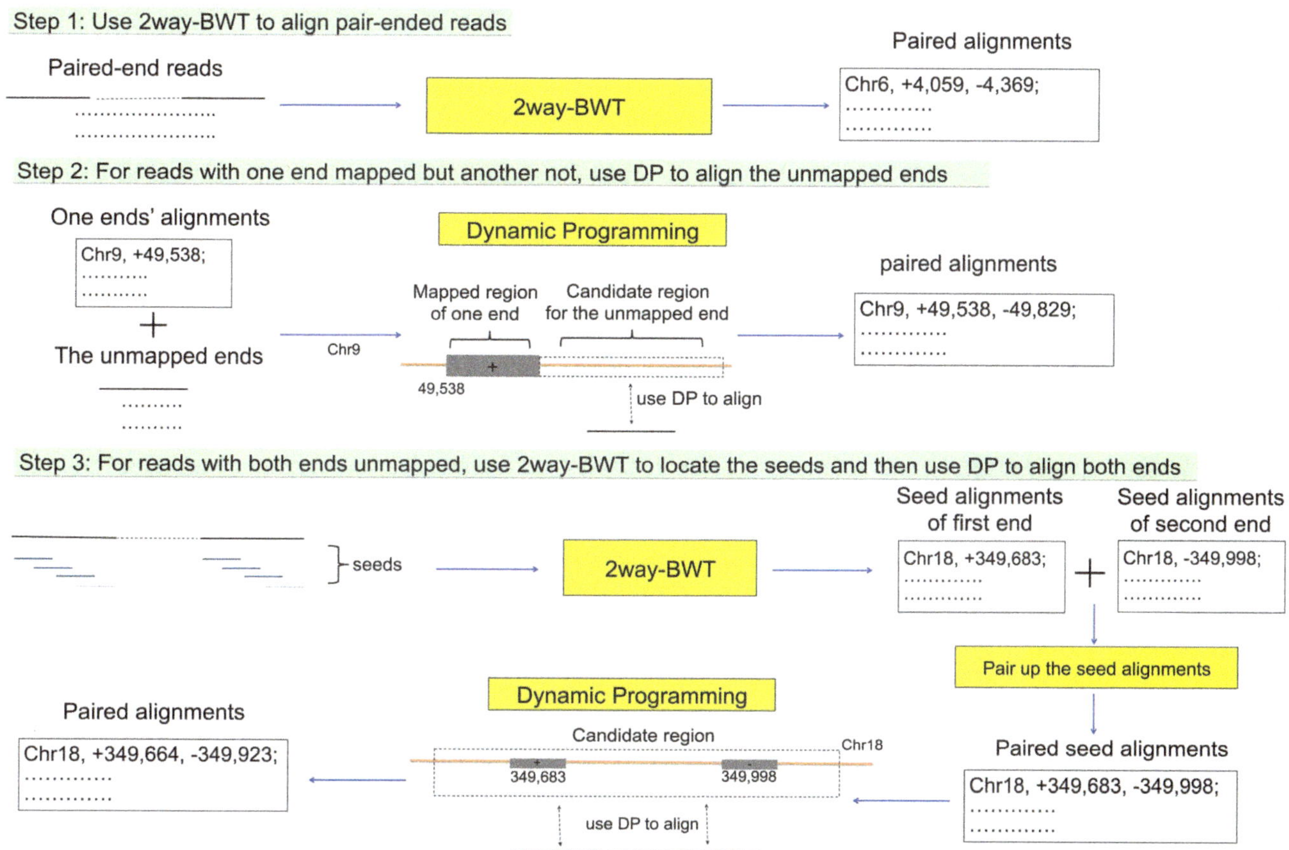

Figure 1. Alignment workflow. For each read (paired-end specifically, single-end is only with step 1 and step 3), the alignment would be decided in at most three steps. In step 1, SOAP3-dp aligns both ends of a read-pair to the reference genome by using GPU version 2way-BWT algorithm (Methods). Pairs with only one end aligned proceed to step 2 for a GPU accelerated dynamic programming (Methods) alignment at candidate regions inferred from the aligned end. Pairs with both ends unaligned in step 1 and those ends failed in step 2 proceed to step 3 to perform a more comprehensive alignment across the whole genome until all seed hits (substrings from the read) are examined or until a sufficient number of alignments are examined.

summarized as receipts in the Supplementary Note. In total, 16 sets of programs and parameters were compared.

In all datasets of simulated reads, SOAP3-dp gives consistent advantage. It is faster and simultaneously has higher sensitivity and lower FDR over all other tools (Table 2 for 100 bp PE, Tables S11–S14 in File S1 for all other simulated datasets, Figure 2). For 100 bp reads, SOAP3-dp with full SA takes 132 seconds to align 6M read pairs, and it is 2.26 to 12.63 times faster than the others (others using the fastest switches). SOAP3-dp's sensitivity is 99.96%, which is 0.13–0.85% higher than the others (others using the sensitive switches), and SOAP3-dp's FDR is 0.34%, which is lower than the others by 0.13–0.85%. Apparently the simulated data is easier to align than the real data due to recombination hotspots with intensified variants in real genome [6]. SOAP3-dp consumes more memory (9.3, 11.9, 17.2 GB for 1/4, 1/2 and full SA in average, respectively) than other software; Bowtie2 has the least (3.5 GB). Nevertheless, workstations and servers nowadays are equipped with at least 16 GB or even 32 GB of memory; SOAP3-dp is designed to take advantage of the available memory to achieve speed.

Mapping quality score is mandatory for most of the popular downstream analysis tools such as GATK and SAMTOOLS. SOAP3-dp uses the same scoring scheme as BWA so as to make its alignment results compatible to the expectations of existing analysis tools. As shown in Figure 3 and Figure S1 in File S1, BWA, Bowtie2, SeqAlto, BarraCUDA, CUSHAW2 and SOAP3-dp provide mapping quality scores that can differentiate different alignments properly, while GEM's scores are too rough, and too many incorrectly aligned reads are given high quality scores, which makes it unsuitable for downstream analysis.

Remarks

Note that GEM can output in a simple format to save time. When compared to SOAP3-dp in its own simple format, GEM and SOAP3-dp can both save about half of their running time; for the 6M simulated paired-end data of length 100 bp, the alignment time of GEM and SOAP3-dp is reduced to 90 seconds and 38 seconds, respectively. Downstream programs for variants calling, if redesigned to utilize these specific formats, could save time.

The simulated dataset is relatively small (6M read pairs), thus when using a large SA, the index loading time of SOAP3-dp dominated the total elapsed time. Considering only the alignment time (time consumption after index loading, including input of raw reads and output of alignment results), SOAP3-dp using the full and 1/2 SA is 12 and 9 seconds faster than 1/4 SA, thus for large real datasets, 1/2 and full SAs are suggested if memory permits.

Different generations of GPU device differ in speed. We compared the performance of SOAP3-dp between the latest GPU "GTX680" and a previous generation "Tesla C2070" using

Table 1. Benchmarking using real reads.

		GPU												CPU																	
Dataset	Volume (Gbp)	SOAP3-dp (Full SA)			SOAP3			BarraCUDA			CUSHAW			BWA			Bowtie2			SeqAlto			GEM			CUSHAW2					
		%	Time (s)	Mem.	%	Time (fold)	Mem.	%	Time (fold)	Mem.	%	Time (fold)	Mem.	%	Time (fold)	Mem.	%	Time (fold)	Mem.	%	Time (fold)	Mem.	%	Time (fold)	Mem.	%	Time (fold)	Mem.			
realYHPE100	12.24	98.12%	1,079	19	−10.60%	2.63	21.1	−6.39%	14.03	4.1	−3.13%	46.79	2.7	−4.14%	16.03	4.9	−3.87%	12.04	3.5	−0.48%	14.25	7.2	−2.86%	3.54	4.5	−1.39%	12.09	3.6			
realYHPE150	56.23	97.16%	6,835	19.6	−28.33%	0.64	22.7	−10.99%	7.22	4.3	−10.11%	24.57	3.1	−8.05%	15.26	5.0	−7.45%	7.82	3.5	−3.65%	17.69	7.2	−5.85%	3.76	5	−6.47%	8.04	3.6			
SRR211279	5.07	97.21%	439	18.4	−6.83%	1.05	20.8	−5.08%	13.42	4.1	−2.22%	54.22	2.1	−2.81%	16.27	4.9	−0.69%	12.00	3.5	−0.48%	17.29	7.2	−2.44%	4.11	4.6	−2.14%	12.03	3.6			

The percentage of reads aligned and time consumption of aligners other than SOAP3-dp are recorded as the difference or ratio based on SOAP3-dp's figures. The '%' column represents 'Properly paired' for PE reads. 'Mem.' represents the peak memory consumption in gigabytes during alignment.

simulated datasets. The alignment time extended about 10% for each dataset (Figure 4) using "Tesla C2070". Furthermore, a large real set of 150 bp paired-end reads was used. The alignment using the "GTX680" consumed 6,835 seconds, which is 4,658 seconds (1.68 times) faster than "Tesla C2070".

Variant Calling Performance

Next, we considered SOAP3-dp's alignment quality for variant calling. The full sets of both PE100 and PE150 were aligned using SOAP3-dp. We used the widely adopted BWA as benchmark. With the alignment results, variants were called using GATK's UnifiedGenotyper [22] and filtered by VariantRecalibrators, with and without GATK's local realignment (see Methods). Before we detail the results on variant detection, it is worth mentioning that BWA, even running in a slower mode to allow a longer gap (one gap up to 50 bp, without "-m" parameter to allow hit entries higher than 2 million due to out of memory error), still cannot catch up the sensitivity of SOAP3-dp in default setting (for a 100 bp read, one gap up to 68 bp, to 23 one-bp gaps, Table 3).

SOAP3-dp's better sensitivity is due to its ability of extensive gapped alignment; the extra reads aligned are crucial for variant detection (in particular, Indels). SOAP3-dp allowed 2.4% and 4.0% more SNPs than BWA, and 6.1% and 9.8% more Indels for the two datasets PE100 and PE150, respectively. Intuitively, longer reads are more favorable for variant detection; this is indeed reflected in SOAP3-dp's performance, but not for BWA (Table 3). We further checked the SNPs detected against dbSNP v135 (an archive of SNPs validated by previous studies); SOAP3-dp has notably 2.1% and 3.6% more SNPs found in dbSNP, confirming a higher sensitivity.

SOAP3-dp allowed more Indels to be detected than BWA, especially more Indels longer than 20 bp (Figures 5a, 5b, Figure S2 in File S1). To validate the novel Indels detected, we randomly selected 50 deletions that SOAP3-dp exclusively detected and are not yet archived in dbSNP v135, and verified them using Fosmid sequencing (see Methods). The Fosmids were sequenced, assembled and then aligned to the reference genome. The 50 deletions were covered by 460 Fosmid sequences. The findings are as follows: 6 deletions were inconclusive due to insufficient coverage of Fosmid sequences, 40 deletions were validated, and 4 rejected, revealing a FDR of 9.1% (Table S15 in File S1, Data S2). SOAP3-dp's ability to allow long gaps without speed penalty provides an unprecedented opportunity to come up with a more comprehensive Indel identification in large-scale genome studies.

With SOAP3-dp's ability to authentically align more reads, more multi-nucleotide polymorphisms (MNP) among the whole genome were identified (Figure S3 in File S1). Notably, GATK's local realignment can eliminate inauthentic alignments and rescue true variants. For SOAP3-dp, the number of MNP increased by 4.1% after realignment; yet for BWA, the number decreased by 6.3% (Table S17 in File S1), indicating that SOAP3-dp initially provided much more reliable alignments and led to more accurate variant calling.

Discussion

SOAP3-dp has been successfully deployed on Amazon EC2, NIH BioWulf and Tianhe-1A computing-cloud. On Amazon EC2, users can access SOAP3-dp's program and a testing dataset by mounting EBS snapshot "snap-154f1c54" named "SOAP3-dp" while creating a GPU instance (Supplementary Note). To test out SOAP3-dp's performance on Amazon EC2, we selected 10 Illumina PE datasets from 1000 genomes project, comprising 131.44 Gbp of raw reads (43.8-fold). The 10 datasets were

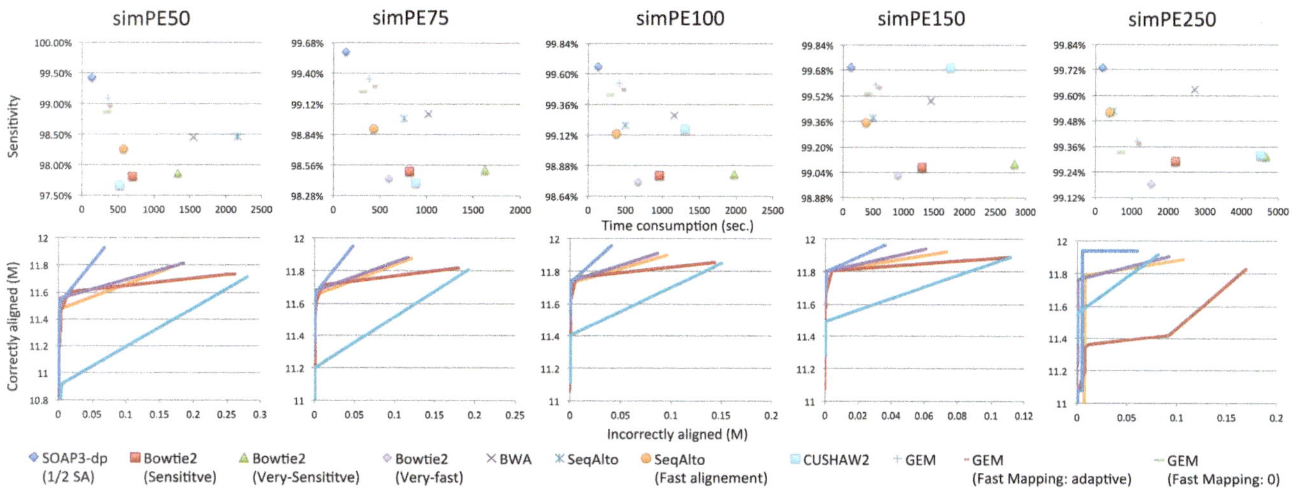

Figure 2. Speed and sensitivity of alignment using simulated paired-end reads. We recorded the number of correct and incorrect alignments stratified by reported mapping quality for each dataset. We then calculated the cumulative number of correct and incorrect alignments from high to low mapping quality. We considered an alignment correct only if the leftmost position was within 50 bp of the position assigned by the simulator on the same strand according to the previous study of Bowtie2 to avoid soft-clipping artifacts.

distributed to the two available Tesla M2050 GPU cards (see Methods and Supplementary Note. Notably, Tesla M2050 is slower than the GTX680 and Tesla C2070 we have used for real and simulated data evaluation) with one copy of the index shared in host memory. Using default parameters and BAM output, the alignment finished in 3.8 hours, yielding a total cost of $7.98, or $0.061 per Gbp reads aligned.

For users' convenience, SOAP3-dp separates the output of reads into three categories: 1) alignments involve gaps and extensive mismatches; 2) few mismatches only and 3) improperly paired or unaligned (file suffix "dpout", "gout", and "unpair" respectively). The separate file scheme fits well with the production environments, where files could be sorted separately in parallel and then merged together, which saves time than sorting a single SAM file. The files could also be concatenated by SAMTOOLS easily.

SOAP3-dp does not enforce a maximum read length. However, read length longer than 500 bp is not recommended while the current version of SOAP3-dp is tailor-made for Illumina reads. A version for longer 454 reads and Ion Torrent reads without performance degradation is our next task.

Overall, SOAP3-dp is an efficient alignment tool that targets the future of genome analysis where reads are longer and the volume is larger. SOAP3-dp is much faster than existing tools while retaining the ability to align more reads correctly. To be flexible, SOAP3-dp outputs both SAM and BAM formats that are compatible with most downstream analysis tools. SOAP3-dp is a free and open-source alignment tool available at http://www.cs.hku.hk/2bwt-tools/soap3-dp/.

Methods

Implementation details of SOAP3-dp

To align a paired-end read, SOAP3-dp proceeds in three steps (Figure 1). In step 1, SOAP3-dp uses GPU-accelerated 2way-BWT [16] to align those reads without gap opening on to the reference using a 3-level stratified alignment pipeline design. In step 2, for those reads with one end mapped but another end unmapped, a candidate region flanking the mapped end is aligned to the unmapped end using GPU-accelerated dynamic programming algorithm. In step 3, for those reads with both ends unmapped as well as reads still unmapped in step 2, seeds (substrings of a read) are extracted at regular intervals along the read and its reverse complement. SOAP3 module aligns these seeds back to the reference genome and enumerates candidate regions to be aligned to the whole read using dynamic programming.

To better illustrate SOAP3-dp's detailed workflow and parameters, we have prepared two sets of slides, which should be read together with the text.

http://bio8.cs.hku.hk/dataset/Workflow.ppsx

1. SOAP3-dp workflow for paired-end alignment
2. SOAP3-dp workflow for single-end alignment

http://bio8.cs.hku.hk/dataset/Parameters.ppsx

1. Optimization of parallel access to the GPU global memory.

SOAP3 makes use of the 2way-BWT indexing technique [23] and involves random access to the indexing data structures in the main memory. The original design of 2way-BWT [5] was based on two-level sampling. The design works well for CPU but not in the highly parallel environment of GPU. The data structures are too large and must be placed into the global memory of GPU, causing serious memory contention among the processors inside the GPU. For this purpose, in SOAP3, the index is redesigned to use one-level sampling instead, which greatly reduces the number of memory accesses by half (for details, one may refer to our previous study [24]).

Apart from reducing the number of global memory accesses, we also optimize the time of individual access to the global memory. This is achieved by *coalescing* simultaneous global memory accesses. To illustrate the idea of coalescing, we first need to explain how the GPU handles threads. GPU threads are grouped into units called *warps* for execution on a streaming multiprocessor (SM). The typical size of a warp is 32 threads. At some point, all threads in the same SM access the global memory. Since the threads are working with different data, it is likely that they access different memory locations at the same time. The GPU architecture is designed in a way that, these memory accesses would be much faster if 1) the memory locations accessed are close to each other (e.g. within a 128-byte segment), and 2) no two threads access the

Table 2. Comparison on 16 sets of programs and parameters using 100 bp paired-end simulated reads.

			GPU						CPU									
6M 100 bp Paired-end reads, 1.2 Gbp bases. 500 bp insert size, 25 bp standard deviation.			**SOAP3-dp[1] (1/4 SA)**	**SOAP3-dp[1] (1/2 SA)**	**SOAP3-dp[1] (Full SA)**	**SOAP3-dp[1] (Full SA, Binary)**	**SOAP3-dp[1] (Full SA, Succinct)**	**SOAP3**	**Bowtie2 (Sensititve)**	**Bowtie2 (Very-Sensititve)**	**Bowtie2 (Very-fast)**	**BWA[2]**	**SeqAlto**	**SeqAlto (Fast alignement)**	**CUSHAW2**	**GEM[3]**	**GEM[3] (Fast Mapping: adaptive)**	**GEM[3] (Fast Mapping: 0)**
Configuration	CPU (thread: core i7-3930k)		4			4	4	4	4	4	4	4	4	4	4	4	4	4
	GPU (device: GTX680)		1			1	1	1	0	0	0	0	0	0	0	0	0	0
Computational	Total Elapsed	sec.	132	137	162	111	112	132	966	1974	672	1154	495	379	1303	416	446	298
Resources		Fold	-	1.04	1.23	0.84	0.85	1.00	7.32	14.95	5.09	8.74	3.75	2.87	9.87	3.15	3.38	2.26
	Loading Index[4]	sec.	32	46	74	74	74	74	38	38	38	53+1+1	96	96	40	40+1	40+1	40+1
	Alignment[5]	sec.	100	91	88	37	38	58	928	1936	634	370+ 369+360	399	283	1263	199+176	238+167	90+167
		Fold	-	0.91	0.88	0.37	0.38	0.58	9.28	19.36	6.34	10.99	3.99	2.83	12.63	3.75	4.05	2.57
	Avg. Memory	GB	9.3	11.9	17.2	17.2	17.2	17.3	3.3	3.3	3.3	3.5	7	6.9	3.6	4.3	4.3	4.3
	Peak Memory	GB	9.7	12.5	18.1	18.1	18.1	19.2	3.5	3.5	3.5	4.8	7.2	7.2	3.6	4.3	4.3	4.3
Alignment	Aligned	#	11,999,827	11,870,740	11,999,763	11,999,936	11,998,226	11,998,804	12,000,000	11,995,872	11,999,975	11,999,763	11,999,484	11,995,422				
Metrics		Diff.	-	−129,087	−64	109	−1,601	−1,023	173	−3,955	148	−64	−343	−4,405				
	Properly Paired	#	11,999,460	11,742,902	11,998,912	11,999,344	11,996,528	11,997,254	11,999,976	11,995,410	11,977,218	11,998,994	11,997,702	11,991,992				
		Diff.	-	−256,558	−548	−116	−2,932	−2,206	516	−4,050	−22,242	−466	−1,758	−7,468				
	Incorrectly Aligned	#	40,561	138,655	143,012	141,373	147,764	85,297	95,672	99,243	99,243	56,514	61,642	61,887				
		Diff.	-	98,094	102,451	100,812	107,203	44,736	55,111	58,682	150,036	15,953	21,081	21,326				
	Sensitivity[6]	%	99.66%	97.77%	98.81%	98.82%	98.75%	99.28%	99.20%	99.14%	99.17%	99.53%	99.48%	99.45%				
		Diff.	-	−1.89%	−0.85%	−0.84%	−0.91%	−0.38%	−0.46%	−0.52%	−0.49%	−0.13%	−0.18%	−0.21%				
	FDR[7]	%	0.34%	1.17%	1.19%	1.18%	1.23%	0.71%	0.80%	0.83%	0.83%	0.47%	0.51%	0.52%				
		Diff.	-	0.83%	0.85%	0.84%	0.89%	0.37%	0.46%	0.49%	0.49%	0.13%	0.18%	0.18%				

[1]Alignment results by the three entries of SOAP3-dp (1/4 SA, 1/2 SA, Full SA) are identical.
[2]The time consumption of BWA is calculated as "align left reads"+"align right reads"+"sampe". The index loading times of "align right reads" and "sampe" modules are 1 second due to the reason that, index files were cached during "align left reads". However, datasets larger than the host memory will flush the cache during alignment.
[3]The alignment time consumption of GEM is calculated as "alignment"+"convert to SAM format". The conversion module was run with 4 threads in consistent with the alignment module.
[4]SOAP3-dp, SOAP3, SeqAlto and GEM aligners explicitly provide index loading time consumption. The index loading time for Bowtie2, CUSHAW2 and BWA are calculated by the total size of index, divided by 100 MB/s, which is the average network file system speed of the testing environment. The index loading time maybe underestimated while the time processing the index was not calculated.
[5]The alignment times were explicitly provided by the aligners (include results processing and input/output time) or calculated by total elapsed time minus estimated index loading time.
[6]Sensitivity is calculated as "Correctly aligned reads"/"All simulated reads". The higher the better.
[7]FDR is calculated as "Incorrectly aligned reads"/"All aligned reads". The lower the better.

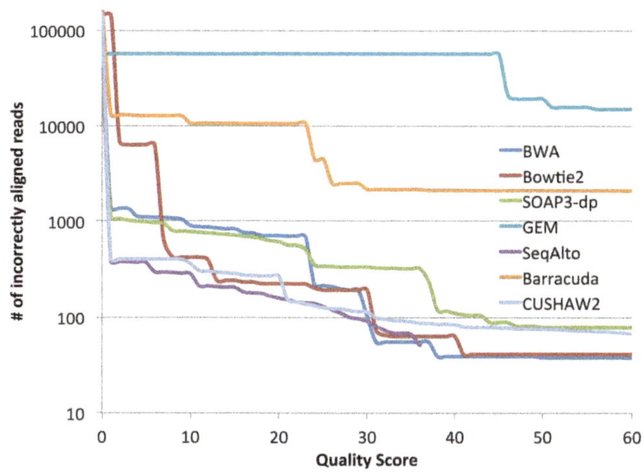

Figure 3. The accumulated number of incorrectly aligned reads categorized at different mapping quality scores by the five aligners.

same memory location. A group of memory accesses is considered to be *coalesced* if they satisfy the above properties.

In SOAP3-dp, we try to coalesce as many memory accesses as possible. The global memory of GPU device holds two large data structures – the 2way-BWT index, and the set of reads. To enable coalesced access. the set of reads is partitioned into groups of 32 (equals the warp size). For each group, the reads are arranged as follows. Let $w_{i,j}$ denote the j-th word of the i-th read in the group ($1<=i<=32$). Instead of storing the reads in the most natural way, i.e. $w_{1,1}, w_{1,2}, \ldots, w_{1,m}, w_{2,1}, w_{2,2}, \ldots, w_{2,m}, w_{3,1}, \ldots$ (where m is the number of words occupied by each read), we rearrange them into: $w_{1,1}, w_{2,1}, \ldots, w_{32,1}, w_{1,2}, w_{2,2}, w_{3,2}, \ldots, w_{32,2}, w_{1,3}, \ldots$. When the threads simultaneously access, say, the first words (more precisely, 4 bytes) of the reads, the memory locations accessed are $w_{1,1}, w_{2,1}, \ldots, w_{32,1}$, 32 words forming a contiguous 128-byte segment. These coalesced accesses could be done in a single memory transaction with only one time of memory address interpretation, achieving excellent memory throughput.

Memory accesses to the BWT index are highly unpredictable, thus coalescing them is difficult. Nevertheless, the BWT index is designed as all the BWT information for matching a base was grouped together in a memory chunk (64 bytes), which could be assessed using one memory transaction.

2. Divergence control and 3-level stratified alignment pipeline.

GPU works in a single-instruction multiple-thread (SIMT) mode. Processors in the same SM must execute the same instruction at one time. When mismatches are allowed, a read can have more than one branch during alignment. Too many diverging branches however would lower the efficiency of GPU drastically, because most processors (with few branches) may become idle and wait for a few others. For this purpose, we derive a suffix array (SA) ranges count based parameter, which can be determined at runtime, whether a read would generate too many branches, and reads are classified into different levels of complexity according to this parameter. The basic idea is that reads of different levels should be aligned separately. In SOAP3-dp's implementation, we have designed three levels of complexity. For the "all best" output parameter, which outputs all alignment results equally the best, if 2 mismatches are allowed, the SA ranges count dividing the three levels are 4 and 32 respectively (it means that, during alignment, a read with SA ranges count exceeding a threshold after extending a base, will be stopped and scheduled for next level, thresholds for other mismatch allowances and output parameters are included in the "definition.h" file in the source code). In particular, we let the GPU handle the first two levels, and use the CPU to take care of the most complicated reads (which account for a small percentage only). Furthermore, to fully utilize both GPU and CPU processing power, SOAP3 overlaps the alignment of complicated reads from the previous batch in CPU with the alignment of the next batch in GPU (as shown in the Figure S4 in File S1).

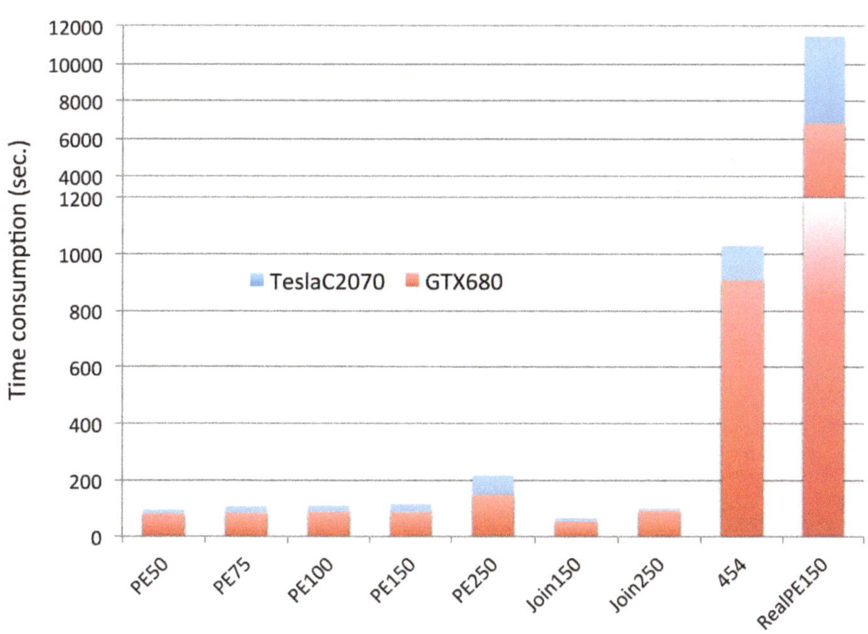

Figure 4. Alignment time consumption of using GPU card "GTX680" and previous generation GPU card "Tesla C2070" respectively.

Table 3. Summary of alignment and variation calling using SOAP3-dp and BWA with different parameters and datasets.

Program/Type	Running time (h)		Reads aligned	Properly Paired	SNP						Indels	
					w/o			w/			w/o	w/
	Aln.	GATK			Total	dbSNP	Homozygous	Total	dbSNP	Homozygous	Total	Total
BWA (Default, PE100)	99	32	93.30%	92.10%	3,996,730	3,926,590	1,620,269	3,922,841	3,876,262	1,609,899	370,231	383,347
BWA (-o 1 -e 50 -L, PE100)	3,363	36	96.74%	94.80%	4,040,813	3,955,380	1,609,729	3,946,627	3,896,737	1,601,049	385,216	417,832
SOAP3-dp (Default, PE100)	6	35	98.03%	96.98%	4,120,165	4,028,414	1,615,376	4,015,725	3,956,371	1,604,106	407,533	406,845
BWA (Default, PE150)	266	46	90.30%	89.17%	3,963,085	3,897,866	1,621,742	3,911,386	3,862,977	1,616,767	350,201	384,684
SOAP3-dp (Default, PE150)	14	92	97.66%	96.54%	4,139,625	4,055,314	1,612,213	4,067,552	4,002,750	1,612,351	409,925	422,427

PE100 and PE150 represent the 100 bp and 150 bp paired-end reads of YH sample. 'w/' and 'w/o' indicates whether the alignments were processed with and without GATK's local realignment respectively. 'dbSNP' is the number of detected SNPs found in dbSNP v135. The running times were rounded to hour.

3. GPU-accelerated dynamic programming.

To perform dynamic programming for aligning a read with a candidate region in genome, Smith-Waterman algorithm is applied. However, a straightforward implementation of the algorithm does not fit well under the GPU environment due to the large number of memory accesses. For this purpose, another implementation is suggested so that the number of memory accesses can be reduced by half. In the following, the straightforward implementation is first described, and then it is shown how to be modified in order to decrease the number of memory accesses. Similar implementations have also been applied in ClustalW [25] and CUDASW++ before [26].

Given a candidate region T (of length m) *and* a read R (of length n), the aim is to find a sub-region T' inside T such that the alignment score between R and T' is maximum.

Let $M(i,j)$ be the maximum alignment score between all suffixes of $T[1\ldots i]$ and $R[1\ldots j]$. The resulting score would be: $\max_{1\leq i\leq m} M(i,n)$.

Let S_{MA}, be the score for match, and let S_{MI}, S_{GO}, S_{GE} be the penalty scores for mismatch, gap opening and gap extension. It is also needed to define $I(i,j)$ as the maximum alignment score all suffixes of $T[\ldots i]$ and $R[1\ldots j]$ under the condition that $R[j]$ is aligned to a space, and $D(i,j)$ as the maximum alignment score between all suffixes of $T[1\ldots i]$ and $R[1\ldots j]$ under the condition that $T[i]$ is aligned to a space.

The recursive formulas are as follows:

$$I(i,j)=\max\begin{cases} M(i,j-1)+S_{GO} \\ I(i,j-1)+S_{GE} \end{cases}$$

$$D(i,j)=\max\begin{cases} M(i-1,j)+S_{GO} \\ I(i-1,j)+S_{GE} \end{cases}$$

$$M(i,j)=\max\begin{cases} M(i-1,j-1)+\delta(R[i],T[j]) \\ I(i,j)+D(i,j) \end{cases}$$

where $\delta(x,y)=S_{MA}$ if $x=y$, or $\delta(x,y)=S_{MI}$ if $x\neq y$.

And the base cases are:

$$I(i,0)=-\infty,\ i=1\ldots m$$

$$D(0,j)=-\infty,\ j=1\ldots n$$

$$M(i,0)=0,\ i=0\ldots m$$

$$M(0,j)=S_{GO}+(j-1)S_{GE},\ j=1\ldots n$$

A straightforward approach of the implementation is as follows:

Smith-Waterman algorithm: Compute I, D and M

```
Initialize tables I, D and M according to the base cases.
For j=1→m do
    For j=1→n do
        I(i,j)←max{M(i,j−1)+S_GO,I(i,j−1)+S_GE}
        D(i,j)←max{M(i−1,j)+S_GO,I(i−1,j)+S_GE}
        M(i,j)←max{M(i−1,j−1)+δ(R[i],T[j]),I(i,j),D{i,j)}
    End for
End for
```

To implement this straightforward approach on GPU, tables I, D and M are created inside the GPU's global memory. The fact that the access of GPU's global memory is much slower than its arithmetic operation affects the efficiency of the algorithm. In every loop, there are 7 table-reading and 3 table-writing operations.

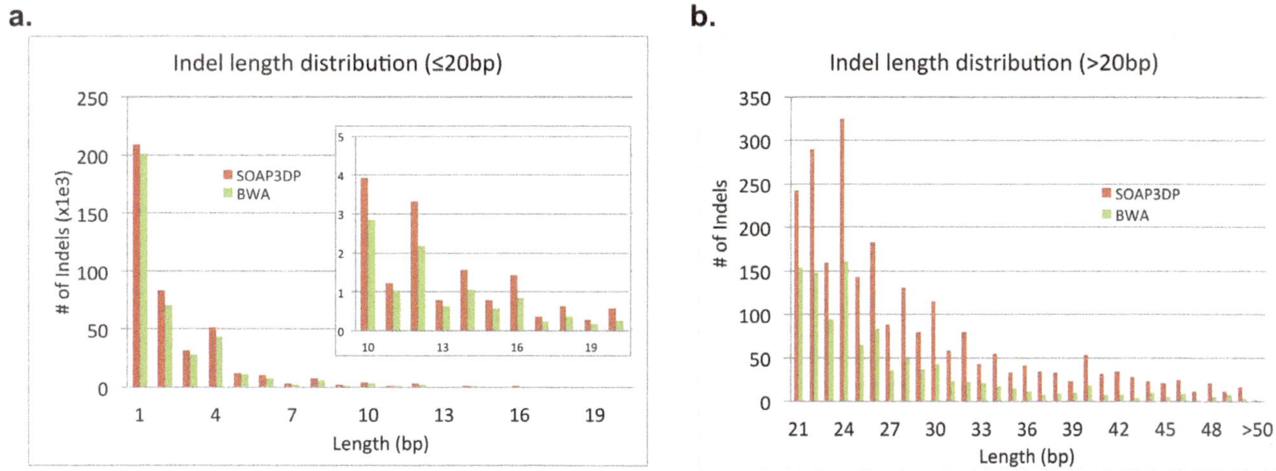

Figure 5. The length distribution of Indels identified by SOAP3-dp and BWA respectively using full set of 100 bp paired-end YH sample reads. a. Indels smaller than or equal to 20 bp, **b.** larger than 20 bp.

It is realized that backtracking could be done with only tables D and M, thus the table I can be eliminated. For this purpose, another approach is suggested as follows, which requires only 2 table-reading and 2 table-writing operations in each loop.

Improved implementation of Smith-Waterman algorithm for GPU:

Compute D and M

Declare register variables: M^u, M^d, V_M, V_I, V_D
(note: M^u refers to $M(i-1, j-1)$ and M^d refers to $M(i-1, j)$)
Initialize tables D and M according to the base cases.
For $j = 1 \rightarrow m$ do
  $M^d \leftarrow 0, V_M \leftarrow 0, V_I \leftarrow -\infty$
  For $j = 1 \rightarrow n$ do
    $M^u \leftarrow M^d$
    $M^d \leftarrow M(i-1,j)$
    $V_I \leftarrow \max\{V_M + S_{GO}, V_I + S_{GE}\}$
    $V_D \leftarrow \max\{M^d + S_{GO}, D(i-1,j) + S_{GE}\}$
    $V_M \leftarrow \max\{M^u + \delta(R[i], T[j]), V_I, V_D\}$
    $D(i,j) \leftarrow V_D$
    $M(i,j) \leftarrow V_M$
  End for
End for

4. Effort Limit for Dynamic Programming.

Reads with seeds that match an exceedingly large amount of places on the genome can spur an excessively large number of dynamic programming problems. For example, a poly-A homopolymer could match over ten thousand loci in the genome. SOAP3-dp avoids executing an excessive number of dynamic programming problems by adopting a ceiling on the number of candidate regions in step 2 and step 3. Candidate regions are scored with number of supporting seeds and sorted descendingly. If the ceiling is set to 30, for example, SOAP3-dp will only perform dynamic programming alignment in the best 30 candidate regions. The ceiling is set in the configuration file, but values higher than the default may strongly affect the performance with limited accuracy improvement.

5. Paired-end alignment.

SOAP3-dp supports alignment of paired-end reads in which both ends of a single DNA fragment are sequenced. The user sets expected minimum and maximum fragment lengths using –v and –u parameters, as well as orientations of the ends in configuration file (typically, Illumina uses Forward-Reverse while SOLiD uses Forward-Forward). A paired-end alignment that matches these expectations is called "properly paired" and an alignment that violates these expectations is "unpaired". If a pair fails to be aligned as properly paired, SOAP3-dp attempts to align each end individually. This is similar to both BWA's and Bowtie2's behavior. When a read pair fails to be aligned properly but both ends could be aligned individually, SOAP3-dp reports these alignments.

In contrast to BWA and SOAP2, which rely on a mapped end to determine a candidate region for further dynamic programming alignment, SOAP3-dp could align those reads with both ends unmapped. This allows read pairs from large period of variation hotspots to be aligned.

6. Scoring functions.

Details discussed in Supplementary Note.

Simulation of single-end and paired-end reads

Mason 0.1 [21] was used to simulate reads using the GRCh37 major build human reference genome, including 22 pairs of autosomes, 2 sex chromosomes and a mitochondrial chromosome. For the paired Illumina-style datasets with read length ranging from 50 bp to 250 bp, Mason was run in 'Illumina' read mode with options -N 6000000 –source-no-N -mp -sq -ll 500 -le 25 -rn 2 -hn 2 –haplotype-snp-rate 0.001 –haplotype-indel-rate 0.0001 –haplotype-no-N -n 100 -pi 0 -pd 0 –no-N'. Each PE set was simulated to contain exactly 12 million reads (6M pairs), which is about 2 times the default batch size of SOAP3-dp.

All simulated datasets are available at http://bio8.cs.hku.hk/dataset/.

Simulation comparison between tools

Executable files for SOAP3-dp v2.3, Bowtie2 v2.0.0-beta4, BWA 0.6.2, SeqAlto basic 0.5-r123, BarraCUDA_r232, CUSHAW-1.0.40, CUSHAW2-v2.1.9, SOAP3_version146 and GEM-core_i3-20121106-022124 were obtained via standard build procedures with default arguments. We indexed the reference genome with each tool's default indexing parameters. SeqAlto uses 22 bp seed length and sub-sampled mode. "Running time" was measured from initial call of the aligner to the completion of SAM-format output. 'Reads aligned' was measured as the number of reads for which the tool found at least one alignment regardless of mapping score. 'Properly paired' was measured as the number of read-pairs aligned with proper read orientation and insert-size range (mean insert-size ±3*standard deviation). 'Peak memory' and 'Average memory' usage was measured by tracking the Linux's *proc* file-system with respective *process id*.

For BWA and BarraCUDA, separate calls of the software modules were required for aligning each end and for processing intermediate alignment results into a final SAM file. 'Running time' was measured for separate modules while 'Peak memory' and 'Average memory' were measured across all modules. For GEM, to be consistent with other tools, a conversion is necessary after alignment to obtain SAM format results with mapping quality. Time consumptions were measured separately and then summed for comparison to other tools. All tools or components were run with 4 threads (except for the alignment module of BarraCUDA, where the CPU thread is constantly 1). Parameters were listed in Supplementary Note as receipts.

The experiments used a single computing node running CentOS v6.3 with an Intel i7-3930k 3.2 Ghz quad-core processor, an Nvidia GTX 680 GPU card with 4 GB non-ECC (Error-correcting code) graphic memory and 64 GB non-ECC memory.

Scripts and command lines to evaluate the authenticity of aligned reads and generate the ROC curves are available at http://bio8.cs.hku.hk/dataset/.

YH data production

Genomic DNA was isolated using standard molecular biology techniques. For each short insert library, 5 μg of DNA was fragmented, end-repaired, A-tailed and ligated to Illumina paired-end adapters. The ligated fragments of 100 bp paired-end reads (PE100) were size-selected at 170 bp and 500 bp on agarose gels, while PE150 were size-selected at 240 bp. All libraries are amplified by LM-PCR to yield the corresponding short insert libraries. PE100 were sequenced using TruSeq v2 while PE150 were sequenced using TruSeq v3 reagent on the Illumina sequencing platform.

Real data comparison

The 100 bp and 150 bp paired-end Illumina HiSeq 2000 reads of YH sample were sequenced and deposited to EBI with study accession number ERP001652. The data are also available at http://yh.genomics.org.cn.

SOAP3-dp uses default parameters. BWA uses both default parameters and "-o 1 –e 50", which allows at most a gap not longer than 50 bp (-m option to elevate the 2M hits limit for each read was not applied due to out of memory error, the option allows more reads to be aligned but consumes much more memory and longer alignment time). The latter option allows more reads to be aligned and more indel signals to be discovered, but would enormously decrease the running speed. Alignments were post-processed by following procedures: 1) local realignment by GATK v2.1, 2) duplication removal by Picard v1.74, 3) base quality score recalibration, 4) variants calling by UnifiedGenotyper and 5) variants quality score recalibration by GATK v1.6. Step 1 is optional according to the experiment while steps 2 to 5 are mandatory. Parameters and known variant sets were set according to the GATK's Best Practice v4 on GATK's website.

SOAP3-dp used a single node with a quad-core Intel Xeon E5570 2.93 Ghz CPU and a GPU while BWA used 10 nodes with the same CPU. To imitate the real production environment, we used Nvidia Tesla C2070 GPU device with 6 G graphic memory and with ECC enabled to perform full YH dataset alignment.

Fosmid sequencing

Fosmid libraries (averagely 40 kbp in size) were constructed according to Kim et al. [27]. In total, ~100 k Fosmid clones were created and every 30 Fosmids were pooled together sharing a barcode for Illumina HiSeq 2000 sequencing. For each pool, one 200 bp and one 500 bp insert size libraries were constructed and sequenced at 20× respectively. If a library had problem of abnormal base content bias or a relatively high base error rate reported by base-calling software, we took it as a non-qualified library and performed sequencing again. Each pool was assembled with SOAPdenovo [28] using 63-mer and other parameters as default. Sequences solved by SOAPdenovo's repeat solving module were remembered and soft-masked in final sequences in order not to obscure the following alignment. The assembled sequences were aligned to the human reference genome using BWASW with default parameters. Most of the Fosmid clones could be assembled to full length. For fragmented Fosmid clone sequences, we further assembled the fragments according to the in-pool linkage information during alignment. We require over 90% of a Fosmid clone sequence to be linearly aligned to only one location in the reference genome. Finally, 460 Fosmid clones were found covering the 50 randomly selected SOAP3-dp specific deletion calls that are not yet archived in dbSNP v135. We define "a Fosmid sequence supporting a deletion" as over 80% of the deleted bases in reference genome cannot be covered by the aligned Fosmid sequence (excluding soft-masked bases), and the identity of the 200 bp alignments flanking the deletion should exceed 90%. While a Fosmid clone can only come from a haploid, we require a heterozygous deletion has at least a Fosmid supporting the deletion, while a homozygous deletion should only have Fosmids supporting the deletion. Heterozygous deletions with lower than 5 spanning Fosmids and without a Fosmid supporting the deletion will be classified as "not clear". Homozygous deletions without a spanning Fosmid will also be classified as "not clear" (Table S15, Data S2).

The assembled Fosmid sequences are available as Data S1 in BAM file format. Raw reads are available upon request.

Experiment on Amazon EC2

The experiment used a single GPU Quadruple Extra Large Instance (cg1.4xlarge, $2.1 per hour) rented from the Amazon Elastic Compute Cloud (EC2) service (http://aws.amazon.com/ec2). The instance has 2 quad-core Intel Xeon X5570 at 2.93 GHz with hyper-threading, 2 Nvidia Tesla M2050 GPU cards with 3 GB ECC graphic-memory per card, 22 gigabytes of physical memory and runs Amazon Linux AMI v2012.09 operating system. Alignments were distributed onto the two GPU cards with two SOAP3-dp processes sharing the same copy of index in host memory. Each process occupies at most 7 threads. 10 sets of Illumina HiSeq 2000 reads generated in 1000 genomes project (Supplementary Note) were downloaded from the Amazon Simple Storage Service (S3). Additional tests have been carried out by NIH biowulf laboratory (http://biowulf.nih.gov/apps/bioinf-gpu.html) and Tianhe-1A super-computing center (http://en.wikipedia.org/wiki/Tianhe-I). While these clouds host CPU and GPU computing nodes and centralized storage system, the ultra-fast SOAP3-dp could be easily integrated into existing pipelines.

Supporting Information

Data S1 Assembled Fosmids sequences in BAM file.

Data S2 Supplementary Table S16. Table S16.xls.

File S1 Supplementary Tables, Figures, Notes and Receipts.

Acknowledgments

We thank the four anonymous reviewers for their professional and constructive suggestions.

Author Contributions

Wrote the paper: TL R. Luo TW. Contributed equally to this work: R. Luo TW JZ CL. Led the project and contributed the main ideas: TL. Co-managed the project: R. Li YL DC HT SY XZ. Wrote the program: JZ TW CL R. Luo EW LL SP. Designed the analysis: R. Luo TW. Conducted the simulation analysis: R. Luo TW. Conducted the real data analysis: R. Luo HL WZ. Assisted in deploying the program: XZ CY SP.

References

1. The 1000 Genomes Project Consortium (2010) A map of human genome variation from population-scale sequencing. Nature 467: 1061–1073.
2. Li H, Ruan J, Durbin R (2008) Mapping short DNA sequencing reads and calling variants using mapping quality scores. Genome Res 18: 1851–1858.
3. Langmead B, Trapnell C, Pop M, Salzberg SL (2009) Ultrafast and memory-efficient alignment of short DNA sequences to the human genome. Genome biology 10: R25.
4. Li H, Durbin R (2009) Fast and accurate short read alignment with Burrows-Wheeler transform. Bioinformatics 25: 1754–1760.
5. Li R, Yu C, Li Y, Lam TW, Yiu SM, et al. (2009) SOAP2: an improved ultrafast tool for short read alignment. Bioinformatics 25: 1966–1967.
6. Li Y, Zheng H, Luo R, Wu H, Zhu H, et al. (2011) Structural variation in two human genomes mapped at single-nucleotide resolution by whole genome de novo assembly. Nat Biotechnol 29: 723–730.
7. Mullaney JM, Mills RE, Pittard WS, Devine SE (2010) Small insertions and deletions (INDELs) in human genomes. Hum Mol Genet 19: R131–136.
8. Mu JC, Jiang H, Kiani A, Mohiyuddin M, Bani Asadi N, et al. (2012) Fast and accurate read alignment for resequencing. Bioinformatics 28: 2366–2373.
9. Liu Y, Schmidt B (2012) Long read alignment based on maximal exact match seeds. Bioinformatics 28: i318–i324.
10. Marco-Sola S, Sammeth M, Guigo R, Ribeca P (2012) The GEM mapper: fast, accurate and versatile alignment by filtration. Nat Methods 9: 1185–1188.
11. Zaharia M, Bolosky WJ, Curtis K, Fox A, Patterson D, et al. (2011) Faster and More Accurate Sequence Alignment with SNAP. arXiv 1111.5572v1.
12. Ewing B, Green P (1998) Base-calling of automated sequencer traces using phred. II. Error probabilities. Genome Res 8: 186–194.
13. Nvidia (2012) Bioinformatics and Life Sciences.
14. Liu Y, Schmidt B, Maskell DL (2012) CUSHAW: a CUDA compatible short read aligner to large genomes based on the Burrows-Wheeler transform. Bioinformatics 28: 1830–1837.
15. Klus P, Lam S, Lyberg D, Cheung MS, Pullan G, et al. (2012) BarraCUDA - a fast short read sequence aligner using graphics processing units. BMC Res Notes 5: 27.
16. Liu CM, Wong T, Wu E, Luo R, Yiu SM, et al. (2012) SOAP3: ultra-fast GPU-based parallel alignment tool for short reads. Bioinformatics 28: 878–879.
17. Wang J, Wang W, Li R, Li Y, Tian G, et al. (2008) The diploid genome sequence of an Asian individual. Nature 456: 60–65.
18. Li H, Durbin R (2010) Fast and accurate long-read alignment with Burrows-Wheeler transform. Bioinformatics 26: 589–595.
19. McKenna A, Hanna M, Banks E, Sivachenko A, Cibulskis K, et al. (2010) The Genome Analysis Toolkit: a MapReduce framework for analyzing next-generation DNA sequencing data. Genome Res 20: 1297–1303.
20. Li H, Handsaker B, Wysoker A, Fennell T, Ruan J, et al. (2009) The Sequence Alignment/Map format and SAMtools. Bioinformatics 25: 2078–2079.
21. Doring A, Weese D, Rausch T, Reinert K (2008) SeqAn an efficient, generic C++ library for sequence analysis. BMC Bioinformatics 9: 11.
22. DePristo MA, Banks E, Poplin R, Garimella KV, Maguire JR, et al. (2011) A framework for variation discovery and genotyping using next-generation DNA sequencing data. Nat Genet 43: 491–498.
23. Lam TW, Li R, Tam A, Wong S, Wu E, et al. High throughput short read alignment via bi-directional BWT; 2009. IEEE. pp. 31–36.
24. Liu C, Lam T, Wong T, Wu E, Yiu S, et al. (2011) SOAP3: GPU-based compressed indexing and ultra-fast parallel alignment of short reads. The 3rd Workshop on Massive Data Algorithmics (MASSIVE 2011). Paris, France.
25. Larkin MA, Blackshields G, Brown NP, Chenna R, McGettigan PA, et al. (2007) Clustal W and Clustal X version 2.0. Bioinformatics 23: 2947–2948.
26. Liu Y, Maskell DL, Schmidt B (2009) CUDASW++: optimizing Smith-Waterman sequence database searches for CUDA-enabled graphics processing units. BMC Res Notes 2: 73.
27. Kim UJ, Shizuya H, de Jong PJ, Birren B, Simon MI (1992) Stable propagation of cosmid sized human DNA inserts in an F factor based vector. Nucleic Acids Res 20: 1083–1085.
28. Li R, Zhu H, Ruan J, Qian W, Fang X, et al. (2010) De novo assembly of human genomes with massively parallel short read sequencing. Genome Res 20: 265–272.

PeSV-Fisher: Identification of Somatic and Non-Somatic Structural Variants Using Next Generation Sequencing Data

Geòrgia Escaramís[1,2,3,4,9], Cristian Tornador[1,2,3,4,9], Laia Bassaganyas[1,2,3,4,9], Raquel Rabionet[1,2,3,4], Jose M. C. Tubio[1,2,3,4,5], Alexander Martínez-Fundichely[1,2,3,4,6], Mario Cáceres[1,2,3,4,6,7], Marta Gut[8], Stephan Ossowski[2,4,9], Xavier Estivill[1,2,3,4]*

1 Genetic Causes of Disease Group, Centre for Genomic Regulation (CRG), Barcelona, Spain, **2** Universitat Pompeu Fabra (UPF), Barcelona, Spain, **3** Centro de Investigación Biomédica en Red en Epidemiología y Salud Pública (CIBERESP), Barcelona, Spain, **4** Hospital del Mar Medical Research Institute (IMIM), Barcelona, Spain, **5** Galician Foundation of Genomic Medicine-SERGAS, Complexo Hospitalitario Universitario de Santiago (CHUS), Santiago de Compostela, Spain, **6** Institut de Biotecnologia i de Biomedicina, Universitat Autònoma de Barcelona, Bellaterra, Spain, **7** Institució Catalana de Recerca i Estudis Avançats (ICREA), Barcelona, Spain, **8** National Center of Genomic Analysis (CNAG-CRG), Barcelona, Spain, **9** Genomic and Epigenomic Variation in Disease Group, Centre for Genomic Regulation (CRG), Barcelona, Spain

Abstract

Next-generation sequencing technologies expedited research to develop efficient computational tools for the identification of structural variants (SVs) and their use to study human diseases. As deeper data is obtained, the existence of higher complexity SVs in some genomes becomes more evident, but the detection and definition of most of these complex rearrangements is still in its infancy. The full characterization of SVs is a key aspect for discovering their biological implications. Here we present a pipeline (*PeSV-Fisher*) for the detection of deletions, gains, intra- and inter-chromosomal translocations, and inversions, at very reasonable computational costs. We further provide comprehensive information on co-localization of SVs in the genome, a crucial aspect for studying their biological consequences. The algorithm uses a combination of methods based on paired-reads and read-depth strategies. *PeSV-Fisher* has been designed with the aim to facilitate identification of somatic variation, and, as such, it is capable of analysing two or more samples simultaneously, producing a list of non-shared variants between samples. We tested *PeSV-Fisher* on available sequencing data, and compared its behaviour to that of frequently deployed tools (BreakDancer and VariationHunter). We have also tested this algorithm on our own sequencing data, obtained from a tumour and a normal blood sample of a patient with chronic lymphocytic leukaemia, on which we have also validated the results by targeted re-sequencing of different kinds of predictions. This allowed us to determine confidence parameters that influence the reliability of breakpoint predictions.

Editor: Jeong-Sun Seo, Seoul National University College of Medicine, Republic of Korea

Funding: This work was supported by AGAUR (Generalitat de Catalunya, 2009 SGR 1502) (X.E.); CIBERESP (Instituto de Salud Carlos III) (G.E.); ESGI (European Commission, 262055_ESGI) (R.R., X.E.), ENGAGE (European Commission, ENGAGE_201413), TECHGENE (European Commission, TECHGENE_223143), and GEUVADIS (European Commission, 261123_GEUVADIS) (X.E.); NOVADIS (Ministerio de Ciencia y Technologia, SAF2008-00357) (X.E.); Galicia Government Xunta de Galicia (Spain) through the project number 10PXIB918057 (J.M.C.T.); MAEC-AEC1 Predoctoral Fellowship (Ministerio de Asuntos Exteriores y Cooperación, Spain) (A.M.F.); and Ramón y Cajal position and grant BFU2007-60930 (Ministerio de Educación y Ciencia) (M.C.). The funders had no role in study design, data collection and analysis, decision to publish, or preparation of the manuscript.

Competing Interests: The authors have declared that no competing interests exist.

* E-mail: xavier.estivill@crg.eu

9 These authors contributed equally to this work.

Introduction

Genomic structural variations (SVs) include copy number variants (CNVs) of genomic segments (typically >100 bp), insertions, and balanced rearrangements, such as inversions and translocations [1,2]. The fraction of the genome affected by SVs is comparatively larger than that accounted for by single nucleotide polymorphisms (SNPs) and other smaller scale variants (Indels) [3]; currently, around 15% of the human genome is considered to fall into copy number variable regions [4]. Thus, their potential role in the contribution to genetic differences between individuals and species, to genetic diseases, and to the somatic differences between normal and cancer cells has become increasingly apparent [1–3,5,6].

Advances in DNA sequencing technologies have enabled the exploration of genomic SVs at a very fine scale, allowing a genome-wide characterization of breakpoints for most classes and sizes of SVs at base-pair resolution. However, the nature and the huge amount of next-generation sequencing (NGS) data pose substantial computational and bioinformatics challenges. Four main strategies exist for the detection of SVs breakpoints from NGS data (reviewed in [7,8]). Briefly, these strategies are: (1) Paired-reads (PR) approaches, based on the alignment of sequence pairs corresponding to both ends of a clone or a DNA fragment, which find clusters of aberrantly mapped pairs of reads that suggest the presence of SVs [9–12]; (2) read-depth (RD) analysis, that detect CNVs by analysing the density of reads mapped to a given interval of the reference genome [13–15]; (3) split-read (SR) [16] and clip-read (CR) [17] analysis, which directly identify sequence reads that contain breakpoints of SVs; and (4) sequence assembly (AS), which enable the fine-scale discovery of SVs, including novel (non-reference) sequence insertions [5,18,19].

Each of the above-mentioned approaches has limits in terms of the type and size of SVs that they are able to detect, as each one has different strengths and weaknesses [7,8]. For example, PR approaches have difficulties in mapping assignments when the read pairs fall into repetitive regions or when there are SNPs or other sequence features mapping close to the breakpoints. Furthermore, the size of insertions that can be detected by PR is limited by the library's insert size mean and standard deviation. Although RD analysis is the only sequencing-based method to accurately predict absolute copy-numbers [13,20], its breakpoint resolution is poor. RD is further hampered by PCR induced coverage biases and is unable to detect copy-number neutral variants such as inversions and balanced translocations. The SR strategy has had a limited application in the analysis of NGS data due to the difficulty of aligning short reads (reliable only in unique genomic regions), algorithmic issues with aligning across large gaps and the need of a higher depth of coverage in order to obtain sufficient split reads matching the breakpoints. Finally, AS approaches require higher computational costs, are very-time consuming, and are still prone to assembly errors.

So far, there is no single SV caller with the capacity to detect the full range of changes, since most of the existing methods for SV prediction use only one type of the described approaches. Although the PR strategy would, *a priori*, be able to detect most types of SV, available methods designed to detect SV by PR focus mainly on the discovery of deletions, insertions and inversions. As an alternative, algorithms like SVMerge [21] propose to integrate multiple existing SVs algorithms, which could complement each other and enhance their capabilities for SVs detection. However, the computational requirements for such a process are very high.

More recent methods have begun to consider both PR and RD signals [22–24], although most of them have focused only on detection of CNVs, with the exception of the recently described GASVPro algorithm that combines these two strategies for the detection of CNVs and inversions [24]. Only the HYDRA [25] and the next-generation VariationHunter [26] algorithms have moved one step further, and offer the capability to predict retrotransposition events. Additionally, two other methods, Pindel and Delly, have considered the combination of PR and SR signals. Pindel [16] uses both strategies for the detection of large deletions and medium-sized insertions, while Delly [27] integrates PR and SR for the characterization of balanced and unbalanced structural variants.

In the last years, the systematic whole-genome sequencing of several cancer samples and germline genomes has revealed a high level of structural genomic complexity in some populations of tumour or germline cells [28–31]. This high genomic complexity consists of a specific pattern of breakpoint accumulation for different types of SVs (deletions, duplications, intra- and inter-chromosomal translocations, and inversions) occurring in relatively small focal regions, involving one or several chromosomes [28]. At the genomic level, this phenomenon demonstrates the importance of the capability to predict the complete spectrum of SV types and to correctly interpret these predictions.

Here we present a novel computational tool to detect and interpret five general types of SVs (deletions, copy-number gains, intra- and inter-chromosomal translocations, and inversions) present in a given genome using NGS data. NGS analysis generates a large amount of SV predictions, and therefore it is essential for computational tools to try to obtain the most reliable set of potential variants in order to facilitate the downstream biological validation. In this sense, our pipeline, called *PeSV-Fisher*, is based on a combinatorial analysis of PR and RD strategies that lead to the detection and correct interpretation of simple and complex structural rearrangements. In addition, for all those breakpoints predicted by PR that are not supported by RD or defined as complex SV, our pipeline includes a module that filters out rearrangement predictions involving multi-copy elements such as segmental duplications (SDs), simple sequence repeats (SSRs) or transposable elements (TEs).

PeSV-Fisher has been designed with the aim of facilitating the identification of somatic variation, and, as such, it is capable of analysing two or more samples simultaneously, producing a list of non-shared variants between tumour and normal samples, although it can also analyse samples individually.

Algorithm Description

PeSV-Fisher works on the two different types of sequence data based on pairs of reads, namely paired-end or mate-pair libraries. Paired-end libraries are constructed with small insert sizes (<600 bp) generated by the fragmentation of genomic DNA into short segments followed by gel-based size selection and sequencing of both ends of the segments. In contrast, mate-pair (jumping) libraries are constructed with larger insert sizes (>2 kb). The DNA is fragmented and adaptors are added at the ends, then it is circularized and randomly sheared, generating smaller fragments; the fragment containing the adaptors is selected, and its ends are sequenced [9]. Computationally, the main difference between the two procedures is the strand order of the resulting aligned pair of reads. However, for readability, from now on we will assume that the data has been generated using paired-end libraries. The pipeline starts from sequence alignment data in the BAM format, a quasi standard widely accepted in the sequencing community and produced by most NGS alignment tools. This way, the users can apply their alignment algorithm of choice, such as MAQ [32], SOAP [33], bowtie2 [34] or BWA [35].

The *PeSV-Fisher* toolkit is a compendium of modules performing the following steps: definition of anomalous read-pairs, clustering procedure and breakpoint prediction, read depth analysis, definition and interpretation of structural variants, and filtering structural variant calls (FinalCountDown). These modules are described in detail in the following subsections.

1.1 Defining Anomalous Read-pairs

First, read-pairs (RPs) are extracted from each BAM file using SAMtools [36]. Unpaired RPs (i.e. one-end anchored RPs) and orphan RPs (i.e. unmapped RPs) are retained apart.

Second, RPs are assigned to four different categories based on the following criteria: (1) RPs mapped in right order and right orientation (i.e., the leftmost read should be aligned in the forward strand and the rightmost read in the reverse strand); (2) RPs aligned in right orientation but wrong order (i.e., the leftmost read is aligned in the reverse strand and the rightmost in the forward strand); (3) RPs mapped in right order but wrong orientation (i.e., both reads mapped in either forward or reverse strands); and (4) RPs mapped to different chromosomes.

Third, category (1) is used to estimate the empirical distribution of the insert size random variable, L. This is used to define the cut-offs to discriminate concordant RPs (those falling into the expected range) from discordant RPs exhibiting a significantly increased or decreased insert size. The empirical distribution of L is calculated based on the locations (*left*, *right*) of the innermost RP positions such that $l = right - left$, where l correspond to the observations of L. Assuming that at least $(1-\alpha) \times 100\%$ of the RPs are concordant, an upper cut-off, *UC*, and a lower cut-off, *LC*, are calculated as the $1-\alpha/2$ and $\alpha/2$ percentiles of the L distribution. If L is normally distributed, when $\alpha = 0.01$, these percentiles correspond to 2.58 standard deviations (2.58σ) from the mean.

Fourth, the type of aberration is defined based on the characteristics of the alignment. Note that this only classifies the aberration, but it does not attempt to define the type of SV yet. RPs in category (1) with $l > UC$ are labelled as *distance*-pairs. This type of RPs are usually considered as indicators of deletions, however, as already described by [26] and as we will show later in this section, other types of SVs might be defined by these mapping aberrations. RPs in this category with $l < LC$, which are considered as indicators of insertions, are discarded. The size of the insertions that could be detected from these RPs is limited by the library's insert size mean and standard deviation, and most PE data does not allow a high-confidence detection of insertions because of this. Detection of novel sequence insertions requires a specific analysis, which is provided by other algorithms [18,25,26]. RPs in category (2) will be called *order*-pairs, and are indicators of intra-chromosomal rearrangements. RPs in category (3) will be called *ori*-pairs, and are usually considered as indicators for putative inversions. However, all these RP categories may also be indicators of more complex SVs, which will be described in detail later in this section. Finally, RPs in category (4) will be called *chrpos* or *chrposori*-pairs, depending on whether both reads of the pair are mapped to opposite strands or to the same strand, respectively, and are considered indicators of different inter-chromosomal rearrangements.

1.2 Clustering Procedure and Breakpoint Predictions

A clustering procedure is carried out to group indicators pointing to the same aberration. Two different types of clustering approaches are used depending on whether (A) the anomalous RPs consist of paired reads where one read is aligned on the positive DNA strand and the other on the negative strand (such as *distance*-, *order*- or *chrpos*-pairs), or (B) both reads are aligned on the same strand (*ori*- or *chrposori*-pairs).

In case (A) any pair (*i,j*) of RPs are clustered together if they meet the following criteria:

$$|right_{+,i} - right_{+,j}| < UC \ \wedge \ |left_{-,i} - left_{-,j}| < UC \ \wedge$$
$$|l_i - l_j| < (UC - LC)$$

where $right_{+,i}$ is the rightmost position of the read aligned on the positive strand corresponding to the *i*th RP (likewise for $left_{-,i}$). Note that if one assumes that L follows a normal distribution, $UC - LC = 2z_{1-\alpha/2}\sigma$, where $z_{1-\alpha/2}$ is the $(1-\alpha/2) \times 100$ percentile of the cumulative standard normal distribution.

In case (B), the criteria are based on the following conditions:

$$|right_{1,+,i} - right_{1,+,j}| < UC \ \wedge \ |right_{2,+,i} - right_{2,+,j}| < UC \ \wedge$$
$$|(right_{1,+,i} + right_{2,+,i}) - (right_{1,+,j} + right_{2,+,j})| < (UC - UL)$$

or

$$|left_{1,-,i} - left_{1,-,j}| < UC \ \wedge \ |left_{2,-,i} - left_{2,-,j}| < UC \ \wedge$$
$$|(left_{1,-,i} + left_{2,-,i}) - (left_{1,-,j} + left_{2,-,j})| < (UC - UL)$$

where $right_{1,+,i}$ is the rightmost position of read 1 of the *i*th RP, pointing to the 1st breakpoint, and $right_{2,+,i}$ is the rightmost position of read 2, pointing to the 2nd breakpoint, both aligned on the forward strand (likewise for $left_{1,-,i}$ and $left_{2,-,i}$ on the reverse strand).

In all cases, any RP that has been included in one cluster is not considered for any other cluster.

Once the clustering procedure is finished, clusters containing at least two RPs are used for breakpoint prediction, and two breakpoints are predicted for each cluster. For reads aligned on the forward strand, putative breakpoints are predicted to fall within the range $[\max\{right_s\}; \min\{right_{s'}\} + UC]$, with $s, s' = 1, \ldots, S$, being S the number of RPs in a cluster. Likewise, putative breakpoints found with reads aligned on the reverse strand are predicted to fall within the range $[\max\{left_s\} - UC; \min\{left_{s'}\}]$.

1.3 Somatic Breakpoints

One of the main objectives in cancer genomic studies is the identification of tumour specific variants. In this sense, *PeSV-Fisher* allows the user to analyse paired samples, i.e. tumour and normal cell samples, simultaneously. If the *somatic* option is activated, the algorithm searches for aberrant clusters present in the tumour sample that are not shared with the normal sample, and only these clusters are taken into account in the subsequent analysis modules. Therefore, when a SV is later predicted, it is automatically defined as a somatic SV.

1.4 Read Depth Analysis

This module uses both RPs with either one- or two-end anchored reads. The algorithm comprises the following steps executed in a hierarchical order:

First, each chromosome is divided into non-overlapping windows of equal size and the GC content in each window is annotated.

Second, every aligned read is assigned to the window where the highest percentage of its base pairs fall. Then, a read-depth count per window, r_i, is calculated.

Third, a GC content normalization is applied to each window i based on the following adjustment: $r_i^* = r_i \times (m/m_{GC})$, where m is the median of read-depth counts across all windows and m_{GC} is the median of read-depth counts across all windows with the same GC content as the actual *i*th window [15]. To liken the data to a normal distribution, we further use the square-root transformation and use the global median as a shift parameter towards 0: $r_i^{**} = \sqrt{r_i^*} - \sqrt{m}$. Based on the formula, values above/below 0 (or a user defined threshold, which we empirically set at $0.4/-0.4$) are interpreted as copy number gains/losses.

Fourth, an adaptation of the Genome Alteration Detection Algorithm (GADA) [37] is used to detect copy number variable genomic segments. This is possible because the adjusted read-depth signal r_i^{**} is analogous to the data properties of the log relative intensities obtained from CGH or SNP array experiments to identify CNV calls. The GADA algorithm exploits the use of piece wise constant vectors to represent CNVs and sparse Bayesian learning to detect the breakpoints. This algorithm has been shown to improve the computational speed among other algorithms, which is desirable given the huge volume of data obtained in NGS-based studies.

1.5 Defining and Interpreting Structural Variants

Finally, the results from the two previous modules are combined to define and interpret four different types of SVs, classified into two main categories: (a) unbalanced SVs, including those rearrangements that are accompanied by significant increase or decrease of the depth-of-coverage (involving deletions and gains of genetic material); and (b) balanced SVs, including those rearrangements with copy number neutral values in depth-of-coverage as described below.

(a) Unbalanced SVs. A deletion is defined by the overlap of genomic regions within breakpoint predictions from distance-pairs with copy number losses from the read depth analysis (Figure 1-a.1).

The definition of a copy number gain depends on the nature of the rearrangement. In the case of tandem duplications, genomic regions within breakpoint predictions from *order*-pairs must overlap with copy number gains in read depth analysis (Figure 1-a.6). A copy number gain of a region within the same chromosome but not in tandem is found combining *distance*- and *order*-pairs clusters with an increase of coverage depth of the copied region (as shown in Figure 1-a.2). If the copied region is inserted in inverted orientation, it is identified by a combination of *ori*-pairs clusters, one comprising *ori*-pairs anchored to the forward strand and the other on the reverse (Figure 1-a.3), and an increase in read depth analysis of the copied region. When the copy is inserted into a different chromosome, the combinatorial process is similar as those just explained, but the clusters combined in these cases are based on *chrpos*- or *chrposori*-pairs (Figure 1-a.4 and 1-a.6). In all cases, *PeSVFisher* defines the copied region and the insertion breakpoint.

(b) Balanced SVs. Inversions are defined by their two breakpoints: breakpoint predictions from clusters of ori-pairs anchored to the forward strand should be equal to breakpoint predictions from clusters of ori-pairs anchored on the reverse strand (Figure 1-b.1).

The definition of neutral intra- and inter-chromosomal translocations is based on the combination of three cluster types, also accompanied by a copy number neutral pattern from the read depth analysis. If a genomic region has been translocated to another location within the same chromosome, two *distance*- and one *order*-cluster should be combined (Figures 1-b.2). When the translocated region is inserted in an inverted orientation, the combination should include *distance*- and *ori*-clusters as shown in Figure 1-b.3. If the region is translocated into another chromosome, the combinatorial process comprises *distance*- and *chrpos*- or *chrposori*-clusters, depending on whether the translocation is inserted directly or in an inverted orientation (Figures 1-b.4 and 1-b.5).

1.6 Filtering Non-defined Structural Variant Calls (FinalCountDown)

The identification of genomic rearrangements based on PR strategies is prone to wrong predictions, especially when reads fall onto repetitive regions of the genome (i.e., simple sequence repeats, segmental duplications, and transposable elements). This is particularly important in the case of balanced SVs, as they are only supported by the PR strategy. Thus, the proportion of false positive predictions could be drastically reduced if those predictions falling within repetitive regions were filtered out. For this purpose, *PeSV-Fisher* includes an optional module called FinalCountDown, which searches for and removes those rearrangements where at least one of the breakpoints falls within repetitive elements of the genome, if these rearrangements are not supported by RD or by a combination of different types of PR aberrant clusters as described in the previous step. The repetitive elements considered in this step are: (1) segmental duplications (SDs), (2) simple sequence repeats (SSRs), or (3) low-divergent transposable elements (TEs) [38] (defined as those transposable element insertion sites that share >90% identity at the nucleotide level with the corresponding family consensus sequence). SSRs, SDs, and TEs GRCh37 coordinates are obtained from the UCSC Genome Bioinformatic Browser (http://genome.ucsc.edu/).

However, since there is evidence that large blocks of sequence homology (or SDs) or shorter common repeat sequences overlap with a relatively high percentage of SV breakpoints [12], the removed rearrangements could still indicate valid structural variants. Therefore, the FinalCountDown module provides one final file with putative rearrangements that passed all filters, and three files with putative rearrangements that did not pass a certain filter (those in SDs; those not in SDs but in SSR; those not in the previous two but with low-divergence from a TE). Finally, the TE-identity of the rearrangements that passed all filters is also provided.

Figure S1 shows a schematic representation of the complete process for the SV detection, interpretation and classification, and the final organization provided by *PeSV-Fisher* for the results is summarized in Figure S2.

Results and Discussion

Targeted Re-sequencing for Evaluating the Performance of *PeSV-Fisher* on Two Individuals with Chronic Lymphocytic Leukaemia

The research presented in this study has been approved by the "Comité Ético de Investigación Clínica" (CEIC, Hospital Clínic de Barcelona). All patients gave informed consent for their participation in the study following the International Cancer Genome Consortium (ICGC) guidelines [39]. All participants provided a written informed consent to take part in this study, which was also approved by CEIC (Hospital Clínic de Barcelona) to enter in the ICGC-Chronic Lymphocytic Leukemia Genome Project (http://www.cllgenome.es/). All patients submitted to mutational screening and clinical validation gave their informed consent in agreement with an Institutional Review Board-approved informed consent for genetic studies. All clinical investigation was conducted according to the declaration of Helsinki.

We first studied the behaviour of *PeSV-Fisher* at fine-scale breakpoint resolution in all the different types of aberrant clusters the tool is able to detect. Our aim was to evaluate the different types of breakpoint predictions by PR strategy rather than the final variant definitions. Given the nature of the genome and of NGS data (such as high percentage of repetitive regions and short read lengths), mapping programs generate a large amount of ambiguous alignments, leading to a similarly large amount of false positive breakpoint calls and thus false positive SV predictions by the PR strategy. Although *PeSV-Fisher* integrates PR and RD strategies to define a SV, balanced SVs (inversions and translocations) can only be detected by combining breakpoints from different aberrant clusters predicted by PR. Therefore, reliable aberrant cluster breakpoints are key to further define a reliable SV, especially in the case of balanced SVs.

The performance of *PeSV-Fisher* was evaluated in terms of true positive (TP) and false negative (FN) rates. To this end, we studied both normal and tumour DNA from one previously published patient with chronic lymphocytic leukaemia, case CLL2, from the Chronic Lymphocytic Leukaemia Genome Project (CLL-GP) [40]. Detailed information about sample collection and processing, biological characteristics of the tumour sample and the whole-

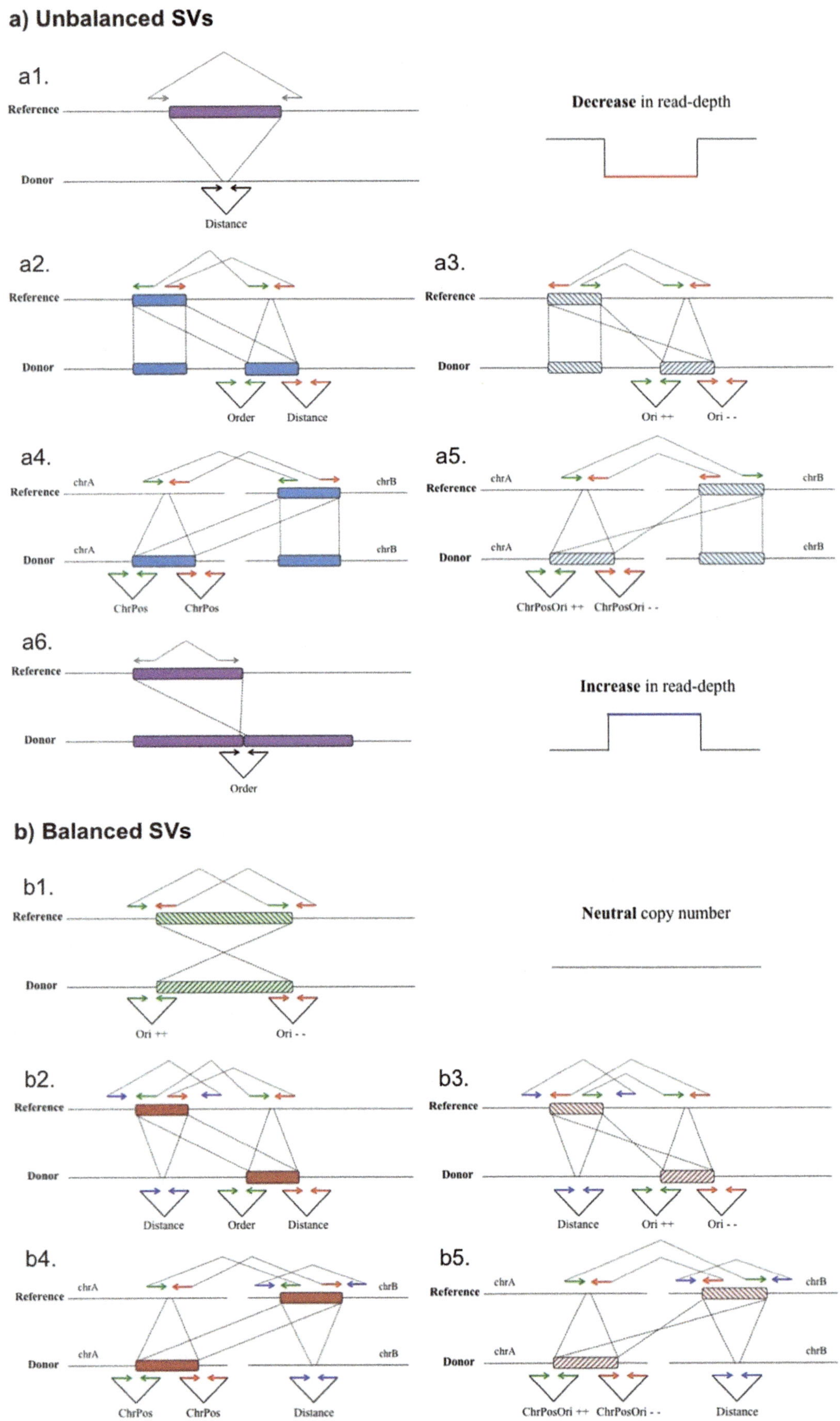
a) Unbalanced SVs
a1.
Reference
Donor
Distance
Decrease in read-depth
a2.
Reference
Donor
Order
Distance
a3.
Reference
Donor
Ori ++
Ori - -
a4.
chrA
chrB
Reference
Donor
ChrPos
ChrPos
a5.
chrA
chrB
Reference
Donor
ChrPosOri ++
ChrPosOri - -
a6.
Reference
Donor
Order
Increase in read-depth
b) Balanced SVs
b1.
Reference
Donor
Ori ++
Ori - -
Neutral copy number
b2.
Reference
Donor
Distance
Order
Distance
b3.
Reference
Donor
Distance
Ori ++
Ori - -
b4.
chrA
chrB
Reference
Donor
ChrPos
ChrPos
Distance
b5.
chrA
chrB
Reference
Donor
ChrPosOri ++
ChrPosOri - -
Distance

Figure 1. Combination of different types of anomalous read-pair alignments together with read depth (RD) pattern to define four different categories of structural variants (SVs). Case a.1 plus a decrease in RD represents a deletion. Cases a.2, a.3, a.4, a5 and a.6 together with an increase in RD represent different types of copy number gains in terms of the co-localization of the copy. Cases a.2 and a.3 represent a copy inserted within the same chromosome, were in case a.3 the copy is inserted in an inverted orientation. Analogous cases a.4 and a.5 represent straight or inverted insertions of copies in another chromosome. Case a.6 represents tandem duplications. Case b1 corresponds to an inversion. Cases b2 and b3 correspond to an intra-chromosomal translocation, but in case b3 the translocated region is inserted in inverted orientation. Similarly cases b4 and b5 correspond to inter-chromosomal translocations.

genome sequencing (WGS) protocol for short-insert library construction are provided in a previous study [40]. Sequencing coverage statistics are shown in Table 1.

To evaluate the TP rate of our method we collected aberrant cluster calls from the tumour sample and constructed different scenarios based on the set of confidence parameters that influence the reliability of breakpoint predictions. These parameters are: phred-scaled quality scores; Q, which evaluate the mappability of the read and the base call accuracy; the number of read-pairs supporting the aberrant cluster; and the length of the detected variant, which is taken into account in all cluster types except for *chrpos*- and *chrposori*-clusters. In particular, we considered two levels for Q: clusters with at least one RP where both reads have $Q \geq 35$ (more than 99.9% of base call accuracy), and clusters with $Q \geq 20$ (99% of base call accuracy) excluding those potentially involved in the first Q level. For the number of RPs supporting each cluster, we considered 4 categories: high [10–50), medium [5–10), low [1–5) and special, in the sense that the number of RPs is significantly larger than the average sequence coverage, ≥ 50. Finally, the different variant lengths categorizations were: high [1 kb –1 Mb), medium [500 bp –1 kb) and special, $\geq 1Mb$. Consequently, we designed 24 scenarios that reflect the spectrum of RP based aberrant clusters obtained by *PeSV-Fisher*.

Breakpoints of the clusters integrated in these scenarios that do not contain any type of the repetitive sequences after the FinalCountDown analysis, were re-analysed by a targeted re-sequencing approach (Supporting Information S1). Note that the criterion to re-sequence only those clusters with "clean" breakpoints is due to the incapability to design specific targets for repetitive sequences. We then applied the PR strategy of *PeSV-Fisher* and an independent split-read analysis using the GEM algorithm [41] on the re-sequencing data in order to establish the TP rates of our tool conditioned on the different reliability parameters.

We captured the two breakpoints of 509 potential SVs, corresponding to 259 *distance*-clusters, 21 *order*-clusters, 49 *ori*-clusters, 104 *chrpos*-clusters, and 76 *chrposori*-clusters. Using the PR strategy we confirmed the two breakpoints of 76% of *distance*-clusters, 67% of *order*-clusters, 84% of *ori*-clusters, 32% of *chrpos*-clusters, and 21% of *chrposori*-clusters, while by split-reads we validated 61%, 10%, 65%, 21%, and 18%, respectively (Figure 2). The efficiency of breakpoint capture was very low in the scenarios where clusters had poor Q values, which could be explained by the presence of some types of repetitive sequences in the breakpoints not included in the FinalCountDown module, or by sequencing/mapping errors. Hence this only allows us to be confident with clusters containing at least one RP with $Q \geq 35$, and to realize the importance of this value for the reliability of a SV even in those rearrangements that do not contain repetitive sequences in the breakpoints.

As expected, the higher the number of RPs pointing to the same alignment aberration, the more confident the results are. This is especially important in *order*-, *chrpos*-, and *chrposori*-clusters (Figure 2). In the category of special number of RPs, which includes those clusters with significantly more RPs than expected by average coverage rate, the re-sequencing approach failed to capture them, mainly because these breakpoints might also fall into specific types of highly repetitive or highly polymorphic regions not included in the FinalCownDown filtering module. Finally, according to our validation results the predicted length of the variant indicated by the anomalous RPs is also important for the reliability.

To evaluate the ability of the *PeSV-Fisher* to detect real complex SVs, we further searched for some examples of SVs that had been defined in sample CLL2 whose breakpoints have been completely validated by our targeted re-sequencing approach. Two interesting examples are further detailed in Table S1: an inter-chromosomal copy-paste event involving two different clusters of *chrposori* supported by a significant increase of RD and an intra-chromosomal cut-paste event formed by two *ori* clusters that overlapped with a *distance* cluster without a significant increase or decrease of RD.

In order to establish FN rates, we extracted breakpoints that were detected only in the tumour sample, i.e. *a priori* somatic breakpoints, and we asked if the breakpoints were also found in the normal sample using the targeted re-sequencing approach. We obtained a set of targets that should be free of variant breakpoints in the normal genome based on the WGS analysis; therefore if we detect the same breakpoint in the normal sample as in the tumour, this is defined as a FN.

Table 1. Whole-genome sequencing data statistics.

		Read length	Insert size			#Reads	Seq. coverage	#Reads	Seq. coverage
			1st percentile	Median	99th percentile	Unpaired+paired-reads		Paired-reads	
CLL2	Tumour cells	~95	112	271	400	1.178.606.714	38.7x	1.138.687.234	37.4x
	Normal cells	~95	78	261	370	1.161.111.680	38.1x	1.120.131.088	36.8x
CLL16	Tumour cells	~95	84	264	405	1.377.406.890	45.2x	1.355.526.562	44.5x
	Normal cells	~95	68	247	402	1.364.685.267	44.8x	1.322.094.298	43.4x
YRI	NA19240	~40	42	186	966	2.328.156.503	32.2x	2.178.558.262	30.3x

Score	Read-Pairs	Length	Distance n	Distance % PR val	Distance % SR val	Order n	Order % PR val	Order % SR val	Orientation n	Orientation % PR val	Orientation % SR val	ChrPos n	ChrPos % PR val	ChrPos % SR val	ChrPosOri n	ChrPosOri % PR val	ChrPosOri % SR val
High Q ≥ 35	High [10 – 50)	High [1kb -1Mb)	74	89,19	72,97	6	83,33	0	15	100	80						
		Medium [500bp – 1kb)	69	86,96	69,57	4	75	25	4	75	50	21	76,19	42,86	11	72,73	54,55
		Special ≥ 1Mb	1	100	100	2	50	0	2	100	50						
	Medium [5 – 10)	High [1kb -1Mb)	27	55,56	37,04	3	66,67	0,00	9	88,89	77,78						
		Medium [500bp – 1kb)	14	57,14	64,29	1	100	100	4	100	75	17	29,41	17,65	15	26,67	20
		Special ≥ 1Mb	0	-	-	0	-	-	1	100	0						
	Low [1 – 5)	High [1kb -1Mb)	27	66,67	37,04	3	0	0	8	62,50	62,50						
		Medium [500bp – 1kb)	8	62,50	62,50	2	100	0	0	-	-	56	16,07	5,36	39	5,13	7,69
		Special ≥ 1Mb	0	-	-	0	-	-	3	33,33	33,33						
	Special ≥ 50	High [1kb -1Mb)	6	100	50	0	-	-	1	100	100						
		Medium [500bp – 1kb)	4	100	50	0	-	-	1	100	0	1	100	0	0	-	-
		Special ≥ 1Mb	0	-	-	0	-	-	0	-	-						
Medium Q ≥ 20	High [10 – 50)	High [1kb -1Mb)	15	73,33	66,67	0	-	-	0	-	-						
		Medium [500bp – 1kb)	0	-	-	0	-	-	0	-	-	0	-	-	0	-	-
		Special ≥ 1Mb	0	-	-	0	-	-	0	-	-						
	Medium [5 – 10)	High [1kb -1Mb)	0	-	-	0	-	-	0	-	-						
		Medium [500bp – 1kb)	0	-	-	0	-	-	0	-	-	2	100	0	0	-	-
		Special ≥ 1Mb	0	-	-	0	-	-	0	-	-						
	Low [1 – 5)	High [1kb -1Mb)	1	0	0	0	-	-	1	0	0						
		Medium [500bp – 1kb)	9	0	44,44	0	-	-	0	-	-	7	0	0	11	18,18	18,18
		Special ≥ 1Mb	0	-	-	0	-	-	0	-	-						
	Special ≥ 50	High [1kb -1Mb)	4	100	75	0	-	-	0	-	-						
		Medium [500bp – 1kb)	0	-	-	0	-	-	0	-	-	0	-	-	0	-	-
		Special ≥ 1Mb	0	-	-	0	-	-	0	-	-						
TOTAL			259	76,45	61,39	21	66,67	9,52	49	83,67	65,31	104	31,73	21,15	76	21,05	18,42

Figure 2. Performance of *PeSV-Fisher* based on the analysis of a high-coverage sequenced blood-cancer genome. Results according to different scenarios based on the set of confidence parameters that influence the reliability of breakpoint predictions of clusters made by anomalous alignment read-pairs. These are phred-scaled quality scores *Q*; the number of read-pairs supporting the aberrant cluster; and the length of the potential variant call. For each scenario and cluster type are represented the number of breakpoints captured by target re-sequecing of *PeSV-Fisher* calls from whole genome sequencing (n), percentage of breakpoints validated by target-re-sequencing using the paired-reads strategy (%PR) and the percentage of breakpoints validated by target-re-sequencing using split-reads analysis (%SR).

Previous analyses of sample CLL2 by WGS showed that the tumour sample contains, excluding antigen receptor loci, only 6 somatic SVs [40], which were all called by *PeSV-Fisher* and validated as somatic by targeted re-sequencing. Thus, in order to establish more significant FN rates, we applied the PR strategy of our tool to another CLL case from the CLL-GP (CLL16, unpublished data). In this case, we detected by whole-genome sequencing and validated by target re-sequencing 42 clusters of potential SVs in the tumour sample, and only 5 (FN = 11.9%) of the putative somatic variants were detected in the normal tissue using the target re-sequencing approach. Information about sample collection and processing, biological characteristics of the tumour sample and the whole-genome sequencing (WGS) and targeted re-sequencing protocols are provided in Supporting Information S1.

Comparative Analyses in a 1000 Genome Project Trio Using Other Approaches

We applied *PeSV-Fisher* to the WGS data from the daughter of the Yoruba high-coverage trio included in the 1000 Genomes Project (1000 GP) dataset (NA19240). For the analysis, we considered different paired-end libraries generated by Illumina with insert sizes ranging from 100 to 600-bp, obtaining ~30x sequencing coverage (Table 1). This same sample is analysed in the publication of Mills *et al.* [5], which reports a complete list of deletion/insertion calls made by different institutions on the high-coverage trios and low coverage sample sets (1000 GP pilot project). From this publication, we obtained the non-redundant set of deletion calls (> = 50 nt) for high coverage trios that were either validated through the use of PCR, assembly of breakpoints (ASM), sequence capture arrays (CAP array) and a high-resolution array-CGH platform (downloaded from: ftp://ftp.1000 genomes.ebi.ac.uk/vol1/ftp/pilot_data/paper_data_sets/a_map_of_human_-variation/trio/sv/(trio.2010_10.deletions.sites.vcf)).

To examine the sensitivity of *PeSV-Fisher* on true deletion calls with well-defined breakpoints, we selected the PCR and/or ASM validated deletions ≥900-bp from the validated deletions list, independently of which algorithm was used for calling, and we checked the overlap with our calls. This 900 bp cut-off was selected based on the variability of the insert sizes in the combined libraries used. Figure 3a shows the overlap of our calls with these validated variants. *PeSV-Fisher* detected 523 distance-clusters overlapping with the validated list, which is similar to those detected by BD (541) or VH (558). Nonetheless, only 374 of these 523 distance-clusters were finally classified as *deletions*, showing a significant decrease of RD and non-overlapping any other PR clusters. Of the remaining 149, in 6 cases the variant definition classified them as other than deletions (two copy-paste and two cut-paste events) and in 28 cases as putative deletions. Unsurprisingly, of these 6 variants not classified as deletions by PeSV-Fisher, 5 contain breakpoints located in highly repetitive regions close to telomeres. Sites with high sequence diversity are prone to generate multiple and similar breakpoint predictions by several types of anomalous RPs, reflecting the complexity of the sequence analysis. This reveals the importance of the combination of two types of strategies like PR and RD to correctly interpret and define the individual status for the high polymorphic loci and the complex genomic events. For the last 115 *distance* calls, the predicted breakpoint ranges for each end of the variant were overlapping and we usually discard these calls as unreliable. This overlap is probably due to the high variability in the libraries insert sizes.

We also compared our performance with that of BreakDancer and VariationHunter, checking the overlap of our calls with those of these two algorithms obtained from a Supplementary Table of Mills *et al.* [5] (downloadable from http://www.nature.com/nature/journal/v470/n7332/extref/nature09708-s5.zip and corresponding to Washington University –WU- and University of Washington –UW- indicators, respectively). We did not compare to algorithms that do not use the PR strategy, as they are conditioned to the analysis of unbalanced SVs, or to data generated by unpublished algorithms. Figure 3b summarizes the overlap of the predicted *defined* deletions made on the Yoruban

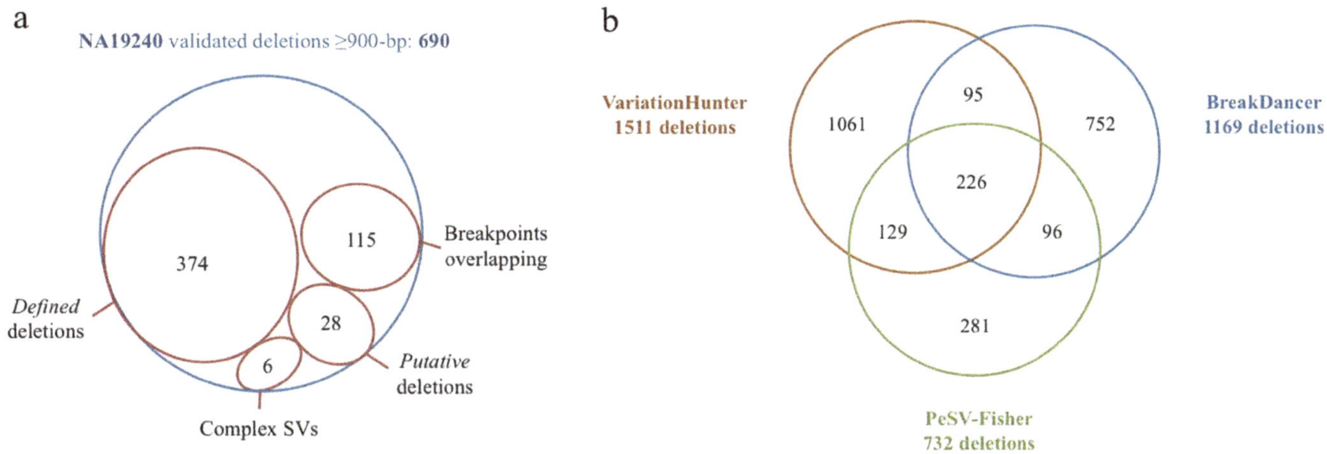

Figure 3. Sensitivity analysis and comparison with results from other SV prediction methods based on the analysis of the Yoruba daughter (NA19240) from the high-coverage trio from the 1000 Genomes Project dataset. (a) 90% based of the overlap of *PeSV-Fisher* calls with the non-redundant set of deletions from the publication of Mills *et al.* [5] that were validated by PCR and/or assembly. *Breakpoints* overlapping indicate *distance* clusters with predicted overlapping breakpoint ranges from each end of the variant; (b) three-way comparison of the deletion calls made by *PeSV-Fisher*, BreakDancer and VariationHunter. The analysis is carried out using a 99% of reciprocal overlap.

daughter NA19240 by the three different algorithms. We considered that two different algorithms were calling the same deletion if there was a 99% reciprocal overlap. With this stringent value we ensured that the overlap measure was symmetric. Each algorithm uniquely called a large number of calls; however a fraction of these calls might be due to false positives or different breakpoint accuracies. Another explanation might be the different size distribution patterns of deletion calls reported by the three algorithms. However, the three algorithms show the expected peak that corresponds to the 6 kb average sizes of full-length LINE-1 human retrotransposons [42] (data not shown).

Computational Considerations

As argued by [43], NGS data imposes challenges on procedures applied for storage and analysis of data. The computer memory is usually a limitation to the analysis and, moreover, the run time scales approximately linearly with the number of reads. A plausible alternative is to search for modular code methods or to split the data in order to divide the process in multiple threads. Thus, the current tendency is to parallelize processes and use large-scale computing clusters with high numbers of cores. However, *PeSV-Fisher* can be launched on either a cluster or a workstation. This is achieved by splitting the data by chromosomes and by making classes with low coupling between them. This approach can be easily parallelized. Furthermore, we paid special attention to memory balancing using sorting strategies that directly manipulate files.

To exemplify the applicability of *PeSV-Fisher* to processing of large datasets we recorded basic timing information for processing of whole-genome paired-end data from medium to high coverage (16x to 45x). PeSV-Fisher was launched in multithreading mode (5 cores for read depth process and 5 cores for the definition of anomalous RPs plus clustering processes by sample) in a workstation with 12 cores and 48GB of memory under a Linux environment. It required approximately an average of 3915 CPU minutes for processing an average of 1.377 million reads generated by an Illumina platform using a median fragment size of 264. Two samples with medium coverage (16x–20x) were completed in parallel mode in approximately 265 minutes and two samples with high coverage (45x) needed 805 minutes. The memory by thread in the medium coverage sample was about 0.1–0.4 GB during the longest part of the execution, but displayed peaks of 1.1–1.4 GB at the sorting step in definition of anomalous RPs plus clustering modules. The memory used for the samples with high coverage was similar but peaks were in the range of 1.7–2.4 GB.

The analysis time for the clustering processes present a high dependency on hard drive speed; however, the analysis time for read depth processing is more dependent on the CPU speed. So, the read depth and definition of anomalous RPs plus clustering modules for each sample can be launched simultaneously for a better performance and resource utilization.

Further computational aspects are given in the Supporting Information S1.

Conclusions

Despite constantly ongoing development of algorithms for the detection of SVs, the complexity of NGS data and of the human genome hampers SV discovery as well as the understanding of the subsequent biological interpretation and consequences.

We presented here our computational tool, *PeSV-Fisher*, for the detection and characterization of SVs using NGS data. We employed a new strategy to correctly interpret SV calls generated by paired-reads approaches, combining different types of aberrantly aligned read-pairs, and a read depth strategy. By doing so, *PeSV-Fisher* is capable of defining five general types of structural rearrangements (deletions, gains, intra- and inter-chromosomal translocations, and inversions), but goes one step further in their definition with the interpretation of co-localization of identified aberrant calls. Additionally, *PeSV-Fisher* keeps information concerning the orientation of aberrantly aligned read-pairs. For example, the positional effect of a gene copy in any region of the genome could be different if this copy is inserted in one orientation or another.

PeSV-Fisher does not take into account the diploid nature of the human genome, although it is an aspect recently being considered in strategies to solve the co-localization of more than two variants at the same locus [26]. However, our tool has been designed especially for the analysis of cancer genomes, where clonal mosaicism, which is defined as the coexistence of cells with two or more distinct genotypes within an individual, has been observed

[44,45]. Interestingly, clonal mosaicism has also been described in aging genomes [44], which means that it not only happens in cancer genomes, but also is a common phenomenon appearing throughout the genome evolution within an individual.

Finally, we should highlight the good performance of the tool in terms of sensitivity and true positive and negative rates, with a reasonable computational cost in a single workstation.

Supporting Information

Figure S1 Workflow of structural variants definition module of *PesV-Fisher*.

Figure S2 Output files organization. *PeSV-Fisher* generates a general *Results* folder containing three subfolders called *clusters, dofc* and *sv,* which contain results from PR strategy, RD strategy and the results from the combination of both strategies, respectively.

Table S1 Examples of complex structural variants defined by *PeSV-Fisher* in the chronic lymphocytic leukemia sample CLL2 and validated by re-sequencing approach.

Supporting Information S1

Acknowledgments

We thank the Chronic Lymphocytic Leukaemia Genome Consortium (led by Elias Campo and Carlos Lopez-Otín) for providing data sequences used in the validation of *PeSV-Fisher.* We are grateful to the Barcelona Supercomputing Center for support in transfer of sequence data used for this project. We also thank Charlotte Hor for her careful reading of the manuscript that improved the English.

Author Contributions

Conceived and designed the experiments: GE LB JMCT XE. Performed the experiments: MG. Analyzed the data: GE LB RR SO. Contributed reagents/materials/analysis tools: MG. Wrote the paper: GE CT LB. Designed the software: GE CT LB JMCT AMF MC. Programmed the software: GE CT RR. Critically revised the manuscript: RR JMCT AMF MC MG SO XE.

References

1. Feuk L, Carson AR, Scherer SW (2006) Structural variation in the human genome. Nat Rev Genet 7: 85–97.
2. Sharp AJ, Cheng Z, Eichler EE (2006) Structural variation of the human genome. Annu Rev Genomics Hum Genet 7: 407–442.
3. Conrad DF, Pinto D, Redon R, Feuk L, Gokcumen O, et al. (2010) Origins and functional impact of copy number variation in the human genome. Nature 464: 704–712.
4. Stankiewicz P, Lupski JR (2010) Structural variation in the human genome and its role in disease. Annu Rev Med 61: 437–455.
5. Mills RE, Walter K, Stewart C, Handsaker RE, Chen K, et al. (2011) Mapping copy number variation by population-scale genome sequencing. Nature 470: 59–65.
6. Zhang F, Gu W, Hurles ME, Lupski JR (2009) Copy number variation in human health, disease, and evolution. Annu Rev Genomics Hum Genet 10: 451–481.
7. Medvedev P, Stanciu M, Brudno M (2009) Computational methods for discovering structural variation with next-generation sequencing. Nat Methods 6: S13–20.
8. Alkan C, Coe BP, Eichler EE (2011) Genome structural variation discovery and genotyping. Nat Rev Genet 12: 363–376.
9. Korbel JO, Urban AE, Affourtit JP, Godwin B, Grubert F, et al. (2007) Paired-end mapping reveals extensive structural variation in the human genome. Science 318: 420–426.
10. Chen K, Wallis JW, McLellan MD, Larson DE, Kalicki JM, et al. (2009) BreakDancer: an algorithm for high-resolution mapping of genomic structural variation. Nat Methods 6: 677–681.
11. Hormozdiari F, Alkan C, Eichler EE, Sahinalp SC (2009) Combinatorial algorithms for structural variation detection in high-throughput sequenced genomes. Genome Res 19: 1270–1278.
12. Kidd JM, Cooper GM, Donahue WF, Hayden HS, Sampas N, et al. (2008) Mapping and sequencing of structural variation from eight human genomes. Nature 453: 56–64.
13. Alkan C, Kidd JM, Marques-Bonet T, Aksay G, Antonacci F, et al. (2009) Personalized copy number and segmental duplication maps using next-generation sequencing. Nat Genet 41: 1061–1067.
14. Abyzov A, Urban AE, Snyder M, Gerstein M (2011) CNVnator: an approach to discover, genotype, and characterize typical and atypical CNVs from family and population genome sequencing. Genome Res 21: 974–984.
15. Yoon S, Xuan Z, Makarov V, Ye K, Sebat J (2009) Sensitive and accurate detection of copy number variants using read depth of coverage. Genome Res 19: 1586–1592.
16. Ye K, Schulz MH, Long Q, Apweiler R, Ning Z (2009) Pindel: a pattern growth approach to detect break points of large deletions and medium sized insertions from paired-end short reads. Bioinformatics 25: 2865–2871.
17. Wang J, Mullighan CG, Easton J, Roberts S, Heatley SL, et al. (2011) CREST maps somatic structural variation in cancer genomes with base-pair resolution. Nat Methods 8: 652–654.
18. Hajirasouliha I, Hormozdiari F, Alkan C, Kidd JM, Birol I, et al. (2010) Detection and characterization of novel sequence insertions using paired-end next-generation sequencing. Bioinformatics 26: 1277–1283.
19. Li R, Zhu H, Ruan J, Qian W, Fang X, et al. (2010) De novo assembly of human genomes with massively parallel short read sequencing. Genome Res 20: 265–272.
20. Sudmant PH, Kitzman JO, Antonacci F, Alkan C, Malig M, et al. (2010) Diversity of human copy number variation and multicopy genes. Science 330: 641–646.
21. Wong K, Keane TM, Stalker J, Adams DJ (2010) Enhanced structural variant and breakpoint detection using SVMerge by integration of multiple detection methods and local assembly. Genome Biol 11: R128.
22. Qi J, Zhao F (2011) inGAP-sv: a novel scheme to identify and visualize structural variation from paired end mapping data. Nucleic Acids Res 39: W567–575.
23. Medvedev P, Fiume M, Dzamba M, Smith T, Brudno M (2010) Detecting copy number variation with mated short reads. Genome Res 20: 1613–1622.
24. Sindi SS, Onal S, Peng L, Wu HT, Raphael BJ (2012) An integrative probabilistic model for identification of structural variation in sequencing data. Genome Biol 13: R22.
25. Quinlan AR, Clark RA, Sokolova S, Leibowitz ML, Zhang Y, et al. (2010) Genome-wide mapping and assembly of structural variant breakpoints in the mouse genome. Genome Res 20: 623–635.
26. Hormozdiari F, Hajirasouliha I, Dao P, Hach F, Yorukoglu D, et al. (2010) Next-generation VariationHunter: combinatorial algorithms for transposon insertion discovery. Bioinformatics 26: i350–357.
27. Rausch T, Zichner T, Schlattl A, Stutz AM, Benes V, et al. (2012) DELLY: structural variant discovery by integrated paired-end and split-read analysis. Bioinformatics 28: i333–i339.
28. Stephens PJ, Greenman CD, Fu B, Yang F, Bignell GR, et al. (2011) Massive genomic rearrangement acquired in a single catastrophic event during cancer development. Cell 144: 27–40.
29. Kloosterman WP, Hoogstraat M, Paling O, Tavakoli-Yaraki M, Renkens I, et al. (2011) Chromothripsis is a common mechanism driving genomic rearrangements in primary and metastatic colorectal cancer. Genome Biol 12: R103.
30. Liu P, Erez A, Nagamani SC, Dhar SU, Kolodziejska KE, et al. (2011) Chromosome catastrophes involve replication mechanisms generating complex genomic rearrangements. Cell 146: 889–903.
31. Kloosterman WP, Guryev V, van Roosmalen M, Duran KJ, de Bruijn E, et al. (2011) Chromothripsis as a mechanism driving complex de novo structural rearrangements in the germline. Hum Mol Genet 20: 1916–1924.
32. Li H, Ruan J, Durbin R (2008) Mapping short DNA sequencing reads and calling variants using mapping quality scores. Genome Res 18: 1851–1858.
33. Li R, Li Y, Kristiansen K, Wang J (2008) SOAP: short oligonucleotide alignment program. Bioinformatics 24: 713–714.
34. Langmead B, Salzberg SL (2012) Fast gapped-read alignment with Bowtie 2. Nat Methods 9: 357–359.
35. Li H, Durbin R (2009) Fast and accurate short read alignment with Burrows-Wheeler transform. Bioinformatics 25: 1754–1760.
36. Li H, Handsaker B, Wysoker A, Fennell T, Ruan J, et al. (2009) The Sequence Alignment/Map format and SAMtools. Bioinformatics 25: 2078–2079.
37. Pique-Regi R, Monso-Varona J, Ortega A, Seeger RC, Triche TJ, et al. (2008) Sparse representation and Bayesian detection of genome copy number alterations from microarray data. Bioinformatics 24: 309–318.
38. Lander ES, Linton LM, Birren B, Nusbaum C, Zody MC, et al. (2001) Initial sequencing and analysis of the human genome. Nature 409: 860–921.

39. Hudson TJ, Anderson W, Artez A, Barker AD, Bell C, et al. (2010) International network of cancer genome projects. Nature 464: 993–998.
40. Puente XS, Pinyol M, Quesada V, Conde L, Ordonez GR, et al. (2011) Whole-genome sequencing identifies recurrent mutations in chronic lymphocytic leukaemia. Nature 475: 101–105.
41. Marco-Sola S, Sammeth M, Guigo R, Ribeca P (2012) The GEM mapper: fast, accurate and versatile alignment by filtration. Nat Methods.
42. Stewart C, Kural D, Stromberg MP, Walker JA, Konkel MK, et al. (2011) A comprehensive map of mobile element insertion polymorphisms in humans. PLoS Genet 7: e1002236.
43. Korbel JO, Abyzov A, Mu XJ, Carriero N, Cayting P, et al. (2009) PEMer: a computational framework with simulation-based error models for inferring genomic structural variants from massive paired-end sequencing data. Genome Biol 10: R23.
44. Jacobs KB, Yeager M, Zhou W, Wacholder S, Wang Z, et al. (2012) Detectable clonal mosaicism and its relationship to aging and cancer. Nat Genet 44: 651–658.
45. Nik-Zainal S, Van Loo P, Wedge DC, Alexandrov LB, Greenman CD, et al. (2012) The life history of 21 breast cancers. Cell 149: 994–1007.

Comparison of Lentiviral and Sleeping Beauty Mediated αβ T Cell Receptor Gene Transfer

Anne-Christine Field[1], Conrad Vink[1], Richard Gabriel[2], Roua Al-Subki[1], Manfred Schmidt[2], Nicholas Goulden[1], Hans Stauss[3], Adrian Thrasher[1], Emma Morris[3], Waseem Qasim[1]*

1 Molecular immunology Unit, Institute of Child Health, University College London, London, United Kingdom, **2** Department of Translational Oncology, National Center for Tumor Diseases and German Cancer Research Center, Heidelberg, Germany, **3** Institute of Immunity & Transplantation, Royal Free Campus University College London, London, United Kingdom

Abstract

Transfer of tumour antigen-specific receptors to T cells requires efficient delivery and integration of transgenes, and currently most clinical studies are using gamma retroviral or lentiviral systems. Whilst important proof-of-principle data has been generated for both chimeric antigen receptors and αβ T cell receptors, the current platforms are costly, time-consuming and relatively inflexible. Alternative, more cost-effective, Sleeping Beauty transposon-based plasmid systems could offer a pathway to accelerated clinical testing of a more diverse repertoire of recombinant high affinity T cell receptors. Nucleofection of hyperactive SB100X transposase-mediated stable transposition of an optimised murine-human chimeric T cell receptor specific for Wilm's tumour antigen from a Sleeping Beauty transposon plasmid. Whilst transfer efficiency was lower than that mediated by lentiviral transduction, cells could be readily enriched and expanded, and mediated effective target cells lysis *in vitro* and *in vivo*. Integration sites of transposed TCR genes in primary T cells were almost randomly distributed, contrasting the predilection of lentiviral vectors for transcriptionally active sites. The results support exploitation of the Sleeping Beauty plasmid based system as a flexible and adaptable platform for accelerated, early-phase assessment of T cell receptor gene therapies.

Editor: Sebastian D. Fugmann, Chang Gung University, Taiwan

Funding: This work was supported by the Children with Cancer, Leukaemia and Lymphoma Research Fund. WQ receives funding support from GOSHCC. AT is a Wellcome Trust Senior Fellow. GOSH/UCL Institute of Child Health receives funding from the Department of Health's NIHR Biomedical Research Centre's funding scheme. The funders had no role in study design, data collection and analysis, decision to publish, or preparation of the manuscript.

Competing Interests: The authors have declared that no competing interests exist.

* E-mail: W.Qasim@ucl.ac.uk

Introduction

Gamma retroviral-mediated transfer of recombinant receptors to redirect T cell immunity against tumour and viral antigens is being extensively investigated in early phase clinical trials. Delivery of chimeric antigen receptors (CAR) against a variety of target antigens, including CD19 in haematological malignancies, GD2 in neuroblastoma and CEA in colonic carcinoma has been reported and reviewed recently [1]. Similar vectors have also been used to transfer conventional recombinant αβ T cell receptors against melanoma antigens [2]. Whilst there were concerns that these vectors may mediate genotoxic side effects in trials involving haematopoietic stem cell modification [3,4], no vector related adverse events have been reported in T cell studies [5]. Recently, HIV-1 derived lentiviral vector based T cell redirection with a CAR against CD19 was reported [6], and these vectors are thought to improve efficiency of gene transfer and yield more functional T cells after *ex-vivo* transduction. Previously, both retroviral [7] and lentiviral [8] modification of T cells expressing recombinant T cell receptors against the Wilm's tumour (WT1) antigen has been described and Phase 1 testing of an enhanced HLA-A2/WT1 peptide specific TCR has been initiated. However, high-costs associated with the manufacture of clinical grade viral vectors have been limiting, and to date only a handful of viral vectors, with limited target specificity have been produced. This limits applicability to individuals with specific HLA types and substitution of existing receptors with more efficacious derivatives is difficult. Alternatively, more readily adaptable gene delivery platforms, capable of mediating stable transgene integration without toxicity, could accelerate clinical testing and extend applicability. Plasmid-based gene transfer systems based on transposable elements are under investigation of a variety of gene therapy applications. The Sleeping Beauty (SB) platform [9] has been adapted to transfer antigen-specific receptors to T cells [10–14], but direct comparisons with lentiviral delivery are lacking. Here we have assessed the ability of an enhanced Sleeping Beauty transposition system, using a hyperactive SB transposase named SB100X [15] to deliver a codon optimised αβ TCR against WT1 antigen. This HLA-A2 restricted receptor has been modified to incorporate murine constant domains and additional disulphide bonds to mitigate against aberrant cross-pairing with endogenous TCR chains. Transduction efficiency, integration sites and redirection of human primary T cells against WT1 antigen was assessed and compared to lentiviral transduction procedures.

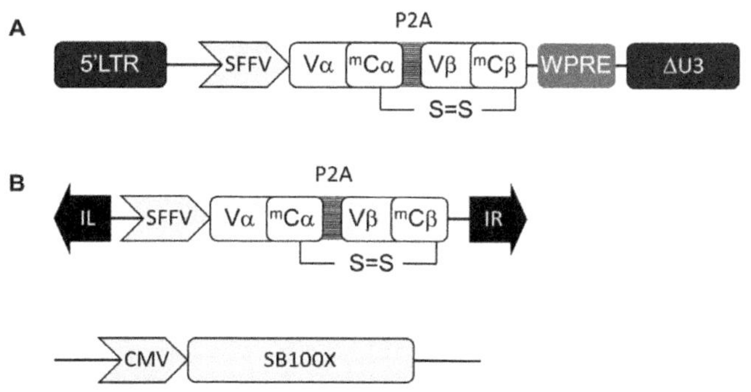

Figure 1. WT1-TCR vector constructs. A. Schematic representation of the pSIN second-generation lentiviral (Lv) vector encoding the codon-optimized, murinized hybrid HLA-A0201–restricted pWT126-specific additional cysteines modified TCR α- and β-chain genes separated by a self-cleaving porcine teschovirus 2A sequence (P2A). LTR indicates long terminal repeat; m, murine SFFV, spleen-forming focus virus; and WPRE, woodchuck hepatitis virus posttranscriptional regulatory element. B. Schematic representation of the plasmid Sleeping Beauty (SB) vector consisting of a transgene expression cassette, encoding the codon-optimized, murinized hybrid HLA-A0201–restricted pWT126-specific additional cysteines modified TCR α- and β-chain genes separated by a self-cleaving porcine teschovirus 2A sequence (P2A), flanked by inverted repeats; and the hyperactive SB transposase SB100X. CMV indicates cytomegalovirus promoter; IL, left inverted repeat; IR, right inverted repeat; m, murine; SFFV, spleen-forming focus virus.

Materials and Methods

Ethics Statement

Informed written consent was obtained from all healthy blood sample donors with institutional ethics committee approval (University College London, UK). All animals were handled in strict accordance with good animal practice as defined by UK Home Office Animal Welfare Legislation, and all animal work was approved by the Institutional Research Ethics Committee (University College London, UK) and performed under Home Office approval (project license number 70/6482).

Cell Lines and Primary T Cells

Human leukaemic immortalized Jurkat cell lines (TCR negative) [16] and TAP (transporter associated with Ag processing) deficient HLA-A0201$^+$ T2 cell line [17] were maintained in RPMI-1640 (Invitrogen, Paisley, UK) supplemented with 10% heat-inactivated foetal calf serum (Sigma-Aldrich, Gillingham, UK), and 1% penicillin/streptomycin (Invitrogen). Human embryonic kidney 293T cells (ATCC CRL-11268^{TM}) were cultured in Dulbecco's modified Eagle's medium (Invitrogen) supplemented with 10% heat-inactivated foetal calf serum and 1% penicillin/streptomycin. Human peripheral blood mononuclear cells (PBMCs) were isolated from healthy HLA-A02 volunteers (with institutional ethics committee approval) by density gradient separation using Ficoll-Paque Plus (GE Healthcare, Little Chalfont, UK) and were cultured in X-VIVO 10 (Lonza, Cologne, Germany) supplemented with human AB serum (10%; Lonza) and the human recombinant common γ-chain cytokine IL2 (Chiron, France).

Antibodies, Cytokines, Peptides and Flow Cytometry

Antibodies for flow cytometry were anti-human CD3 APC, CD8 FITC, anti-mouse TCRβ chain (mCβ APC) (BD Biosciences, Cowley, UK); anti-human Vβ2.1 PE and pWT126 tetramer PE (Beckman Coulter, High Wycombe, UK). The antibodies used for cytokine capture assay were IFNγ catch reagent and anti-human IFNγ PE (Miltenyi Biotech, Germany). The concentration of the human recombinant IL2 was 100 U/ml in all culture conditions. The peptides used in this study were the HLA-A0201-binding peptides pWT126 (RMFPNAPYL) and pWT235 (CMTWNQMNL), which were synthesized by ProImmune as described previously [18]. Flow cytometric analysis was performed using a LSR II flow cytometer (BDBiosciences) and the data were analysed using FlowJo version 7 software (TreeStar, Ashland, OR, USA).

Lentiviral and Sleeping Beauty Constructs

A self-inactivating lentiviral vector, encoding the SFFV (spleen focus forming virus) LTR promoter and the HIV-1 central polypurine tract sequence and mutated Woodchuck post transcriptional regulatory element (WPRE) [19] was generated to include a codon-optimized hybrid HLA-A0201-restricted WT1-TCR genes [8]. The receptors comprise murine constant region with additional disulphide bonds and human variable (Vα1.5 and Vβ2.1) region sequences. The α and β chains are linked by a P2A self-cleaving peptide sequence and the TCR recognises WT126 peptide, RMFPNAPYL in the context of HLA-A0201 [8]. The SB transposon plasmid pT2/HB and hyperactive transposase plasmid pCMV-SB100X have been previously described [15,20,21] and were modified to encode an identical SFFV promoter WT1-TCR cassette.

Lentiviral Vector Production

Lentiviral vector stocks were produced by transient co-transfection of subconfluent HEK293T cells using polyethylenimine (Sigma-Aldrich) with helper plasmids encoding vesicular stomatitis virus (VSV) envelope plasmid (pMDG), and gag/pol packaging plasmid (pCMVΔ8.74) as previously described [22]. Viral supernatant was harvested 48 and 72 hours later, 0.22 μM filtered, and concentrated by ultracentrifugation before cryopreservation. Vector stocks were titred on TCR negative Jurkat T cells by flow cytometry.

Lentiviral Vector Transduction and Sleeping Beauty Nucleofection

PBMCs were activated using magnetic beads conjugated to antibodies to CD3 and CD28 (Invitrogen, Stockholm, Sweden), at a ratio of 1:1 for 48 hours in X-VIVO 10 medium supplemented with human AB serum and 100 U/ml IL2 [22]. The cells were then exposed to lentiviral infection at a multiplicity of infection of 20 or nucleofected with DNA supercoiled plasmids (5 μg each) encoding the SB transposon and the SB transposase, using an Amaxa Nucleofector II device and Human T Cell Nucleofector Kit and protocol for Stimulated Human T cells (Lonza). After 48–72 hours after expression of TCR transgenes was analysed by flow cytometry.

Expansion and Purification of TCR-modified T Cells

TCR modified T cells were either assessed immediately or expanded weekly by antigen-specific stimulation in 24-well plates as follows: 5×10^5 TCR-transduced cells were co-cultured with 2×10^5 T2 cells pulsed with pWT126 peptide as stimulators and 2×10^6 irradiated autologous feeder PBMCs, in the presence of 100 U/ml IL2. In some experiments, WT1-TCR T cells were expanded and then enriched by sorted by magnetic bead selection using MiniMacs columns following two stage incubation with an

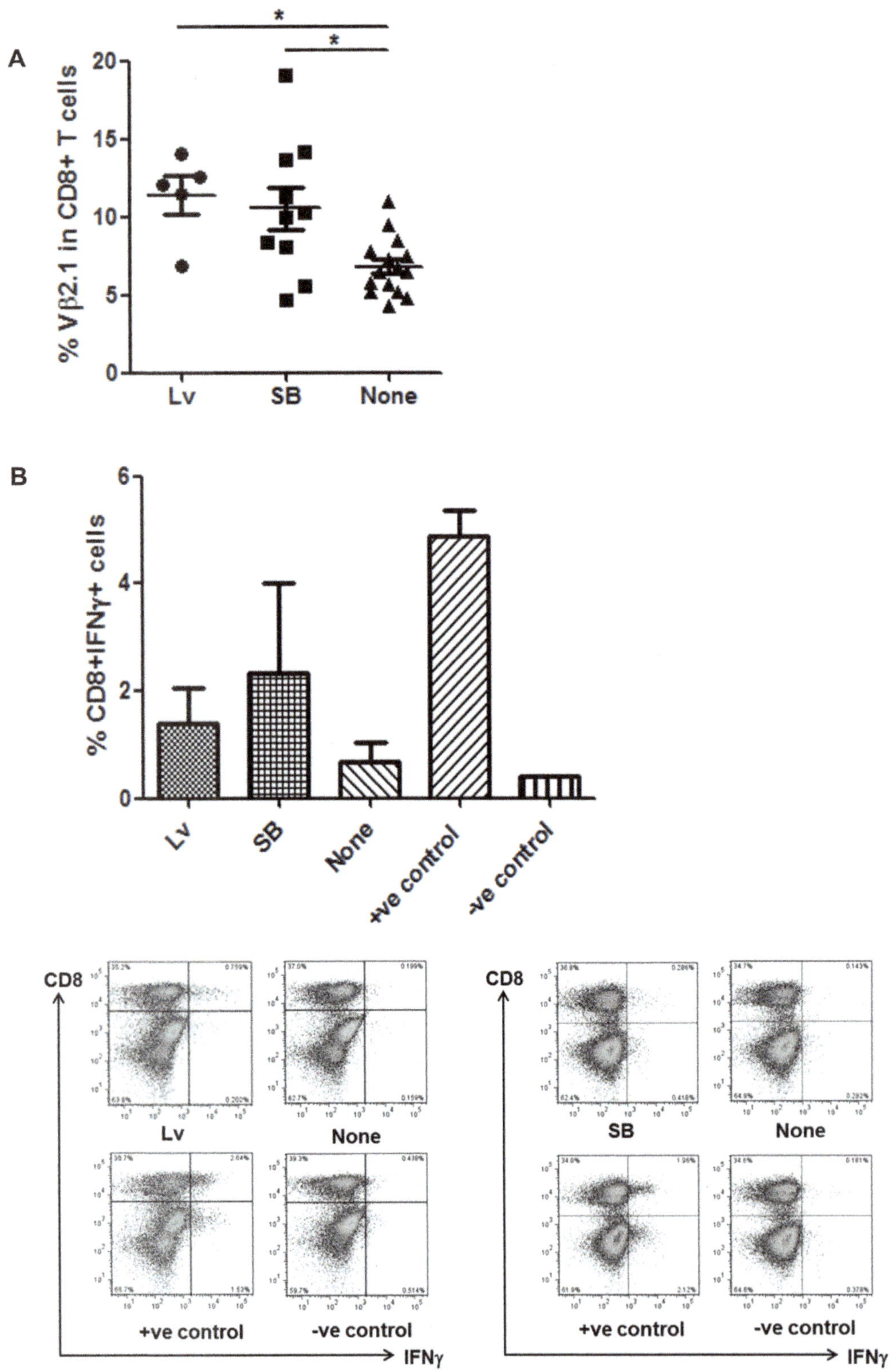

Figure 2. WT1-TCR expression and function in human primary T cells. A. WT1-TCR cell surface expression. Percentages of Vβ2.1 positive cells in CD8+ T cells for each HLA-A2+ donor are shown after WT1-TCR lentiviral (Lv) or Sleeping Beauty (SB) vector gene transfers. CD3/CD28 microbeads-activated PBMCs were transduced at a multiplicity of infection of 20 or nucleofected with 5 μgs transposon and transposase plasmids. *$p<0.05$. B. Antigen-specific effector function. Mean percentages of IFNγ positive cells in CD8+ T cells from HLA-A2+ donors assayed by MACS Cytokine Secretion Assay of human primary CD8+ T cells, after WT1-TCR lentiviral (Lv) or Sleeping Beauty (SB) vector gene transfer, restimulated with either the specific WT1 peptide, anti-CD3 Ab (positive control) or irrelevant peptide (negative control), and autologous PBMCs (top). Representative dot plots of IFNγ secretion are also shown after WT1-TCR Lv (bottom left) or SB (bottom right) gene transfer.

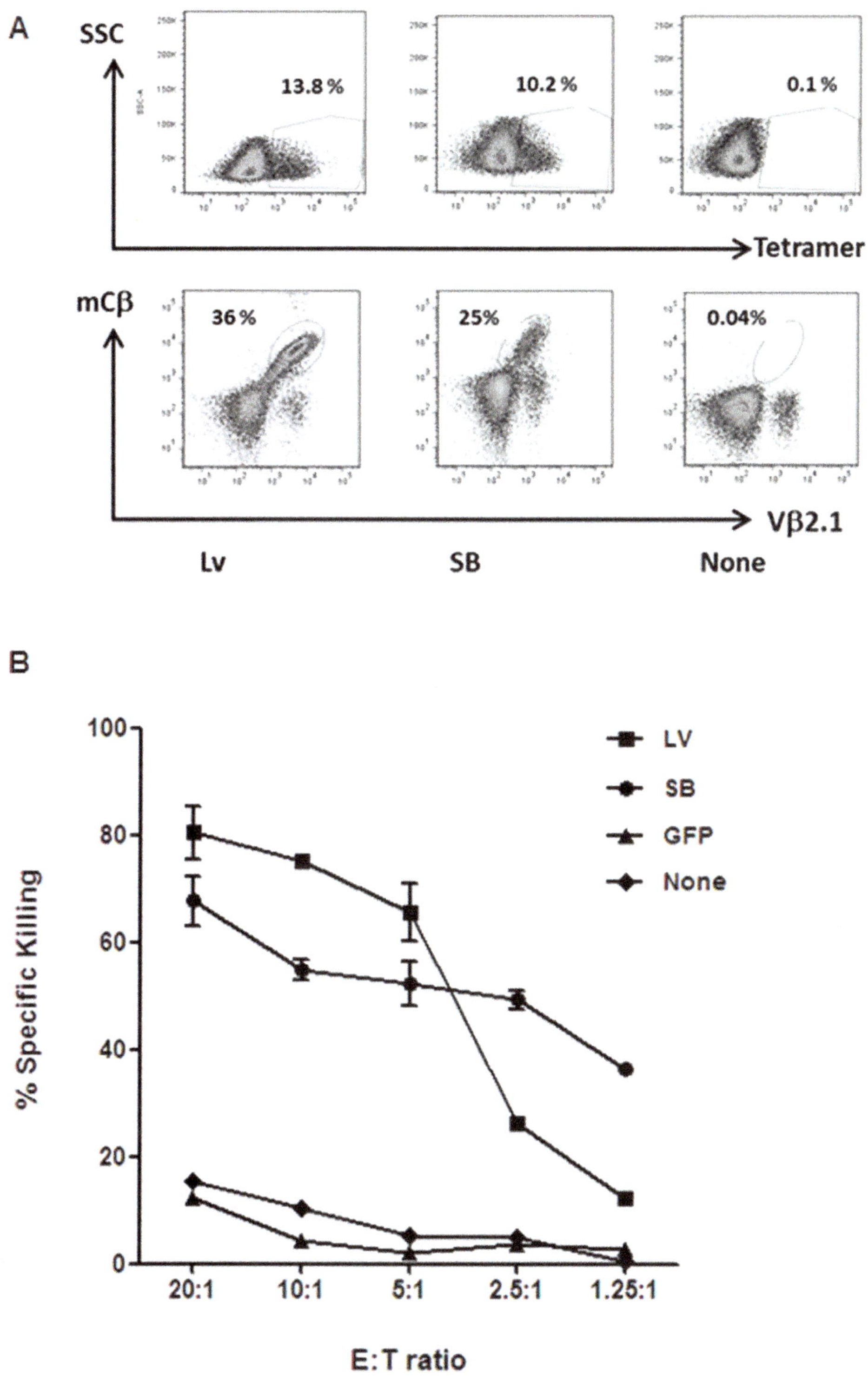

Figure 3. Antigen-specific responses of human primary T cells. A. Murine Cβ/human Vβ2.1 expression and WT1 tetramer binding are shown in representative human primary lymphocyte dot plots, after SFFV-driven WT1-TCR lentiviral vector (Lv) or Sleeping Beauty (SB) gene transfer and in unmodified controls. CD3/CD28 microbeads-activated PBMCs were transduced at a multiplicity of infection of 20 or nucleofected with 5 µgs transposon and transposase plasmids and expanded with T2 cells coated with specific peptide pWT126 and irradiated autologous PBMCs as feeder cells. B. *In vitro* cytotoxicity of donor T cells against specific peptide-loaded T2 target cells, Lv versus SB. Percentages of cytolytic activity of T cells, after WT1-TCR lentiviral (Lv) or Sleeping Beauty (SB) vector gene transfers, was tested in a ^{51}Cr release CTL assay. CD3/CD28 microbeads-activated PBMCs were transduced at a multiplicity of infection of 20 or nucleofected with 5 µgs transposon and transposase plasmids and expanded with the specific WT1 peptide and autologous irradiated PBMCs. Both WT1-TCR-modified donor T cells recognized and killed their specific targets (2 way ANOVA, $p<0.001$ at all E:T ratios compared to GFP-modified or unmodified cells).

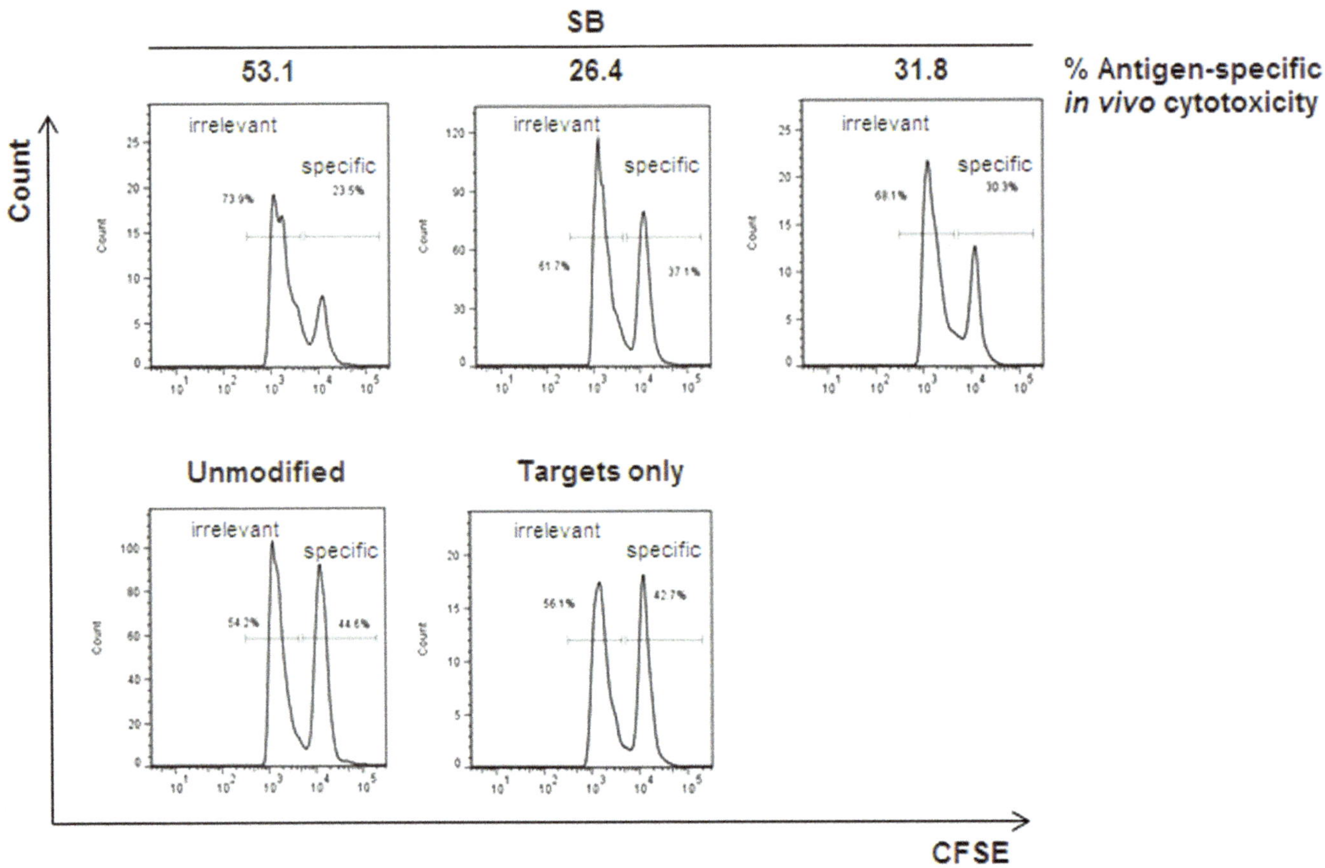

Figure 4. ***In vivo* cytotoxicity of murine primary T cells.** *In vivo* cytotoxicity of modified mouse splenocytes against CFSE-labelled peptide-loaded mouse target B cells. A2K^b Tg mice, 2 days after intravenous injection of syngeneic effector cells modified with Sleeping Beauty (SB) WT1-TCR vector system, were again intravenously injected with a 1:1 mix of relevant: irrelevant peptide-loaded A2K^b Tg target B cells, differentially labelled with CFSE (specific WT126 peptide, 1.5 μM CFSE, and irrelevant WT235 peptide for WT1-TCR, 0.15 μM CFSE). Eighteen hours later, splenocytes of injected animals were harvested and analysed by FACS to identify CFSE-labelled cells. Control A2K^b Tg mice were injected with only CFSE-labelled peptide-loaded target B cells or with unmodified splenocytes as effectors. Antigen-specific cytotoxicity was calculated as [1- (number of relevant peptide-loaded targets in experimental mice/number of irrelevant peptide-loaded targets in experimental mice)/(number of relevant peptide-loaded targets in control mice/number of irrelevant peptide-loaded targets in control mice)] ×100.

anti-murine TCRβ chain APC antibody, followed by anti-APC microbeads (Miltenyi Biotech).

Cytotoxicity Assays

In vitro ^{51}Cr release cytotoxicty assays were performed with 2×10^6 T2 cells pre-incubated at 37°C for 1 hour in 200 μl assay medium (RPMI 1640 containing 5% heat-inactivated FCS) with 100 μM specific peptide as target cells. Peptide-coated T2 cells were then labelled with ^{51}Cr for 1 hour, washed, and added to serial 2-fold dilutions of expanded WT1-TCR modified PBMCs as effector cells in round-bottom 96-well plates to obtain a total volume of 200 μl/well. Assay plates were incubated at 37°C, 5% CO2, and after 4 hours, 50 μl supernatant was harvested and radioactivity counted. The specific killing was calculated by the equation: (experimental release-spontaneous release)/(maximum release-spontaneous release) ×100%.

In vivo cytotoxicity assays were performed with 0.7×10^6 WT1-TCR modified splenocytes from C57Bl/6 mice as effector cells, which were activated with Concanavalin A 2 μg/ml (Sigma-Aldrich) and IL2 100 U/ml for 3 days in RPMI 1640 containing 10% heat-inactivated FCS, 0.1 mM 2-mercaptoethanol (Invitrogen) and 0.1 mM sodium pyruvate (Lonza) and nucleofected with the SB transposon and the SB transposase DNA plasmids, using an Amaxa Nucleofector II device and Mouse T Cell Nucleofector Kit and protocol (Lonza). The effector cells were injected intravenously into A2Kb mice without conditioning and left 2 days to equilibrate. Splenocytes from female A2Kb Tg mice were prepared as target cells by selecting B cells (B Cell Isolation Kit; Miltenyi Biotec), peptide-loaded with 100 μM of either specific peptide (WT126p) or an irrelevant HLA-A0201-presented peptide before labelling with 1.5 μM or 0.15 μM CFSE, respectively, for 5 minutes at 37°C. Labelled cells were mixed at a 1:1 ratio, relevant: irrelevant targets and a total of 10×10^6 mixed cells were injected into recipient A2K^b mice previously injected with effector cells. Eighteen hours later, splenocytes of injected animals were analyzed by flow cytometry to identify CFSE-labelled cells. Control untreated A2K^b Tg mice were injected with labelled target cells only and in vivo cytotoxicity was calculated as previously described. Percentage antigen-specific cytotoxicity was determined using the following formula: [1- (mean number of relevant peptide-loaded targets in experimental mice/mean number of irrelevant peptide-loaded targets in experimental mice)/(mean number of relevant peptide-loaded targets in control mice/mean number of irrelevant peptide-loaded targets in control mice)] ×100.

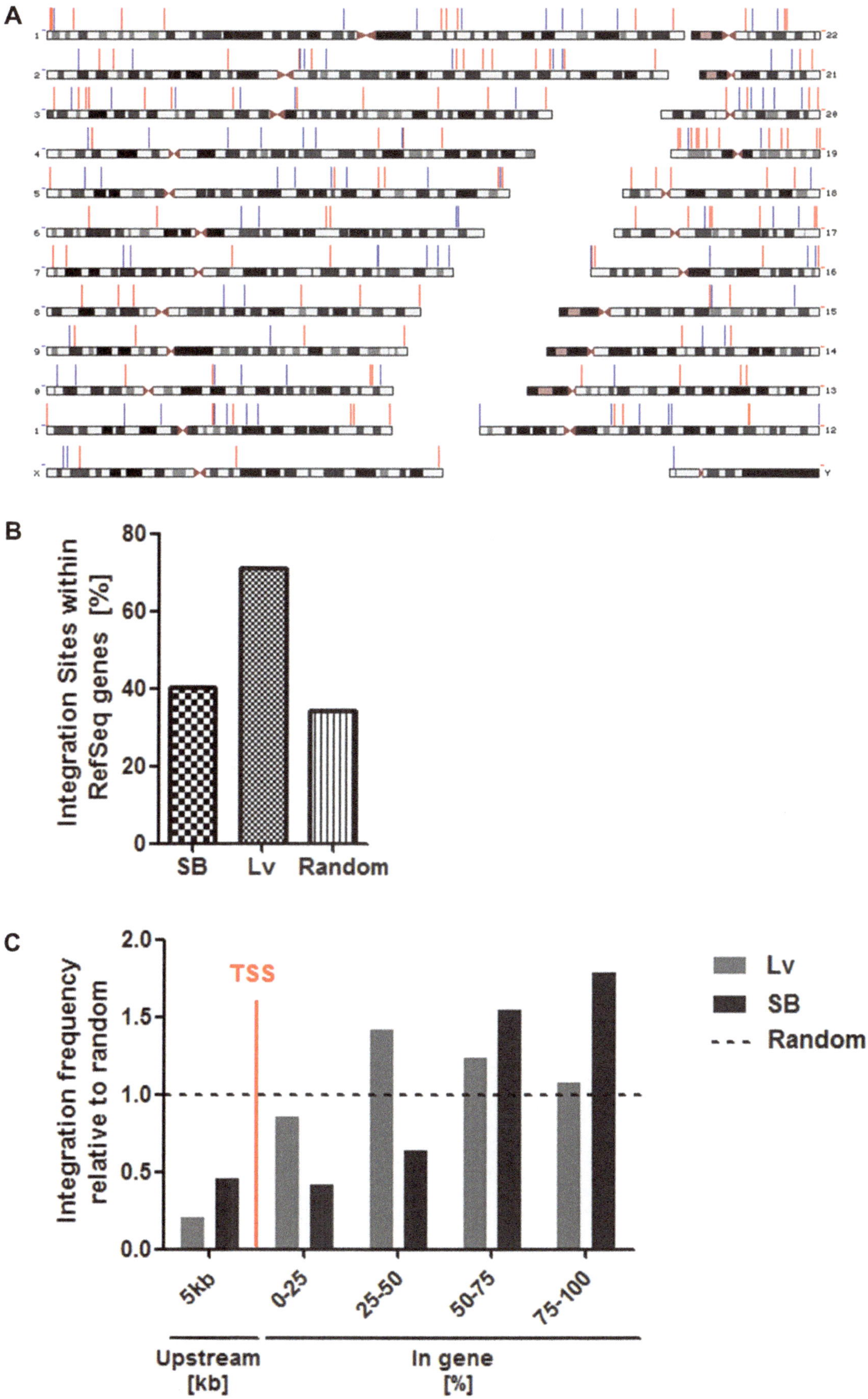
A
B
80
60
40
20
0
Integration Sites within RefSeq genes [%]
SB
Lv
Random
C
2.0
1.5
1.0
0.5
0.0
TSS
Integration frequency relative to random
5kb
0-25
25-50
50-75
75-100
Upstream [kb]
In gene [%]
Lv
SB
Random

Figure 5. Integration profiles in human primary T cells by LM-PCR. CD3/CD28 microbeads-activated PBMCs were transduced at a multiplicity of infection of 20 (Lv) or nucleofected with both 5 µgs transposon and transposase plasmids (SB). Both vectors encoded the identical WT1-TCR transgene. Genomic DNA was extracted and lentivirus-chromosome or transposon-chromosome junctions were recovered by ligation-mediated PCR and sequenced. A. Genome-wide mapping of vector integrations at the chromosome level. Sequences were mapped to the University of California at Santa Cruz (UCSC) human genome by BLAT search and integration sites were depicted relative to chromosomes using the UCSC Genome Graphs tool. Lv in Blue, SB in Red. B. Frequency of integration sites of vectors within RefSeq genes. 1,000 random integration sites were generated by bioinformatics, as already described [21]. C. Proximity of integration sites to transcription start sites (TSS). RefSeq genes containing integration sites were divided by length into 4 equally sized regions and 1 upstream region (0–5 kb), and the proportion of integration sites within each region was counted. To allow statistical comparison of integration preferences with average genomic content, 1,000 random chromosomal sites were generated by multiplying the total length of the genome by a random number between 0 and 1 and converting this value to a chromosomal coordinate. Vector integration frequencies are expressed relative to the proportion of random sites within each region.

C57Bl/6 mice were purchased from Charles River (Margate, UK). HLA-A0201K^b transgenic (A2K^b Tg) mice on a C57Bl/6 background were a kind gift from Theobald M. (University Medical Center, Utrecht, The Netherlands) and were bred and maintained in the Comparative Biology Unit of University College London (Royal Free Hospital, London, UK) under pathogen-free conditions.

Ligation-mediated PCR (LM-PCR)

To recover integration sites, DNA from WT1-TCR modified PBMCs by lentiviral transduction or SB nucleofection was extracted by Proteinase K digestion. 1 µg of DNA was digested with NlaIII to produce vector-chromosome junction fragments, ligated to excess linker and PCR was performed with GoTaq polymerase (Promega, Madison, WI) using the linker first-round and second-round primers as previously described [20,21]. PCR products were shotgun-cloned into the Topo-TA vector (Invitrogen) and sequenced (UCL, London, UK) using the second round primers. LM-PCR products were 454 pyrosequenced as previously described [21].Raw sequence reads obtained after sequencing were trimmed and aligned to the human genome (NCBI, hg19) using HISAP [23]. Integration sites were considered to be valid if vector-genome junction sequence was present and the flanking genomic region had a unique sequence match of at least 95% after alignment to the human genome (University of California at Santa Cruz, RefSeq genes and RepeatMasker). Chromosome graphs were generated using the University of California at Santa Cruz Genome Graphs tool.

Statistics

Where shown, error bars represent the standard error of the mean of the replicates indicated, and differences were found to be significant by application of non-parametric Anova tests.

Results

1. Sleeping Beauty Mediated αβ TCR Transfer to Primary T Cells Compared to Lentiviral Delivery

Sleeping Beauty transposon-mediated TCR delivery was compared against lentiviral vector-mediated gene transfer in primary human lymphocytes using identical promoter/transgene expression cassettes. These comprised an optimised human-murine chimeric WT1 αβTCR encoding Vα1.5 and Vβ2.1 chains [8] in combination with the spleen-forming focus virus (SFFV) long terminal repeat promoter (Figure 1A, B). Sleeping Beauty transposons were delivered in combination with a functionally enhanced, hyperactive, Sleeping Beauty transposase variant, SB100× (Figure 1B). Preliminary experiments in TCR-deficient Jurkat T cells allowed optimisation of transposon/SB100X plasmid ratios, and confirmed robust and durable expression following both lentiviral and SB delivery.

Freshly isolated human peripheral blood lymphocytes (PBL) from healthy HLA-A2 donors were stimulated for 48 hours with anti-CD3/CD28 microbeads and human IL-2 then exposed to a single round of lentivirus infection or nucleofected with transposon/transposase DNA plasmids. Flow cytometry following staining with anti-Vβ2.1 antibody identified primary T cells expressing cell surface WT1-TCR and revealed comparable gene transfer. These investigations using multiple healthy volunteer donor lymphocytes revealed inter-individual variability, with mean levels of SB mediated gene transfer in CD8$^+$ T cells of 10.5% (n = 10) compared to 11.4% (n = 5) following lentiviral transduction (Figure 2A). These engineered CD8$^+$ T cells were pulsed with HLA-A2 restricted WT1 peptide and antigen-specific responses were confirmed on the basis of IFNγ release (Figure 2B). Assuming that WT1 specific responses were only mediated by WT1-TCR expressing cells, events captured by flow cytometry within the CD8+IFNγ+ gate (1.4% for LV, 2.3% for SB) accounting for 28% and 57% of the modified populations respectively. Furthermore, WT1-TCR-modified T cell populations could be expanded and were then subjected to challenge by peptide loaded target cells. Co-staining for murine TCRβ chain epitopes distinguished recombinant receptor expression from endogenous Vβ2.1 expression and HLA-A2/pWT126 multimer staining confirmed WT1-specific TCR expression (Figure 3A). Human TAP-deficient T2 target cells loaded with pWT126 peptide were specifically lysed by both lentiviral and SB modified T cells, in contrast to cells presenting non-relevant peptides or cultures challenging control (eGFP) modified T cells (Figure 3B).

2. *In vivo* Functional Assessment of SB Modified T Cells

Assessments of TCR redirected human T cells in immunodeficient mice injected with human WT1 expressing tumour cells have previously demonstrated the potency of retroviral modified T cells in vivo, including T cell homing and tumour eradication. However, these models are technically challenging with variable reproducibility. Alternatively A2K^b transgenic mice have been used to assess the function of T cells derived from murine stem cells modified using lentiviral vectors to express HLA-A2/WT1 specific TCRs [24]. Mature murine T cells resist lentiviral transduction through the expression of innate restriction factors [25] but we successfully adopted the model for confirmation that SB modified murine T cells expressing human TCR specific for HLA-A2/WT1 are functional in vivo. Activated splenocytes from C57Bl/6 mice were nucleofected with the SB transposon system expressing WT1-TCR and were injected into unconditioned recipient A2K^b transgenic mice. Two days later, target B cells from female A2K^b Tg mice were peptide-loaded with relevant (pWT126) or irrelevant HLA-A2-restricted peptides and labelled with CFSE. Labelled target with relevant or non-relevant targets were mixed at a 1:1 ratio and were injected in the same recipient mice. Eighteen hours later, splenocytes of injected animals were analysed by flow cytometry to identify CFSE-labelled target cells.

Control (untreated) A2K^b Tg mice were injected with only labelled target cells (Figure 4). In vivo, target specific cytotoxicity ranged from 26–53%, confirming antigen-specific function of SB modified T cells.

3. Sleeping Beauty Transposition Sites in Primary Human T Cells

Lentiviral integration sites in T cells have been well characterised in subjects with HIV-1 infection and in gene engineered T cells and there is a preference for integration in transcriptionally active regions, though in contrast to gamma-retroviral vectors, no bias towards transcription start sites (TSS). Sleeping Beauty transposons target TA dinucleotides and are thought not to target gene coding sequences. Integration sites ($n \geq 150$) of lentiviral vectors and SB transposons expressing WT1-TCR transgenes were detected by high throughput sequencing. Overall, 70.8% of lentiviral integrations were found within genes compared to 40.2% of SB100X/transposon-mediated transgene insertions, which was more comparable to anticipated frequencies of 34.2% if random integration had occurred (Figure 5B). When integrations within genes were mapped relative to their position within the gene or upstream region (Figure 5C), no preference toward TSS of RefSeq genes was observed along the length of the gene and no significant variation in integration pattern was observed along the length of the gene. The bias towards RefSeq genes can be expressed as a ratio of frequencies in genes above random, and this yielded values of 2.05 for lentiviral vectors and 1.17 for the SB platform. Thus, in primary human lymphocytes plasmid-based SB delivery of αβTCR gene expression cassettes appears less likely to perturb endogenous gene function.

Discussion

Wilms' Tumour antigen 1 (WT1) is a well characterised tumour antigen overexpressed by haematological and certain non-haematological malignancies. A phase 1 study of gamma retrovirally modified T cells to treat relapsed leukaemia has recently been initiated to investigate the safety and efficacy of a recombinant αβ TCR specific for HLA-A2/WT1 peptide antigen (ClinicalTrials.-gov; NCT01621724). As with other T cell modification therapies requiring stable transgene integration, a major hurdle has been the production and validation of clinical grade vector stocks. In the case of gamma-retroviral vectors, efficient stable packaging cell lines have facilitated the production of large volumes of vector stocks, and yet the very high levels of investment required have meant only a handful of candidate antigen-specific receptors have been adopted for translational studies. Recent advances support a switch to lentiviral vector platforms [22,26], but GMP production of high titre lentiviral stocks is problematic and is hampered by the lack of stable producer lines. Rather, lentiviral stocks are produced by transient transfection, using high quality plasmids and then require extensive and costly downstream validation and release testing. The availability of an alternative plasmid based system for direct modification of T cells would allow accelerated assessment of different TCRs and revised CAR configurations. In this regard transposon-based systems are attractive and proposals for initial clinical testing of a nucleofection-based protocol for transfer of a CD19 specific CAR are well developed [14,27,28]. The enhanced SB100X mutant has been evaluated for transfer of a similar single chain receptor, but information on delivery of a αβ TCR configuration has been lacking. We have demonstrated stable gene transfer by SB100X-mediated transposition, and have determined that integration sites in primary human T cells are similar to those reported for human cell lines, and are not skewed towards gene rich, transcriptionally active loci. We speculate that this may reduce the risk of genotoxicity, albeit noting that there have been no vectors-mediated adverse events in any gamma retroviral or lentiviral studies targeting human T cells even when integrants close to protoncogenes have been detected. We speculate that this relates to the differentiated nature of T cells, in contrast to haematopoietic stem cells where risks of transformation may be greater. There have been concerns that background integration of transposase encoding plasmid may mediate remobilisation of transposed sequences, and this could also be a risk factor for mutagenesis. Strategies to mitigate against such eventualities include the use of mRNA rather than plasmid for transposase delivery [11], or the incorporation of a linked suicide gene for elimination of cells in case of adverse effects [22]. Whilst transposition may be less efficient than lentiviral-mediated gene transfer, we have demonstrated the feasibility of enriching cells by magnetic bead selection and expanding antigen-specific populations using stimulator cells and feeder layers if high numbers of cells are required [14,29]. Alternatively, it is entirely possible that the early reinfusion of fewer SB modified T cells, cultured for a shorter period and retaining functional properties may be sufficient to mediate beneficial in vivo effects.

In summary, SB-mediated transfer of antigen-specific T cell receptor redirects effector function and offers a flexible pathway for rapid evaluation of different receptor configurations. Further optimisation and adaptation to GMP conditions will allow therapeutic applications to be explored in early phase clinical studies.

Acknowledgments

We are grateful to Zultan Ivics for providing the SB100X plasmid.

Author Contributions

Conceived and designed the experiments: ACF NG HS AT EM WQ. Performed the experiments: ACF CV RA RG MS. Analyzed the data: ACF CV RA RG MS. Contributed reagents/materials/analysis tools: RG MS HS EM WQ. Wrote the paper: ACF WQ.

References

1. Jena B, Dotti G, Cooper LJ (2010) Redirecting T-cell specificity by introducing a tumor-specific chimeric antigen receptor. Blood 116: 1035–1044. blood-2010-01-043737 [pii];10.1182/blood-2010-01-043737 [doi].
2. Morgan RA, Dudley ME, Wunderlich JR, Hughes MS, Yang JC, et al. (2006) Cancer regression in patients after transfer of genetically engineered lymphocytes. Science 314: 126–129.
3. Hacein-Bey-Abina S, Von Kalle C, Schmidt M, Le Deist F, Wulffraat N, et al. (2003) A serious adverse event after successful gene therapy for X-linked severe combined immunodeficiency. N Engl J Med 348: 255–256.
4. Ott MG, Schmidt M, Schwarzwaelder K, Stein S, Siler U, et al. (2006) Correction of X-linked chronic granulomatous disease by gene therapy, augmented by insertional activation of MDS1-EVI1, PRDM16 or SETBP1. Nat Med 12: 401–409.
5. Scholler J, Brady TL, Binder-Scholl G, Hwang WT, Plesa G, et al. (2012) Decade-long safety and function of retroviral-modified chimeric antigen receptor T cells. Sci Transl Med 4: 132ra53. 4/132/132ra53 [pii];10.1126/scitranslmed.3003761 [doi].
6. Porter DL, Levine BL, Kalos M, Bagg A, June CH (2011) Chimeric Antigen Receptor-Modified T Cells in Chronic Lymphoid Leukemia. N Engl J Med. 10.1056/NEJMoa1103849 [doi].
7. Xue SA, Gao L, Thomas S, Hart DP, Xue JZ, et al. (2010) Development of a Wilms' tumor antigen-specific T-cell receptor for clinical trials: engineered patient's T cells can eliminate autologous leukemia blasts in NOD/SCID mice. Haematologica 95: 126–134. haematol.2009.006486 [pii];10.3324/haematol.2009.006486 [doi].

8. Perro M, Tsang J, Xue SA, Escors D, Cesco-Gaspere M, et al. (2010) Generation of multi-functional antigen-specific human T-cells by lentiviral TCR gene transfer. Gene Ther 17: 721–732. gt20104 [pii];10.1038/gt.2010.4 [doi].
9. Ivics Z, Hackett PB, Plasterk RH, Izsvak Z (1997) Molecular Reconstruction of Sleeping Beauty, a Tc1-like Transposon from Fish, and Its Transposition in Human Cells. Cell 91: 501–510.
10. Huang X, Guo H, Tammana S, Jung YC, Mellgren E, et al. (2010) Gene transfer efficiency and genome-wide integration profiling of Sleeping Beauty, Tol2, and piggyBac transposons in human primary T cells. Mol Ther 18: 1803–1813. mt2010141 [pii];10.1038/mt.2010.141 [doi].
11. Huang X, Haley K, Wong M, Guo H, Lu C, et al. (2010) Unexpectedly high copy number of random integration but low frequency of persistent expression of the Sleeping Beauty transposase after trans delivery in primary human T cells. Hum Gene Ther 21: 1577–1590. 10.1089/hum.2009.138 [doi].
12. Huang X, Wilber AC, Bao L, Tuong D, Tolar J, et al. (2006) Stable gene transfer and expression in human primary T cells by the Sleeping Beauty transposon system. Blood 107: 483–491.
13. Jin Z, Maiti S, Huls H, Singh H, Olivares S, et al. (2011) The hyperactive Sleeping Beauty transposase SB100X improves the genetic modification of T cells to express a chimeric antigen receptor. Gene Ther. gt201140 [pii];10.1038/gt.2011.40 [doi].
14. Huls MH, Figliola MJ, Dawson MJ, Olivares S, Kebriaei P, et al. (2013) Clinical application of sleeping beauty and artificial antigen presenting cells to genetically modify T cells from peripheral and umbilical cord blood. J Vis Exp. 50070 [pii];10.3791/50070 [doi].
15. Mates L, Chuah MK, Belay E, Jerchow B, Manoj N, et al. (2009) Molecular evolution of a novel hyperactive Sleeping Beauty transposase enables robust stable gene transfer in vertebrates. Nat Genet 41: 753–761.
16. Mommaas B, van Halteren AG, Pool J, van d, V, Wieles B, et al. (2005) Adult and cord blood T cells can acquire HA-1 specificity through HA-1 T-cell receptor gene transfer. Haematologica 90: 1415–1421.
17. Demars R, Chang CC, Shaw S, Reitnauer PJ, Sondel PM (1984) Homozygous deletions that simultaneously eliminate expressions of class I and class II antigens of EBV-transformed B-lymphoblastoid cells. I. Reduced proliferative responses of autologous and allogeneic T cells to mutant cells that have decreased expression of class II antigens. Hum Immunol 11: 77–97. 0198-8859(84)90047-8 [pii].
18. Xue SA, Gao L, Hart D, Gillmore R, Qasim W, et al. (2005) Elimination of human leukemia cells in NOD/SCID mice by WT1-TCR gene-transduced human T cells. Blood 106: 3062–3067. 2005-01-0146 [pii];10.1182/blood-2005-01-0146 [doi].
19. Demaison C, Parsley K, Brouns G, Scherr M, Battmer K, et al. (2002) High-level transduction and gene expression in hematopoietic repopulating cells using a human imunodeficiency virus type 1-based lentiviral vector containing an internal spleen focus forming virus promoter. Hum Gene Ther 13: 803–813.
20. Cui Z, Geurts AM, Liu G, Kaufman CD, Hackett PB (2002) Structure-Function Analysis of the Inverted Terminal Repeats of the Sleeping Beauty Transposon. Journal of Molecular Biology 318: 1221–1235.
21. Vink CA, Gaspar HB, Gabriel R, Schmidt M, McIvor RS, et al. (2009) Sleeping beauty transposition from nonintegrating lentivirus. Mol Ther 17: 1197–1204.
22. Qasim W, Mackey T, Sinclair J, Chatziandreou I, Kinnon C, et al. (2007) Lentiviral vectors for T-cell suicide gene therapy: preservation of T-cell effector function after cytokine-mediated transduction. Mol Ther 15: 355–360.
23. Arens A, Appelt JU, Bartholomae CC, Gabriel R, Paruzynski A, et al. (2012) Bioinformatic clonality analysis of next-generation sequencing-derived viral vector integration sites. Hum Gene Ther Methods 23: 111–118. 10.1089/hgtb.2011.219 [pii];10.1089/hgtb.2011.219 [doi].
24. Pospori C, Xue SA, Holler A, Voisine C, Perro M, et al. (2011) Specificity for the tumor-associated self-antigen WT1 drives the development of fully functional memory T cells in the absence of vaccination. Blood 117: 6813–6824. blood-2010-08-304568 [pii];10.1182/blood-2010-08-304568 [doi].
25. Noser JA, Towers GJ, Sakuma R, Dumont JM, Collins MK, et al. (2006) Cyclosporine increases human immunodeficiency virus type 1 vector transduction of primary mouse cells. J Virol 80: 7769–7774.
26. Cavalieri S, Cazzaniga S, Geuna M, Magnani Z, Bordignon C, et al. (2003) Human T lymphocytes transduced by lentiviral vectors in the absence of TCR activation maintain an intact immune competence. Blood 102: 497–505.
27. Vandendriessche T, Chuah MK (2009) Moving gene therapy forward with mobile DNA. Hum Gene Ther 20: 1559–1561. 10.1089/hum.2009.1109 [doi].
28. Maiti SN, Huls H, Singh H, Dawson M, Figliola M, et al. (2013) Sleeping beauty system to redirect T-cell specificity for human applications. J Immunother 36: 112–123. 10.1097/CJI.0b013e3182811ce9 [doi].
29. Singh H, Manuri PR, Olivares S, Dara N, Dawson MJ, et al. (2008) Redirecting specificity of T-cell populations for CD19 using the Sleeping Beauty system. Cancer Res 68: 2961–2971. 68/8/2961 [pii];10.1158/0008-5472.CAN-07-5600 [doi].

PERMISSIONS

All chapters in this book were first published in PLOS ONE, by The Public Library of Science; hereby published with permission under the Creative Commons Attribution License or equivalent. Every chapter published in this book has been scrutinized by our experts. Their significance has been extensively debated. The topics covered herein carry significant findings which will fuel the growth of the discipline. They may even be implemented as practical applications or may be referred to as a beginning point for another development.

The contributors of this book come from diverse backgrounds, making this book a truly international effort. This book will bring forth new frontiers with its revolutionizing research information and detailed analysis of the nascent developments around the world.

We would like to thank all the contributing authors for lending their expertise to make the book truly unique. They have played a crucial role in the development of this book. Without their invaluable contributions this book wouldn't have been possible. They have made vital efforts to compile up to date information on the varied aspects of this subject to make this book a valuable addition to the collection of many professionals and students.

This book was conceptualized with the vision of imparting up-to-date information and advanced data in this field. To ensure the same, a matchless editorial board was set up. Every individual on the board went through rigorous rounds of assessment to prove their worth. After which they invested a large part of their time researching and compiling the most relevant data for our readers.

The editorial board has been involved in producing this book since its inception. They have spent rigorous hours researching and exploring the diverse topics which have resulted in the successful publishing of this book. They have passed on their knowledge of decades through this book. To expedite this challenging task, the publisher supported the team at every step. A small team of assistant editors was also appointed to further simplify the editing procedure and attain best results for the readers.

Apart from the editorial board, the designing team has also invested a significant amount of their time in understanding the subject and creating the most relevant covers. They scrutinized every image to scout for the most suitable representation of the subject and create an appropriate cover for the book.

The publishing team has been an ardent support to the editorial, designing and production team. Their endless efforts to recruit the best for this project, has resulted in the accomplishment of this book. They are a veteran in the field of academics and their pool of knowledge is as vast as their experience in printing. Their expertise and guidance has proved useful at every step. Their uncompromising quality standards have made this book an exceptional effort. Their encouragement from time to time has been an inspiration for everyone.

The publisher and the editorial board hope that this book will prove to be a valuable piece of knowledge for researchers, students, practitioners and scholars across the globe.

LIST OF CONTRIBUTORS

Wenhua Zhou and Liming Ying
Molecular Medicine, National Heart and Lung Institute, Imperial College London, London, United Kingdom

Kogularamanan Suntharalingam and Ramon Vilar
Department of Chemistry, Imperial College London, London, United Kingdom

Nigel J. Brand
Harefield Heart Science Centre, National Heart and Lung Institute, Imperial College London, Middlesex, United Kingdom

Paul J. R. Barton
Harefield Heart Science Centre, National Heart and Lung Institute, Imperial College London, Middlesex, United Kingdom
NIHR Cardiovascular Biomedical Research Unit, Royal Brompton and Harefield NHS Trust, London, United Kingdom

Lifan Zhang, Jennifer J. Michal, Xiang Zhou, Bo Ding, Michael V. Dodson and Zhihua Jiang
Department of Animal Sciences, Washington State University, Pullman, Washington, United States of America

Michelle R. Mousel and Gregory S. Lewis
USDA/ARS US Sheep Experiment Station, Dubois, Idaho, United States of America

Xiaolin Wu
Department of Dairy Science, University of Wisconsin-Madison, Madison, Wisconsin, United States of America

Nermin K. El-Halawany
Cell Biology Department, Division of Genetic Engineering and Biotechnology, National Research Center, Dokki, Gueza, Egypt

Wanjun Gu, Xiaofei Wang, Chuanying Zhai and Xueying Xie
Research Center of Learning Sciences, Southeast University, Nanjing, Jiangsu, China

Tong Zhou
Institute for Personalized Respiratory Medicine, The University of Illinois at Chicago, Chicago, Illinois, United States of America
Section of Pulmonary, Critical Care, Sleep & Allergy, Department of Medicine, The University of Illinois at Chicago, Chicago, Illinois, United States of America

Qingguo Wang
Department of Biomedical Informatics, Vanderbilt University School of Medicine, Nashville, Tennessee, United States of America

Peilin Jia
Department of Biomedical Informatics, Vanderbilt University School of Medicine, Nashville, Tennessee, United States of America
Center for Quantitative Sciences, Vanderbilt University School of Medicine, Nashville, Tennessee, United States of America

Zhongming Zhao
Department of Biomedical Informatics, Vanderbilt University School of Medicine, Nashville, Tennessee, United States of America
Department of Psychiatry, Vanderbilt University School of Medicine, Nashville, Tennessee, United States of America
Department of Cancer Biology, Vanderbilt University School of Medicine, Nashville, Tennessee, United States of America
Center for Quantitative Sciences, Vanderbilt University School of Medicine, Nashville, Tennessee, United States of America

Li Huang, Shiqiang Li, Xueshan Xiao, Xiaoyun Jia, Wenmin Sun, Yang Gao, Panfeng Wang, Xiangming Guo and Qingjiong Zhang
State Key Laboratory of Ophthalmology, Zhongshan Ophthalmic Center, Sun Yat-sen University, Guangzhou, Guangdong, China

Qingyan Zhang, Liping Guan, Jianguo Zhang, Zhihong Zhu, Ye Yin and Jun Wang
BGI-Shenzhen, Shenzhen, Guangdong, China

Aakrosh Ratan, Webb Miller and Stephan C. Schuster
Center for Comparative Genomics and Bioinformatics, Pennsylvania State University, University Park, Pennsylvania, United States of America

Joseph Guillory, Jeremy Stinson and Somasekar Seshagiri
Department of Molecular Biology, Genentech Inc., South San Francisco, California, United States of America

Elisa Tassano and Giorgio Gimelli
Laboratorio di Citogenetica, Istituto G. Gaslini, Genova, Italy

Jens Buttgereit
Experimental and Clinical Research Center (ECRC), a joint institution of the Max Delbrück Center for Molecular Medicine (MDC) and the Charité Medical Faculty, Berlin, Germany
Max Delbrück Center for Molecular Medicine (MDC), Campus Berlin-Buch, Berlin, Germany

Michael Bader
Max Delbrück Center for Molecular Medicine (MDC), Campus Berlin-Buch, Berlin, Germany

Margherita Lerone and Maria Teresa Divizia
Laboratorio di Genetica Molecolare, Istituto G. Gaslini, Genova, Italy

Renata Bocciardi
Dipartimento di Neuroscienze, Riabilitazione, Oftalmologia, Genetica e Scienze Materno-Infantili, (DiNOGMI), and Laboratorio di Genetica Molecolare, Istituto G. Gaslini, Genova, Italy

Frédérique Sloan-Béna and Stefania Gimelli
Service of Genetic Medicine, University Hospitals of Geneva, Geneva, Switzerland

Flavia Napoli and Giovanna Pala
Dipartimento di Pediatria, Istituto G. Gaslini, Universita` di Genova, Genova, Italy

George W. Cook, Miriam K. Konkel, Jerilyn A. Walker, Matthew G. Bourgeois and Mark A. Batzer
Department of Biological Sciences, Louisiana State University, Baton Rouge, Louisiana, United States of America

Mitchell L. Fullerton
Department of Bioengineering, Clemson University, Clemson, South Carolina, United States of America

John T. Fussell
Electrochemical Materials, Louisiana Business and Technology Center, Baton Rouge, Louisiana, United States of America

Heath D. Herbold
Albemarle Corporation, Pasadena, Texas, United States of America
Eric C. Larsen and Sharmila Banerjee-Basu
MindSpec, McLean, Virginia, United States of America

Idan Menashe
MindSpec, McLean, Virginia, United States of America
Department of Public Health, Faculty of Health Sciences, Ben Gurion University of the Negev, Beer-Sheva, Israel

Maria O'Rourke, P. Noel Barrett and Ian Livey
Vaccine R&D, Baxter Bioscience, Orth/Donau, Austria

Lara Lusa
Faculty of Medicine, Institute for Biostatistics and Medical Informatics, Ljubljana, Slovenia

Andreas Traweger
Vaccine R&D, Baxter Bioscience, Orth/Donau, Austria
Paracelsus Medical University, Spinal Cord Injury and Tissue Regeneration Center Salzburg
Institute of Tendon and Bone Regeneration, Salzburg, Austria

Dasa Stupica, Vera Maraspin and Franc Strle
Department of Infectious Diseases, University Medical Centre Ljubljana, Ljubljana, Slovenia

Edward Wijaya and Kiyoshi Asai
Graduate School of Frontier Sciences, University of Tokyo, Kashiwa, Japan
Computational Biology Research Center, National Institute of Advanced Industrial Science and Technology (AIST), Tokyo, Japan

Martin C. Frith and Paul Horton
Computational Biology Research Center, National Institute of Advanced Industrial Science and Technology (AIST), Tokyo, Japan
Sanne Vendelbosch, Martin de Boer, Remko A. T. W. Gouw, Cynthia K. Y. Ho, Judy Geissler, Michael J. Moorhouse, Dirk Roos and Timo K. van den Berg
Sanquin Research and Landsteiner Laboratory, Department of Blood Cell Research, Amsterdam, The Netherlands

Wendy T. N. Swelsen and Neubury M. Lardy
Sanquin Blood Supply Foundation, Department of Immunogenetics, Amsterdam, The Netherlands

Taco W. Kuijpers
Sanquin Research and Landsteiner Laboratory, Department of Blood Cell Research, Amsterdam, The Netherlands
Emma Children's Hospital, Department of Pediatric Hematology, Immunology and Infectious Diseases, Academic Medical Center, Amsterdam, The Netherlands

Andrew R. Wood, Michael N. Weedon and Timothy M. Frayling
Genetics of Complex traits, Institute of Biomedical and Clinical Sciences, Peninsula College of Medicine and Dentistry, University of Exeter, Exeter, United Kingdom

John R. B. Perry
Genetics of Complex traits, Institute of Biomedical and Clinical Sciences, Peninsula College of Medicine and Dentistry, University of Exeter, Exeter, United Kingdom
Wellcome Trust Centre for Human Genetics, University of Oxford, Oxford, United Kingdom
Department of Twin Research and Genetic Epidemiology, King's College London, London, United Kingdom

Tim D. Spector
Department of Twin Research and Genetic Epidemiology, King's College London, London, United Kingdom

Toshiko Tanaka and Luigi Ferrucci
Longitudinal Studies Section, Clinical Research Branch, Gerontology Research Center, National Institute on Aging, Baltimore, Maryland, United States of America

Michael A. Nalls and Andrew B. Singleton
Laboratory of Neurogenetics, National Institute of Aging, Bethesda, Maryland, United States of America

Dena G. Hernandez and J. Raphael Gibbs
Laboratory of Neurogenetics, National Institute of Aging, Bethesda, Maryland, United States of America
Department of Molecular Neuroscience and Reta Lila Laboratories, Institute of Neurology, UCL, London, United Kingdom

Hou-Feng Zheng
Departments of Medicine, Human Genetics, Epidemiology and Biostatistics, Jewish General Hospital, McGill University, Montreal, Quebec, Canada

David Melzer
Epidemiology and Public Health, Institute of Biomedical and Clinical Sciences, Peninsula College of Medicine and Dentistry, University of Exeter, Exeter, United Kingdom

J. Brent Richards
Department of Twin Research and Genetic Epidemiology, King's College London, London, United Kingdom
Departments of Medicine, Human Genetics, Epidemiology and Biostatistics, Jewish General Hospital, McGill University, Montreal, Quebec, Canada

Stefania Bandinelli
Tuscany Regional Health Agency, Florence, Italy, I.O.T. and Department of Medical and Surgical Critical Care, University of Florence, Florence, Italy
Geriatric Unit, Azienda Sanitaria di Firenze, Florence, Italy

Arianna Moiani
Division of Genetics and Cell Biology, Istituto Scientifico H. San Raffaele, Milan, Italy

Annarita Miccio
Department of Life Sciences, University of Modena and Reggio Emilia, Modena, Italy

Ermanno Rizzi, Marco Severgnini and Gianluca De Bellis
Institute for Biomedical Technologies, Consiglio Nazionale delle Ricerche, Milan, Italy

Danilo Pellin
CUSSB, Vita-Salute San Raffaele University, Milan, Italy

Julia Debora Suerth and Christopher Baum
Institute of Experimental Hematology, Hannover Medical School, Hannover, Germany

Fulvio Mavilio
Department of Life Sciences, University of Modena and Reggio Emilia, Modena, Italy
Genethon, Evry, France

Ruibang Luo, Thomas Wong, Chi-Man Liu, Edward Wu, Lap-Kei Lee, David W. Cheung, Hing-Fung Ting, Siu-Ming Yiu and Tak-Wah Lam
HKU-BGI Bioinformatics Algorithms and Core Technology Research Laboratory & Department of Computer Science, University of Hong Kong, Hong Kong, China

Xiaoqian Zhu and Shaoliang Peng
School of Computer Science, National University of Defense Technology, Changsha, Hunan, China

Haoxiang Lin, Wenjuan Zhu, Chang Yu and Yingrui Li
BGI-Shenzhen, Shenzhen, Guangdong, China

Ruiqiang Li
Peking-Tsinghua Center for Life Sciences, Biodynamic Optical Imaging Center and School of Life Sciences, Peking University, Beijing, China

Jianqiao Zhu
HKU-BGI Bioinformatics Algorithms and Core Technology Research Laboratory & Department of Computer Science, University of Hong Kong, Hong Kong, China
Department of Computer Sciences, University of Wisconsin-Madison, Madison, Wisconsin, United States of America

Geòrgia Escaramís, Cristian Tornador, Laia Bassaganyas, Raquel Rabionet and Xavier Estivill
Genetic Causes of Disease Group, Centre for Genomic Regulation (CRG), Barcelona, Spain
Universitat Pompeu Fabra (UPF), Barcelona, Spain
Centro de Investigación Biomédica en Red en Epidemiología y Salud Pública (CIBERESP), Barcelona, Spain
Hospital del Mar Medical Research Institute (IMIM), Barcelona, Spain

Jose M. C. Tubio
Genetic Causes of Disease Group, Centre for Genomic Regulation (CRG), Barcelona, Spain
Universitat Pompeu Fabra (UPF), Barcelona, Spain
Centro de Investigación Biomédica en Red en Epidemiología y Salud Pública (CIBERESP), Barcelona, Spain
Hospital del Mar Medical Research Institute (IMIM), Barcelona, Spain
Galician Foundation of Genomic Medicine-SERGAS, Complexo Hospitalitario Universitario de Santiago (CHUS), Santiago de Compostela, Spain

Alexander Martínez-Fundichely
Genetic Causes of Disease Group, Centre for Genomic Regulation (CRG), Barcelona, Spain
Universitat Pompeu Fabra (UPF), Barcelona, Spain
Centro de Investigación Biomédica en Red en Epidemiología y Salud Pública (CIBERESP), Barcelona, Spain
Hospital del Mar Medical Research Institute (IMIM), Barcelona, Spain
Institut de Biotecnologia i de Biomedicina, Universitat Autònoma de Barcelona, Bellaterra, Spain

Mario Cáceres
Genetic Causes of Disease Group, Centre for Genomic Regulation (CRG), Barcelona, Spain
Universitat Pompeu Fabra (UPF), Barcelona, Spain
Centro de Investigación Biomédica en Red en Epidemiología y Salud Pública (CIBERESP), Barcelona, Spain
Hospital del Mar Medical Research Institute (IMIM), Barcelona, Spain
Institut de Biotecnologia i de Biomedicina, Universitat Autònoma de Barcelona, Bellaterra, Spain
Institució Catalana de Recerca i Estudis Avanc¸ats (ICREA), Barcelona, Spain

Marta Gut
National Center of Genomic Analysis (CNAG-CRG), Barcelona, Spain

Stephan Ossowski
Universitat Pompeu Fabra (UPF), Barcelona, Spain
Hospital del Mar Medical Research Institute (IMIM), Barcelona, Spain
Genomic and Epigenomic Variation in Disease Group, Centre for Genomic Regulation (CRG), Barcelona, Spain

Anne-Christine Field, Conrad Vink, Roua Al-Subki, Nicholas Goulden, Adrian Thrasher and Waseem Qasim
Molecular immunology Unit, Institute of Child Health, University College London, London, United Kingdom

Richard Gabriel and Manfred Schmidt
Department of Translational Oncology, National Center for Tumor Diseases and German Cancer Research Center, Heidelberg, Germany

Hans Stauss and Emma Morris
Institute of Immunity & Transplantation, Royal Free Campus University College London, London, United Kingdom

Index